W9-CRB-305

Fundamentals of the Physical Theory of Diffraction

BICENTENNIAL
1807
⊛WILEY
2007
BICENTENNIAL

THE WILEY BICENTENNIAL—KNOWLEDGE FOR GENERATIONS

*E*ach generation has its unique needs and aspirations. When Charles Wiley first opened his small printing shop in lower Manhattan in 1807, it was a generation of boundless potential searching for an identity. And we were there, helping to define a new American literary tradition. Over half a century later, in the midst of the Second Industrial Revolution, it was a generation focused on building the future. Once again, we were there, supplying the critical scientific, technical, and engineering knowledge that helped frame the world. Throughout the 20th Century, and into the new millennium, nations began to reach out beyond their own borders and a new international community was born. Wiley was there, expanding its operations around the world to enable a global exchange of ideas, opinions, and know-how.

For 200 years, Wiley has been an integral part of each generation's journey, enabling the flow of information and understanding necessary to meet their needs and fulfill their aspirations. Today, bold new technologies are changing the way we live and learn. Wiley will be there, providing you the must-have knowledge you need to imagine new worlds, new possibilities, and new opportunities.

Generations come and go, but you can always count on Wiley to provide you the knowledge you need, when and where you need it!

WILLIAM J. PESCE
PRESIDENT AND CHIEF EXECUTIVE OFFICER

PETER BOOTH WILEY
CHAIRMAN OF THE BOARD

Fundamentals of the Physical Theory of Diffraction

Pyotr Ya. Ufimtsev

WILEY-INTERSCIENCE
A JOHN WILEY & SONS, INC., PUBLICATION

Published by John Wiley & Sons, Inc., Hoboken, New Jersey
Published simultaneously in Canada

For general information on our other products and services or for technical support, please contact our Customer Care Department within the United States at (800) 762-2974, outside the United States at (317) 572-3993 or fax (317) 572-4002.

Wiley also publishes its books in a variety of electronic formats. Some content that appears in print may not be available in electronic formats. For more information about Wiley products, visit our web site at www.wiley.com.

Library of Congress Cataloging-in-Publication Data is available.
ISBN 978-0-470-09771-X

Printed in the United States of America

10 9 8 7 6 5 4 3 2 1

Contents

Foreword   xi

Preface   xv

Acknowledgments   xvii

Introduction   1

1. Basic Notions in Acoustic and Electromagnetic Diffraction Problems   5

   1.1   Formulation of the Diffraction Problem   5
   1.2   Scattered Field in the Far Zone   7
   1.3   Physical Optics   11

       1.3.1   Definition of the Physical Optics   11
       1.3.2   Total Scattering Cross-Section   13
       1.3.3   Optical Theorem   15
       1.3.4   Introducing the Notion of "Shadow Radiation"   16
       1.3.5   Shadow Contour Theorem and theTotal Scattering
               Cross-Section   21
       1.3.6   Summary of Properties of Physical
               Optics Approximation   24

   1.4   Nonuniform Component of Induced Surface Field   25
   1.5   Electromagnetic Waves   27
   Problems   31

2. Wedge Diffraction: Exact Solution and Asymptotics   33

   2.1   Classical Solutions   33
   2.2   Transition to the Plane Wave Excitation   38
   2.3   Conversion of the Series Solution to the Sommerfeld Integrals   40
   2.4   The Sommerfeld Ray Asymptotics   44
   2.5   The Pauli Asymptotics   47
   2.6   Uniform Asymptotics: Extension of the Pauli Technique   51
   2.7   Comments on Alternative Asymptotics   55
   Problems   56

3. Wedge Diffraction: The Physical Optics Field   **59**

3.1   Original PO Integrals   59
3.2   Conversion of the PO Integrals to the Canonical Form   61
3.3   Ray Asymptotics for the PO Diffracted Field   67
Problems   68

4. Wedge Diffraction: Radiation by the Nonuniform Component of Surface Sources   **71**

4.1   Integrals and Asymptotics   71
4.2   Integral Form of Functions $f^{(1)}$ and $g^{(1)}$ 76
4.3   Oblique Incidence of a Plane Wave at a Wedge   78
Problems   82

5. First-Order Diffraction at Strips and Polygonal Cylinders   **83**

5.1   Diffraction at a Strip   83

5.1.1   Physical Optics Part of the Scattered Field   85
5.1.2   Total Scattered Field   87
5.1.3   Numerical Analysis of the Scattered Field   92
5.1.4   First-Order PTD with Truncated Scattering Sources $j_h^{(1)}$ 95

5.2   Diffraction at a Triangular Cylinder   99

5.2.1   Symmetric Scattering: PO Approximation   101
5.2.2   Backscattering: PO Approximation   102
5.2.3   Symmetric Scattering: First-Order PTD Approximation   104
5.2.4   Backscattering: First-Order PTD Approximation   107
5.2.5   Numerical Analysis of the Scattered Field   110

Problems   112

6. Axially Symmetric Scattering of Acoustic Waves at Bodies of Revolution   **115**

6.1   Diffraction at a Canonical Conic Surface   115

6.1.1   Integrals for the Scattered Field   117
6.1.2   Ray Asymptotics   118
6.1.3   Focal Fields   124
6.1.4   Bessel Interpolations for the Field $u_{s,h}^{(1)}$ 125

6.2   Scattering at a Disk   126

6.2.1   Physical Optics Approximation   127
6.2.2   Field Generated by Nonuniform Scattering Sources   130
6.2.3   Total Scattered Field   132

6.3   Scattering at Cones: Focal Field   134

      6.3.1   Asymptotic Approximations for the Field   134
      6.3.2   Numerical Analysis of Backscattering   138

6.4   Bodies of Revolution with Nonzero Gaussian Curvature:
      Backscattered Focal Fields   141

      6.4.1   PO Approximation   143
      6.4.2   Total Backscattered Focal Field: First-Order PTD
            Asymptotics   145
      6.4.3   Backscattering from Paraboloids   145
      6.4.4   Backscattering from Spherical Segments   151

6.5   Bodies of Revolution with Nonzero Gaussian Curvature:
      Axially Symmetric Bistatic Scattering   155

      6.5.1   Ray Asymptotics for the PO Field   156
      6.5.2   Bessel Interpolations for the PO Field in the
            Region $\pi - \omega \leq \vartheta \leq \pi$ 159
      6.5.3   Bessel Interpolations for the PTD Field in the
            Region $\pi - \omega \leq \vartheta \leq \pi$ 160
      6.5.4   Asymptotics for the PTD Field in the Region
            $2\omega < \vartheta \leq \pi - \omega$ away from the GO Boundary $\vartheta = 2\omega$ 161
      6.5.5   Uniform Approximations for the PO Field in the Ray Region
            $2\omega \leq \vartheta \leq \pi - \omega$ Including the GO Boundary $\vartheta = 2\omega$ 161
      6.5.6   Approximation for the PO Field in the Shadow
            Region for Reflected Rays   165

Problems   166

7. Elementary Acoustic and Electromagnetic Edge Waves   **169**

7.1   Elementary Strips on a Canonical Wedge   170
7.2   Integrals for $j_{s,h}^{(1)}$ on Elementary Strips   171
7.3   Triple Integrals for Elementary Edge Waves   175
7.4   Transformation of Triple Integrals into One-Dimensional Integrals   178
7.5   General Asymptotics for Elementary Edge Waves   183
7.6   Analytic Properties of Elementary Edge Waves   187
7.7   Numerical Calculations of Elementary Edge Waves   191
7.8   Electromagnetic Elementary Edge Waves   194
7.9   Improved Theory of Elementary Edge Waves   198

      7.9.1   Acoustic EEWs   199
      7.9.2   Electromagnetic EEWs   203

7.10   Some References Related to Elementary Edge Waves   209
Problems   210

8. Ray and Caustics Asymptotics for Edge Diffracted Waves   **213**

8.1   Ray Asymptotics   213

   8.1.1   Acoustic Waves   213
   8.1.2   Electromagnetic Waves   218
   8.1.3   Comments on Ray Asymptotics   219

8.2   Caustic Asymptotics   220

   8.2.1   Acoustic Waves   221
   8.2.2   Electromagnetic Waves   225

Problems   226

9. Multiple Diffraction of Edge Waves: Grazing Incidence and Slope Diffraction   **229**

9.1   Statement of the Problem and Related References   229
9.2   Grazing Diffraction   230

   9.2.1   Acoustic Waves   230
   9.2.2   Electromagnetic Waves   234

9.3   Slope Diffraction in the Configuration of Figure 9.1   236

   9.3.1   Acoustic Waves   236
   9.3.2   Electromagnetic Waves   238

9.4   Slope Diffraction: General Case   240

   9.4.1   Acoustic Waves   240
   9.4.2   Electromagnetic Waves   243

Problems   245

10. Diffraction Interaction of Neighboring Edges on a Ruled Surface   **247**

10.1   Diffraction at an Acoustically Hard Surface   248
10.2   Diffraction at an Acoustically Soft Surface   250
10.3   Diffraction of Electromagnetic Waves   252
Problems   254

11. Focusing of Multiple Acoustic Edge Waves Diffracted at a Convex Body of Revolution with a Flat Base   **255**

11.1   Statement of the Problem and its Characteristic Features   255
11.2   Multiple Hard Diffraction   256
11.3   Multiple Soft Diffraction   258
Problems   260

12. Focusing of Multiple Edge Waves Diffracted at a Disk   **261**

12.1   Multiple Hard Diffraction   261
12.2   Multiple Soft Diffraction   264
12.3   Multiple Diffraction of Electromagnetic Waves   267
Problems   268

13. Backscattering at a Finite-Length Cylinder   **269**

13.1   Acoustic Waves   269

    13.1.1   PO Approximation   269
    13.1.2   Backscattering Produced by the Nonuniform
           Component $j^{(1)}$ 273
    13.1.3   Total Backscattered Field   277

13.2   Electromagnetic Waves   279

    13.2.1   E-Polarization   279
    13.2.2   H-Polarization   283

Problems   285

14. Bistatic Scattering at a Finite-Length Cylinder   **287**

14.1   Acoustic Waves   287

    14.1.1   PO Approximation   287
    14.1.2   Shadow Radiation as a Part of the Physical Optics Field   289
    14.1.3   PTD for Bistatic Scattering at a Hard Cylinder   290
    14.1.4   Beams and Rays of the Scattered Field   297
    14.1.5   Refined Asymptotics for the Specular Beam   300

14.2   Electromagnetic Waves   304

    14.2.1   *E*-Polarization   304
    14.2.2   *H*-Polarization   306
    14.2.3   Refined Asymptotics for the Specular Beam Reflected
           from the Lateral Surface   308

Problems   311

Conclusion   **313**

References   **315**

Index   **323**

Foreword

Ideas have consequences. Great ideas have far-reaching consequences.

The physical theory of diffraction (PTD) that Professor Ufimtsev introduced in the 1950s—a methodology for approximate evaluation at high enough frequency of the scattering from a body, especially a body of complicated shape—has proven to be a truly great idea.

The first form of PTD developed by Professor Ufimtsev, the vector form applicable to electromagnetic scattering from three-dimensional bodies, has played a key role in the development of modern low-radar-reflectivity weapons systems such as the Lockheed F-117 Stealth Fighter and the Northrop B-2 Stealth Bomber, functioning both as a design tool and as a conceptual framework. These systems in turn have revolutionized the conduct of large-scale government-versus-government warfare and thus have helped to shape history.

Ben Rich, who oversaw the F-117 project as head of Lockheed's fabled Skunk Works, refers to Professor Ufimtsev's work as "the Rosetta Stone breakthrough for stealth technology." At Northrop, where I worked on the B-2 project, we were so enthusiastic about PTD that a co-worker and I sometimes broke into choruses of "Go, Ufimtsev" to the tune of "On, Wisconsin." At both Lockheed and Northrop we referred to PTD as "industrial-strength" diffraction theory to distinguish it from the approach to diffraction then being favored in the universities, which was not well enough developed to handle the problems of stealth design.

Like many good theories PTD is much easier to apply than to explain. But let us now nevertheless examine the inner workings of PTD and seek to understand why it is such a useful approach. First of all, PTD is based on two important principles which it will be convenient to refer to here as the *physical principle* and the *geometrical principle*.

The physical principle shows how the scattered field at a point outside a scattering body can be determined from an integral of appropriate field quantities over the surface of the body. In acoustics these quantities are the pressure at a hard surface, the normal velocity at a soft surface, both at an impedance boundary or the surface of a penetrable body. In electromagnetics they are the tangential magnetic field at the surface of a perfect conductor, the tangential magnetic and electric fields at an impedance boundary or the surface of a penetrable body.

The geometrical principle states that at high enough frequency, when the wavelength is small enough compared to the critical dimensions of the scattering body, the surface integrals can be evaluated asymptotically to yield a description of the total field outside the body in terms of geometrical rays, including diffracted rays. The change in field amplitude along a ray can be calculated geometrically by tracing the divergence and convergence of ray bundles except in the regions surrounding (a) a geometrical shadow boundary, for which ray tracing predicts a field discontinuity across the boundary, and (b) a caustic, that is, a locus where adjacent geometrical rays meet or cross (such as, in the simplest case, a focal point), at which ray tracing predicts an infinite field. The correct value for the field in these regions, which shrink as frequency increases, can be found by using uniform asymptotic techniques to evaluate the surface integrals.

One of the important features of PTD is this ability to calculate the field accurately in shadow boundary and caustic regions. It is especially important in low observables design because we are often interested in far-field scattering of a plane wave from a body with straight or slightly curved edges, a configuration for which parts of the far-field region lie in caustic regions.

The other major advantages of PTD arise from the way the surface fields are handled. There is a *uniform* part which is *defined everywhere on the surface* and a *nonuniform* part that serves as a correction term.

For electromagnetics the uniform part is usually, though not always, given by the physical optics (PO) approximation, namely that the surface fields at a point are the same as if the point lay on an infinite plane surface tangent to the actual body at the point and with the same boundary conditions as at the point. For acoustics the uniform part is usually given by the analogous approximation. Because this acoustics approximation does not have a firmly established name and because other investigators have set the precedent, Professor Ufimtsev uses the terminology PO in both electromagnetics and acoustics throughout this book. Much of Chapter 1 is devoted to PO and its implications.

The nonuniform fields for a non-penetrable body, for example a hard body in acoustics or a perfect conductor in electromagnetics, tend to be strongest near a diffracting feature such as an edge where two faces of a faceted surface meet, and these fields often diminish rapidly with distance from the feature. It should be emphasized here that this desirable behavior is a consequence of the judicious choice of the uniform part.

The nonuniform surface fields are determined using the results of simpler scattering problems, often called *canonical problems*. Consider again, for example, an edge on a faceted surface. Let the body be a perfect conductor and the edge be straight with the wedge angle formed by the two faces constant along its length, let the illuminating field be a plane wave, and let us choose the PO fields as the uniform part. Then the canonical problem is diffraction of an appropriately oriented plane wave from an infinitely long wedge with perfectly conducting flat faces (even if the faces on the body of interest are not flat). This problem reduces to two scalar two-dimensional problems, one for incident electric field normal to the edge, the other for incident magnetic field normal to the edge, and exact solutions exist for these problems. The vector surface fields can be constructed from the two scalar solutions, and the nonuniform surface fields associated with the edge are then found by subtracting the physical optics fields of the canonical problem from the full solution.

There now arises the problem of reconciling the uniform part and the nonuniform part, which is defined on a surface that may not exactly match the body surface. Professor Ufimtsev addresses this in Chapter 7, where he reduces the nonuniform part to a continuous array of *elementary edge waves* concentrated along the edge. These elementary edge waves are sources of diffracted rays and have a directivity pattern that is related to the canonical problem. In the parlance of engineering they would be called *diffraction coefficients*.

The nonuniform contribution to the field diffracted from the edge is now given by an integral of the elementary edge waves over the length of the edge. But, when we asymptotically evaluate the integral for the physical optics diffraction from a face, we see that it reduces to an integral along the illuminated part of the face perimeter plus possibly other localized terms (such as a specular reflection contribution). Thus there are edge diffraction contributions from the uniform part of the surface field on both the faces that meet at the edge (if both are illuminated) as well as from the nonuniform part, and these three terms give the total edge diffraction. Furthermore, it turns out that each element of the edge produces diffraction in essentially all directions.

We can now, from this investigation of how the surface fields are modeled, extract these additional important features of PTD:

1. PTD can find accurately the reflection and diffraction from a body of complicated shape without having to match the entire body to canonical problems, just the regions that give rise to diffraction;
2. PTD minimizes the difficulty of reconciling the geometries of the body and of the canonical problem;
3. PTD yields diffracted rays in all directions from each element of a linear diffracting feature rather than just in directions on the well-known diffraction cone.

The third point is extremely important in low observables work, where the off-cone rays can sometimes yield the strongest fields in a region.

This book presents a thorough development of the fundamentals of PTD for both the scalar and vector cases as applied to acoustics and electromagnetics, including important aspects of the theory only recently developed by Professor Ufimtsev. For acoustics it is of course the scalar theory that is of interest. For electromagnetics both scalar and vector theory should be of interest. Canonical problems are often two-dimensional, and two-dimensional problems can be reduced to scalar form.

Emphasis in the book is on nonpenetrable bodies with "classical" boundary conditions at the surface: The Dirichlet and Neumann problems of applied mathematics; the corresponding soft and hard boundary problems of acoustics; and the perfect conductor problem of electromagnetics.

PTD is, however, in principle readily extended to the cases of a body with an impedance boundary condition at its surface and of a penetrable but opaque body and has in fact been used extensively for such bodies, though much of the work is classified, proprietary, or otherwise restricted. The extension to translucent and transparent bodies is more challenging, not because of any shortcoming of PTD but because it can be necessary to deal with such complicated phenomena as diffracted waves that travel through the body and are then refracted out of the body.

Much has been said and written about the relative merits of the two major modern approaches to diffraction theory, PTD on the one hand and, on the other hand, Professor Joseph Keller's geometrical theory of diffraction (GTD) and its modified versions, the uniform theory of diffraction (UTD) developed at ohio state university and the similar uniform asymptotic theory of diffraction (UAT).

Both approaches are valid, each yields a ray description of the field (PTD as an end result, GTD as a starting point), each has its advantages, and the two have now been cross-fertilizing each other for half a century. The work of the next generation, I fervently hope, will be to mold these approaches and other contributions together into a single modern theory of diffraction from bodies.

By his detailed exposition of the fundamentals of PTD in this present volume, Professor Ufimtsev has not only produced a work of great contemporary value but also a compendium that can be extremely useful in this reconciliation process.

KENNETH M. MITZNER

November 2006

Preface

The physical theory of diffraction (PTD) is a high-frequency asymptotic technique for the investigation of antennas and scattering problems. This monograph presents the first complete and comprehensive description of the modern PTD based on the concept of elementary edge waves (EEWs). Its subject is the diffraction of acoustic and electromagnetic waves by perfectly reflecting objects located in a homogeneous lossless medium.

The basic idea of PTD is that: The diffracted field is considered as the radiation generated by the scattering sources (currents) induced on the objects. The so-called *uniform* and *nonuniform* scattering sources are introduced in PTD. Uniform sources are defined as sources induced on the infinite plane tangent to the object at a source point. Nonuniform sources are caused by any deviation of the scattering surface from the tangent plane. For large convex objects with sharp edges, the basic contributions to the scattered field are produced by the uniform sources and by those nonuniform sources that concentrate near edges (often called *fringe* sources).

The integration of uniform sources leads to the physical optics (PO) approximation for the scattered field. The PTD is the natural extension of the PO approximation, taking into account the additional field created by the nonuniform/fringe sources.

This book provides high-frequency asymptotics for fringe scattering sources and for the scattered field in the far zone. Scattering characteristics are calculated for a variety of objects, such as strips, polygonal cylinders, cones, bodies of revolution with nonzero Gaussian curvature (including paraboloids and spherical segments), and finite circular cylinders with flat bases.

The title of the book underlines the fact that a great deal of attention is to be given to scattering physics. The derived analytic expressions clearly explain the physical structure of the scattered field and describe, in detail, all of the reflected and diffracted rays and beams, as well as the fields in the vicinity of caustics and foci. Also, a new fundamental component of the field, the so-called *shadow radiation*, is introduced. It is shown that this component contains *half* of the total scattered power. The physical manifestations of the shadow radiation are the well-known phenomena of Fresnel diffraction and forward scattering.

Plotted numeric results supplement the theory and provide visualizations of the individual contributions of different parts of the scattering objects to the total diffracted field. Detailed comments explain all critical steps in the analytic and numeric calculations to facilitate their examination and utilization by readers. All chapters are followed by problems for independent investigations, which will be helpful in studying PTD, especially for students.

This book is intended for researchers working on antennas and scattering problems in industry and university laboratories. It can also be useful for teaching a variety of university courses, that include topics on high-frequency asymptotic techniques in diffraction theory. University instructors and graduate students will benefit from this book as well.

PYOTR YA. UFIMTSEV

Los Angeles, California
June 2006

Acknowledgments

The work on this book was partially sponsored by the Center of Aerospace Research and Education in the University of California at Irvine. I highly appreciate the support by director of this center, Dr. Satya N. Atluri.

Many thanks go to Dr. A.V. Kaptsov for his professional advice, which greatly helped in my work with FORTRAN and SIGMA-PLOT programs.

During the preparation of this book I often appealed to my sons Ivan and Vladimir with requests to check and improve my English and to fix arising computer problems. I am thankful for their assistance.

Thanks are also due to J. V. Jull, K. M. Mitzner, Y. Rahmat-Samii, and A. J. Terzuoli, Jr., for their reviews of the manuscript and valuable comments.

This book includes, in revised form, materials from certain articles I wrote for the journals *Zhurnal Tekhnicheskoi Fiziki* (Russia), *Journal of the Acoustical Society of America* (USA), *Annals of Telecommunications* (France), and *Electromagnetics* (USA). I thank the editorial boards of the journals for their permission to use these materials.

P. YA. U.

Introduction

The physical theory of diffraction (PTD) is an asymptotic high-frequency technique that originated in earlier work by this author (Ufimtsev, 1957, 1958a,b,c, 1961). The results of the initial journal publications on PTD were summarized in a monograph (Ufimtsev, 1962), which became a bibliographical rarity a long time ago. To acquaint a new generation of readers with the original form of PTD, some sections of this monograph were updated and included in a more recent book (Ufimtsev, 2003). The selected topics of the modern form of PTD have been published in concise form in the articles by Butorin and Ufimtsev (1986), Butorin et al. (1987), Ufimtsev (1989, 1991), and Ufimtsev and Rahmat-Samii (1995).

This book presents the first complete and comprehensive description of the modern PTD based on the concept of elementary edge waves (EEWs). The theory is developed for acoustic and electromagnetic waves scattered by perfectly reflecting objects.

For acoustic waves, *soft* (Dirichlet) or *hard* (Neumann) boundary conditions are imposed on scattering objects located in a homogeneous nonviscous medium. Absence of viscosity is justified for a fluid (such as air and water) in the linear approximation (Kinsler et al., 1982; Pierce, 1994).

In diffraction problems for electromagnetic waves, the scattering objects are considered as perfectly conducting bodies located in a vacuum. Assumption of infinite conductivity is acceptable for metallic objects detected by radar. The boundary condition related to electromagnetic waves states that on the surface of perfectly conducting bodies, the tangential component of the electric vector is equal to zero (Balanis, 1989).

The diffraction theory of acoustic waves is scalar, and it is simpler than the vector theory of electromagnetic waves. Because of this, we investigate first in detail an acoustic diffraction problem and then briefly present its electromagnetic version referring to similar elements in acoustic theory. This facilitates the study of electromagnetic problems. Notice also that from the mathematical point of view, all two-dimensional (2-D) diffraction problems have identical solutions for acoustic and electromagnetic waves. These problems are considered for acoustic waves. The relationships between acoustic and electromagnetic diffracted waves are emphasized throughout the book.

Fundamentals of the Physical Theory of Diffraction. By Pyotr Ya. Ufimtsev
Copyright © 2007 John Wiley & Sons, Inc.

They are also formulated in the text boxes placed at the very beginning of most chapters and sections.

Notice that PTD has found various applications. Some related references are collected at the end of the book in the Section "Additional References Related to the PTD Concept: Applications, Modifications, and Developments". In particular, PTD was successfully used in the design of the American stealth-fighter F-117 and stealth-bomber B-2 (Browne, 1991a, Browne, 1991b, Rich, 1994, Rich and Janos, 1994; see also the foreword written by Mitzner for the Ufimtsev book, 2003). The present book contains only original results obtained by the author (some of them in collaboration with colleagues).

The distinctive feature of PTD is that it belongs to the class of source-based theories. The scattered/diffracted field is considered as radiation by surface sources, which are induced (due to diffraction) on the scattering objects by the incident waves. In the case of electromagnetic waves and metallic scattering objects, these sources are the surface electric charges and currents. In the case of acoustic waves, these sources are the surface distributions of the "acoustic pressure" on rigid objects, or the surface distributions of the "fluid velocity" on soft (pressure-release) objects. The advantage of this approach, as compared to the ray-based techniques, is that it allows the calculation of the scattered field everywhere, including the diffraction regions, such as foci and caustics, where the diffracted field does not have a ray structure.

The central and original idea of PTD is the separation of surface sources into the so-called uniform and nonuniform components. This separation is a flexible procedure, based on an appropriate choice of canonical diffraction problems (Ufimtsev, 1998). In the present book (except Section 7.9), the *uniform* component is defined as the scattering sources induced on the infinite plane tangent to the object at a source point. In the case of incident waves with a ray structure, this component is determined according to the geometrical optics (geometrical acoustics) for electromagnetic (acoustic) waves. The field found by the integration of the uniform component is considered as a high-frequency approximation for the scattered field. In acoustic diffraction problems, this approximation is interpreted as the extended Kirchhoff approximation (KA). In electromagnetic diffraction problems, it is known as the physical optics (PO) approach. In the present book we use the term *physical optics* for both electromagnetic and acoustic waves, just as in the work by (Bowman et al. 1987, p. 29).

The PTD is the natural extension of PO and takes into account the additional field generated by the *nonuniform* component, which has a diffraction nature and is caused by any deviation of the scattering surface from an infinite tangent plane. Another definition of the uniform and nonuniform scattering sources is introduced in Section 7.9. Here, the uniform component is defined as the field induced on the half-plane tangential to the illuminated face of the scattering edge (and to the edge itself). The nonuniform component is the difference between the exact field on the tangential wedge and this new uniform component. This type of separation of the surface field allows the formulation of the advanced version of PTD, which is free from the so-called grazing singularity (Section 7.9).

The *localization principle* related to the behavior of high-frequency diffracted fields is used to determine the asymptotic approximations for the nonuniform component. In particular, according to this principle, the nonuniform sources induced in the vicinity of sharp curved edges are asymptotically identical to the nonuniform sources induced on a tangential wedge near the tangency point.

Thus, the wedge diffraction is the basic canonical problem for the investigation of edge waves, and it is studied in detail in this book. Exact and asymptotic expressions for the 2-D edge waves are derived in Chapters 2, 3, and 4. These results are then used in Chapter 5 to construct simple asymptotic expressions for the field diffracted at strips and polygonal cylinders.

Notice that 2-D diffraction problems for acoustically soft (hard) scattering objects are equivalent to the electromagnetic problems where the electric vector $\vec{E}$ (magnetic vector $\vec{H}$) is parallel to the generatrix of perfectly conducting objects. Due to this equivalence, some results obtained in Ufimtsev (1962) for 2-D electromagnetic problems are transferable for acoustic problems, with proper re-definitions of physical quantities. For the same reason, the asymptotics derived in Chapter 5 for acoustic waves are also valid for electromagnetic waves diffracted at perfectly conducting strips and trilateral cylinders.

A new physical interpretation of the classical physical optics is introduced in Chapter 1. The scattered PO field is separated into the *reflected field* and the *shadow radiation*. The first part contains the ordinary reflected rays and beams, and dominates in the geometrical optics region. The shadow radiation is equivalent to the field scattered at a black body (of the same shape and size as the actual scattering object), and dominates in the vicinity of the shadow region (Fig. 1.4, Fig. 14.6). Manifestations of the shadow radiation are the well-known phenomena *Fresnel diffraction* and *forward scattering*.

The Shadow Contour Theorem is established in Section 1.3.5, which states that different objects with identical shadow boundaries on their surfaces generate identical shadow radiation. This theorem significantly facilitates the approximate estimation of scattering at complex objects. It is also shown here that the shadow radiation contains *half* of the total power scattered by perfectly reflecting objects. Thus, the new formulation of the PO field elucidates the scattering physics and explains the nature of the fundamental diffraction law according to which the total scattering cross-section of large (compared to the wavelength) perfectly reflecting objects equals double the transverse area of geometrical optics shadow zone behind the object.

A significant part of this book is devoted to the theory of *elementary edge waves* and to its applications. The elementary edge wave is a wave radiated by surface sources, induced in the vicinity of an infinitesimal element of the edge. High-frequency asymptotic expressions are found for the elementary edge waves and allow one to investigate the diffraction at arbitrary curved edges with large radii of curvature (as compared to the wavelength).

Elementary edge waves can also be interpreted as the *elementary edge-diffracted rays*. The PO field too can be understood as the linear superposition of the other type of *elementary rays*. Because of this, PTD can be considered as a ray theory on the level of elementary rays. Even in the diffraction regions such as geometrical optics boundaries, foci, and caustics, the wave field can be represented in terms of elementary rays. The ordinary reflected and diffracted rays are found in PTD by the asymptotic evaluation of the field integrals and can be interpreted as the beams of elementary rays generated in the vicinity of the stationary points. Such a possible interpretation of PTD goes back to the intuitive Huygens principle, which was rigorously formulated by Helmholtz in terms of elementary spherical waves/rays (Bakker and Copson, 1939).

The general theory of elementary waves is applied in this book to the solution of a variety of diffraction problems. Backscattering and bistatic scattering at bodies of revolution are considered in Chapter 6. Ray and caustic asymptotics are derived in Chapter 8. Slope and multiple diffraction at large objects are investigated in Chapters 9 and 10. The results of these chapters are utilized in Chapters 11 and 12 to analyze the focusing of multiple edge waves on the symmetry axis of the bodies of revolution. The example of the disk diffraction problem (whose exact asymptotic solution is known), establishes that PTD provides correct expressions for the first term in the total asymptotic expansion for each multiple edge-diffracted wave.

Chapters 13 and 14 derive the PTD asymptotics for the field scattered at a finite cylinder under oblique incidence of a plane wave. Together with the numerical results illustrated in the figures, they explain the physical structure of the scattered field. New fine features of the theory are emphasized here. They concern the necessity to calculate the high-order terms in the PO field as well as radiation by the nonuniform component of the scattering sources caused by the smooth bending of the cylindrical surface.

The theory developed in the book can find various applications. Among them are the problems associated with the design of microwave antennas, estimation of scattering cross-sections, identification of scattering objects, propagation of waves in urban environments, and so on. In combination with numerical methods, it can be used for the development of efficient hybrid techniques for the investigation of complex diffraction problems. This book can also be useful for teaching a variety of university courses, including topics on high-frequency asymptotic techniques in diffraction theory. The problems following each chapter are intended for independent investigation and will be helpful in studying PTD, especially for students.

The International System of Units (SI) and the time dependence of $\exp(-i\omega t)$ for the wave fields and sources are used in this book.

ACRONYMS

GO	Geometrical optics
GA	Geometrical acoustics
GTD	Geometrical theory of diffraction
EEW	Elementary edge wave
KA	Kirchhoff approximation
PO	Physical optics
PTD	Physical theory of diffraction

Chapter 1

Basic Notions in Acoustic and Electromagnetic Diffraction Problems

1.1 FORMULATION OF THE DIFFRACTION PROBLEM

This book develops the Physical Theory of Diffraction (PTD) for both acoustic and electromagnetic waves diffracted at perfectly reflecting objects.

> In the case of two-dimensional (2-D) problems, this theory is valid for both *electromagnetic* and *acoustic* waves.

First we present the theoretical fundamentals for acoustic waves and then for electromagnetic waves. In the linear approximation, the velocity potential u of harmonic acoustic waves satisfies the wave equation (Kinsler et al., 1982; Pierce, 1994)

$$\nabla^2 u + k^2 u = I. \tag{1.1}$$

Here $k = 2\pi/\lambda = \omega/c$ is the wave number, λ is the wavelength, ω the angular frequency, c the speed of sound, and I the source strength characteristic. The time dependence is assumed to be in the form $\exp(-i\omega t)$, and is suppressed below. The acoustic pressure p and the velocity v of fluid particles, caused by sound waves, are determined through the velocity potential (Kinsler et al., 1982; Pierce, 1994)

$$p = -\rho \frac{\partial u}{\partial t} = i\omega\rho u, \qquad \vec{v} = \nabla u \tag{1.2}$$

where ρ is the mass density of a fluid. The power flux density of sound waves, which is the analog of the Poynting vector for electromagnetic waves, equals

$$\vec{P} = p\vec{v} = p\nabla u. \tag{1.3}$$

Fundamentals of the Physical Theory of Diffraction. By Pyotr Ya. Ufimtsev
Copyright © 2007 John Wiley & Sons, Inc.

Its value averaged over the period of oscillations $T = 2\pi/\omega$ equals

$$\vec{P}_{av} = \frac{1}{2}\mathrm{Re}(p^*\vec{v}). \tag{1.4}$$

Here and everywhere below, the superscript asterisk is used for complex conjugate quantities.

Two types of boundary conditions are imposed on the surface of perfectly reflecting objects: the Dirichlet condition

$$u = 0 \quad \text{or} \quad p = 0 \quad \text{(soft)} \tag{1.5}$$

for objects with a soft (pressure-release) surface, and the Neumann condition

$$\frac{\partial u}{\partial n} = \hat{n} \cdot \nabla' u = 0 \quad \text{(hard)} \tag{1.6}$$

for objects with a hard (rigid) surface. Here u is the total field that is the *sum of the incident and scattered waves*. The symbol $\hat{n}$ stands for a unit outward vector, which is normal to the scattering surface S (Fig. 1.1). The gradient operator ∇' is applied to coordinates of the integration /source point Q.

To complete the formulation of the diffraction problem, and to ensure the uniqueness of its solution, the above wave equation and the boundary conditions are supplemented by the Sommerfeld radiation condition for the scattered field,

$$\lim r \left(\frac{\partial u}{\partial r} - iku \right) = 0 \quad \text{with } r \to \infty, \tag{1.7}$$

where r is the distance from the scattering object to the observation point.

In the International System (SI) of units, the quantities introduced above have the following dimensions

$$[r] = \mathrm{m}, \quad [t] = \mathrm{sec}, \quad [\vec{v}] = [c] = \frac{\mathrm{m}}{\mathrm{sec}}, \quad [\omega] = \frac{1}{\mathrm{sec}},$$

$$[\rho] = \frac{\mathrm{kg}}{\mathrm{m}^3}, \quad [p] = \frac{\mathrm{kg}}{\mathrm{m} \cdot \mathrm{sec}^2}, \quad [u] = \frac{\mathrm{m}^2}{\mathrm{sec}}, \quad [\vec{P}] = \frac{\mathrm{kg}}{\mathrm{sec}^3}. \tag{1.8}$$

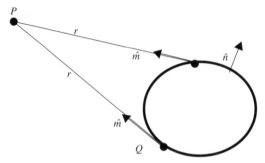

Figure 1.1   Scattering surface S. Here r is the distance between the observation point P (which can be in the far zone) and the integration point Q (on the surface of the scatterer), the unit vector $\hat{m}$ is directed from the point Q to the point P.

The standard denotations are used here: m for meter, kg for kilogram, and sec for second. The pressure unit is called Pascal (Pa) and $1\,Pa = 1$ Newton/$1\,m^2$. The SI unit of the power flux density is $1\,Watt/1\,m^2 = 1\,Joule/(1\ sec \cdot 1\,m^2)$.

In the scattering problems, which admit the electromagnetic interpretation, the quantity u plays the role of the electric field intensity ($[E] = $ Volt/m) or magnetic field intensity ($[H] = $ Ampere/m), depending on the polarization of electromagnetic waves. Their power flux density is called the *Poynting* vector and is defined as

$$\vec{P} = \vec{E} \times \vec{H} \qquad \text{and} \qquad \vec{P}_{av} = \frac{1}{2}\text{Re}[\vec{E} \times \vec{H}^*]. \tag{1.9}$$

1.2 SCATTERED FIELD IN THE FAR ZONE

The scattered field is determined by the Helmholtz integral expressions (Bakker and Copson, 1939):

$$u_s = -\frac{1}{4\pi}\int_S \frac{\partial u}{\partial n}\frac{e^{ikr}}{r}\,ds, \qquad u_h = \frac{1}{4\pi}\int_S u\frac{\partial}{\partial n}\frac{e^{ikr}}{r}\,ds, \tag{1.10}$$

where the integrals are taken over the scattering surface S. The function u_s describes the field scattered by an acoustically soft object, and the function u_h relates to the field scattered by an acoustically hard object. The field quantities u and $\partial u/\partial n$ in the integrands belong to the total field on the object surface, that is, to the sum of the incident and scattered fields. These quantities represent the surface sources of the scattered field induced by the incident wave. We denote them by symbols

$$j_s = \frac{\partial u}{\partial n}, \qquad j_h = u, \tag{1.11}$$

similar to those used for induced sources/currents in the electromagnetic version of PTD (Ufimtsev, 2003). The quantity e^{ikr}/r in Equation (1.10) represents the Green function of a homogeneous medium, that is, the fundamental solution of the wave equation, and $\hat{n}$ is a unit outward vector normal to the surface S.

In the far field, where $r \gg kd^2$ (d is the characteristic linear dimension of the object), the field expressions (1.10) can be simplified. We choose the origin of the coordinate system somewhere inside the object, as shown in Figure 1.2. Under the conditions $R \gg r'$, $R \gg kd^2$, we have

$$r \approx R - r'\cos\Omega, \qquad \frac{e^{ikr}}{r} \approx \frac{e^{ikR}}{R}e^{-ikr'\cos\Omega} \tag{1.12}$$

with

$$\cos\Omega = \cos\vartheta\cos\vartheta' + \sin\vartheta\sin\vartheta'\cos(\varphi - \varphi'). \tag{1.13}$$

In addition,

$$\frac{\partial}{\partial n}\frac{e^{ikr}}{r} = \nabla'\frac{e^{ikr}}{r}\cdot\hat{n} = -\left(ik - \frac{1}{r}\right)\frac{e^{ikr}}{r}\nabla r\cdot\hat{n}, \qquad \text{with } \nabla r = \hat{m}, \tag{1.14}$$

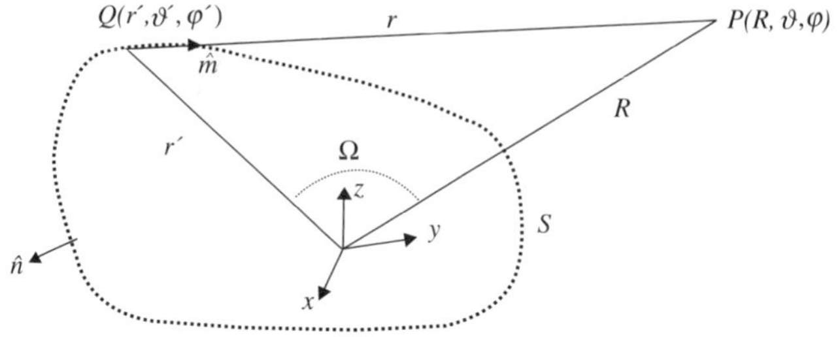

Figure 1.2 S is the surface of the scattering object; Q is the integration point (with the spherical coordinates r', ϑ', φ') on the surface S; P is the observation point (with the coordinates R, ϑ, φ) in the far zone; Ω is the angle between the directions from the origin to the integration and observation points.

or in view of Equations (1.12)

$$\frac{\partial}{\partial n} \frac{e^{ikr}}{r} \approx -ik \frac{e^{ikR}}{R} e^{-ikr' \cos \Omega} \cdot (\hat{m} \cdot \hat{n}). \tag{1.15}$$

Finally, we obtain the following approximations for the field in the far-away point P:

$$u_s = -\frac{1}{4\pi} \frac{e^{ikR}}{R} \int_S j_s e^{-ikr' \cos \Omega} \, ds, \tag{1.16}$$

$$u_h = -\frac{ik}{4\pi} \frac{e^{ikR}}{R} \int_S j_h e^{-ikr' \cos \Omega} (\hat{m} \cdot \hat{n}) ds \tag{1.17}$$

where $\hat{m}$ and $\hat{n}$ are unit vectors. In this book, we develop asymptotic approximations first for the surface sources $j_{s,h}$ and then for the scattered field (1.16), (1.17).

Expressions (1.16) and (1.17) can be written in the generic form

$$u_{s,h} = u_0 \Phi_{s,h} \frac{e^{ikR}}{R} \tag{1.18}$$

where the functions

$$\Phi_s = -\frac{1}{4\pi u_0} \int_S j_s e^{-ikr' \cos \Omega} \, ds, \qquad \Phi_h = -\frac{ik}{4\pi u_0} \int_S j_h e^{-ikr' \cos \Omega} (\hat{m} \cdot \hat{n}) ds$$

$$\tag{1.19}$$

represent the directivity patterns of the scattered field, and u_0 is the complex amplitude of the incident wave at the origin of the coordinates ($R = 0$). Notice that in the vicinity of the scattering object located in the far zone from the source Q, the incident wave can be approximated by the equivalent plane wave with the amplitude u_0.

According to Equations (1.2), (1.3) and (1.4), the power flux density of the scattered field is determined by

$$\vec{P}^{sc} = i\omega\rho u \nabla u. \tag{1.20}$$

In the far field,

$$\nabla u \approx iku \cdot \hat{R}, \qquad \text{with } \hat{R} = \nabla R. \tag{1.21}$$

Therefore, the power flux density averaged over the oscillation period $T = 2\pi/\omega$ equals

$$\vec{P}^{sc}_{av} = \frac{1}{2}\text{Re}[p^*\vec{v}] = \frac{1}{2}\text{Re}[(i\omega\rho u)^*(iku)] \cdot \hat{R} = \frac{1}{2}k^2 Z |u|^2 \cdot \hat{R}, \tag{1.22}$$

where

$$Z = \rho c \tag{1.23}$$

is the characteristic impedance of the medium.

Usually, the far field is characterized by the *bistatic cross-section* σ introduced through the relation

$$P^{sc}_{av} = \frac{\sigma \cdot P^{inc}_{av}}{4\pi R^2}, \tag{1.24}$$

where

$$P^{inc}_{av} = \frac{1}{2}k^2 Z |u^{inc}|^2 \tag{1.25}$$

is the power flux density of the incident wave. This definition suggests this interpretation given in the following paragraphs.

The bistatic cross-section is the area σ of a hypothetical plate perpendicular to the direction of the incident wave. This plate intercepts the incident power $P^{inc}_{av} \cdot \sigma$ and distributes it uniformly into the whole surrounding space *with the power flux density that is equal to the actual one scattered by the object in the direction of observation*. Because the scattered power depends on the direction of scattering, the scattering cross-section σ is a function of this direction. The term bistatic means that the direction of scattering can be arbitrary. In the particular case when the scattering direction coincides with the direction to the source of the incident wave, the quantity σ is called the *backscattering* cross-section or *monostatic* cross-section. Thus, according to Equations (1.22) and (1.24)

$$\sigma = 4\pi R^2 \frac{P^{sc}_{av}}{P^{inc}_{av}} = 4\pi R^2 \frac{|u^{sc}|^2}{|u^{inc}|^2}. \tag{1.26}$$

In the directions where the field scattered from a *smooth convex* surface has a ray structure, the bistatic cross-section is predicted by Geometrical Optics (Geometrical

Acoustics) and equals

$$\sigma = \pi \rho_1 \rho_2. \tag{1.27}$$

Here ρ_1 and ρ_2 are principal radii of curvature of the scattering surface at the *reflection point*. It is also assumed that this surface is perfectly reflecting (soft or hard). Two interesting features of this quantity should be emphasized.

First, the expression (1.27) is universal. It is applicable both for acoustic and electromagnetic waves. The reason for this is that the ray structure does not depend on the nature of the waves, and it is totally determined by the geometry of the scattering surface. If the geometry is the same, the divergence of reflected rays will be the same for both acoustic and electromagnetic rays. Also, the modulus of reflection coefficient for any perfectly reflecting surfaces (soft or hard for acoustic waves, or perfectly conducting for electromagnetic waves) equals unity. However, just these two factors, the ray divergence and the reflection coefficient, totally determine the amplitude of reflected rays, and eventually the bistatic cross-section.

Equation (1.27) can be generalized for imperfect reflecting surfaces:

$$\sigma = |\mathcal{R}|^2 \pi \rho_1 \rho_2, \tag{1.28}$$

where $\mathcal{R}$ is the reflection coefficient, which can be different for acoustic and electromagnetic waves.

The second interesting and not obvious feature of Equation (1.27) is the following. This expression does not depend on the angle between the incident and reflected rays at the same reflection point (Fig. 1.3). In other words, it is constant for any bistatic angles, including zero angle related to backscattering. This property of scattering from perfectly reflecting objects follows from the theory of Fock (1965) as it was shown in Ufimtsev (1999).

The theory of Fock (1965) is more general. It is also valid for imperfect scattering surfaces characterized by the reflection coefficient $\mathcal{R}$. In this case, Fock's theory leads straight to Equation (1.28), where $\mathcal{R}$ depends on the bistatic angle as well as on the boundary conditions.

Figure 1.3   Scattering from the same reflection point (at the same reflecting object) for different bistatic angles. Bistatic cross-section σ of this perfectly reflecting object is constant for all of these angles and equals the monostatic cross-section.

1.3 PHYSICAL OPTICS

This high-frequency approach is widely used in acoustic and electromagnetic diffraction problems.

1.3.1 Definition of the Physical Optics

Physical Optics (PO) was suggested by Macdonald (1912), and since then it has been successfully applied in the theory of diffraction. In particular it is often used in the analysis of electromagnetic waves scattered from large metallic objects. Basic features of this approach in the study of electromagnetic diffraction are exposed in the article by Ufimtsev (1999). The scalar version of PO is applicable for acoustic waves and it is known in acoustics as the extended Kirchhoff approximation (Brill and Gaunaurd, 1993; Menounou et al., 2000; Moser et al., 1993). Physical Optics is a constituent part of the Physical Theory of Diffraction developed in the present book. According to this approximation, the field induced on the surface of the object is determined by Geometrical Optics (Geometrical Acoustics).

The physics behind this is as follows. Geometrical Optics describes a wave field in the limiting case when a wavelength tends to zero. With respect to such a small wavelength, the scattering surface at the reflection point can be considered approximately as a tangential plane. Therefore, the surface field induced at the tangential infinite plane is a good high-frequency approximation for true scattering sources induced on a large scattering object. Two such planes P_1 and P_2 tangent at the points Q_1 and Q_2 are shown in Figure 1.4. These points are located at the "illuminated" side of the object. Notice that according to Geometrical Optics, the field equals zero in the shadow region, including the points on the object surface.

Thus, the reflection from a tangential plane is an appropriate canonical ("fundamental") problem. Its exact solution can be easily found using Geometrical Optics, as well as by image theory. The total field generated by an external source above an infinite reflecting plane (Fig. 1.5) is the sum of the incident field and the reflected field, which can be interpreted as the field created by the image source. On the acoustically soft plane, the total field is zero as a result of the boundary condition (1.5), but its normal derivative equals

$$\frac{\partial u_s}{\partial n} = 2\frac{\partial u_s^{inc}}{\partial n} \qquad (1.29)$$

due to the exact solution of this problem. On the acoustically hard plane, the normal derivative of the total field is zero as a result of the boundary condition (1.6), and the field itself equals

$$u_h = 2u_h^{inc}, \qquad (1.30)$$

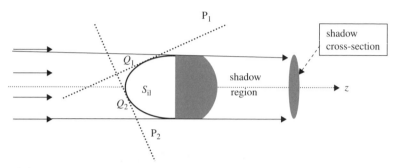

Figure 1.4   Surface fields induced on the scattering objects at the points Q_1 and Q_2 are asymptotically identical to the fields induced at the tangential planes P_1 and P_2, respectively. S_{il} is the illuminated part of the object surface. A dark plate behind the object displays the cross-section of the geometrical shadow region.

as follows from the solution of this reflection problem. In the general PTD, these quantities are interpreted as the *uniform components* of induced sources on a smooth convex scattering surface:

$$j_s^{(0)} = 2\frac{\partial u^{inc}}{\partial n}, \qquad j_h^{(0)} = 2u^{inc}. \tag{1.31}$$

These expressions define the induced sources only on the "illuminated" part of the scattering object. On the shadowed part, these components are set to zero.

By substituting Equation (1.31) into Equation (1.10) one obtains expressions for the scattered field at any distance from the scatterer (Fig. 1.4):

$$u_s^{PO} \equiv u_s^{(0)} = -\frac{1}{4\pi}\int_{S_{il}} j_s^{(0)}\frac{e^{ikr}}{r}\,ds,$$

$$u_h^{PO} \equiv u_h^{(0)} = \frac{1}{4\pi}\int_{S_{il}} j_h^{(0)}\frac{\partial}{\partial n}\frac{e^{ikr}}{r}\,ds. \tag{1.32}$$

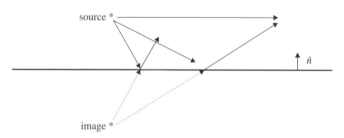

Figure 1.5   Reflection from an infinite plane.

These expressions represent the *scalar Physical Optics* approximation, also known in Acoustics as the *extended Kirchhoff Approximation* (KA). In the present book we use the term Physical Optics for both acoustic and electromagnetic waves. The symbols $u_{\mathrm{s}}^{(0)}$ and $u_{\mathrm{h}}^{(0)}$ are introduced in Equations (1.32) to emphasize that these fields are generated by the *uniform* component of the induced surface sources. Thus, Physical Optics, which deals with these uniform components, is a constituent part of the general PTD.

The Physical Optics of Equations (1.32) possesses a special property related to the field scattered in the direction to the source of the incident wave. According to Equations (1.16) and (1.17) the PO far field is determined as

$$u_{\mathrm{s}}^{(0)} = -\frac{1}{2\pi}\frac{e^{ikR}}{R}\int_{S_{\mathrm{il}}}\frac{\partial u^{\mathrm{inc}}}{\partial n}e^{-ikr'\cos\Omega}\,ds \tag{1.33}$$

and

$$u_{\mathrm{h}}^{(0)} = -\frac{ik}{2\pi}\frac{e^{ikR}}{R}\int_{S_{\mathrm{il}}}u^{\mathrm{inc}}e^{-ikr'\cos\Omega}(\hat{m}\cdot\hat{n})\,ds. \tag{1.34}$$

The field incident on the scattering object (being at the large distance from the source) can be represented in the form

$$u^{\mathrm{inc}} = \mathrm{const}\,e^{ik\phi^i}. \tag{1.35}$$

The unit vector $\nabla\phi^i = \hat{k}^i$ indicates the direction of the incident wave, and the unit vector $\hat{m} = \nabla r = \hat{k}^{\mathrm{s}}$ shows the direction of scattering. In the case of backscattering, the equality $\hat{k}^{\mathrm{s}} = -\hat{k}^i$ is valid. Note also that

$$\frac{\partial u^{\mathrm{inc}}}{\partial n} = \nabla u^{\mathrm{inc}}\cdot\hat{n} = iku^{\mathrm{inc}}(\nabla\phi^i\cdot\hat{n}) = iku^{\mathrm{inc}}(\hat{k}^i\cdot\hat{n}). \tag{1.36}$$

The substitution of Equation (1.36) and the quantity $(\hat{m}\cdot\hat{n}) = -(\hat{k}^i\cdot\hat{n})$ into Equations (1.33) and (1.38) leads to the equation

$$u_{\mathrm{s}}^{(0)} = -u_{\mathrm{h}}^{(0)} = -\frac{ik}{2\pi}\frac{e^{ikR}}{R}\int_{S_{\mathrm{il}}}u^{\mathrm{inc}}e^{-ikr'\cos\Omega}(\hat{k}^i\cdot\hat{n})\,ds. \tag{1.37}$$

Hence, in the frame of the Physical Optics approximation, the *backscattered fields* created by the *soft* and *hard* objects (of the same shape and size) have *equal magnitudes* and differ only in *sign*.

1.3.2 Total Scattering Cross-Section

The power flux density of the scattered waves is defined by Equation (1.22). By the integration of this quantity over the object surface, one can find the total power scattered in all directions. In the PO approximation, the total power scattered from an acoustically soft object equals

$$P^{\text{tot}} = \frac{1}{2}\text{Re}\int_{S_{\text{il}}} (p_s^{\text{sc}})^* (\vec{v}_s^{\text{tot}} \cdot \hat{n}) ds, \tag{1.38}$$

where

$$\vec{v}_s^{\text{tot}} \cdot \hat{n} = 2\frac{\partial u^{\text{inc}}}{\partial n}, \tag{1.39}$$

and in accordance with the boundary condition (1.5), $p_s^{\text{sc}} = -p^{\text{inc}} = -i\omega\rho u^{\text{inc}}$. The incident wave in the vicinity of the scattering object can be approximated by the plane wave (Fig. 1.4)

$$u^{\text{inc}} = u_0 e^{ikz}. \tag{1.40}$$

Then

$$\vec{v}_s^{\text{tot}} \cdot \hat{n} = 2\frac{\partial u^{\text{inc}}}{\partial n} = 2iku_0 e^{ikz}(\hat{z} \cdot \hat{n}), \qquad (p_s^{\text{sc}})^* = i\omega\rho u_0^* e^{-ikz} \tag{1.41}$$

and

$$P^{\text{tot}} = -k\omega\rho|u_0|^2 \int_{S_{\text{il}}} (\hat{z} \cdot \hat{n}) ds = k^2 Z A |u_0|^2, \tag{1.42}$$

where A is the area of the object's projection on the plane perpendicular to the direction of propagation or, in other words, the area of the shadow cross-section (Fig. 1.4). In view of Equation (1.25), the power flux density of the incident wave equals

$$P_{\text{av}}^{\text{inc}} = \frac{1}{2}k^2 Z |u_0|^2. \tag{1.43}$$

The total cross-section is defined by the ratio

$$\sigma^{\text{tot}} = P^{\text{tot}}/P_{\text{av}}^{\text{inc}} \tag{1.44}$$

and equals

$$\sigma^{\text{tot}} = 2A. \tag{1.45}$$

This result is also valid for hard objects and for perfectly conducting objects, which scatter electromagnetic waves. It can be easily verified for hard objects. Indeed in this case,

$$P^{\text{tot}} = \frac{1}{2}\text{Re}\int_{S_{\text{il}}} (p_{\text{h}}^{\text{tot}})^{*}(\vec{v}_{\text{h}}^{\text{sc}} \cdot \hat{n})ds \tag{1.46}$$

and

$$p_{\text{h}}^{\text{tot}} = 2p^{\text{inc}} = 2i\,\omega\rho u_0 e^{ikz}, \qquad (\vec{v}_{\text{h}}^{\text{sc}} \cdot \hat{n}) = -(\vec{v}^{\text{inc}} \cdot \hat{n}) = -iku_0 e^{ikz}(\hat{z} \cdot \hat{n}). \tag{1.47}$$

The substitution of Equation (1.47) into (1.46) leads to Equations (1.42) and to (1.45).

1.3.3 Optical Theorem

There is a specific connection between the total scattering cross-section and the far scattered field in the shadow/forward direction. In the PO approximation, the far-field expressions (1.18) and (1.19) take the form

$$u_{\text{s}}^{\text{PO}} = u_0 \Phi_{\text{s}}^{\text{PO}} \frac{e^{ikR}}{R}, \qquad u_{\text{h}}^{\text{PO}} = u_0 \Phi_{\text{h}}^{\text{PO}} \frac{e^{ikR}}{R}, \tag{1.48}$$

where

$$\Phi_{\text{s}}^{\text{PO}} = -\frac{1}{2\pi u_0}\int_{S_{\text{il}}} \frac{\partial u^{\text{inc}}}{\partial n} e^{-ikr'\cos\Omega}\,ds, \tag{1.49}$$

$$\Phi_{\text{h}}^{\text{PO}} = -\frac{ik}{2\pi u_0}\int_{S_{\text{il}}} u^{\text{inc}} e^{-ikr'\cos\Omega}(\hat{m} \cdot \hat{n})ds \tag{1.50}$$

with $\cos\Omega$ defined in Equation (1.13). The incident wave (1.40) propagates in the z-direction. For the observation point in the forward direction, we have $\vartheta = 0$, $\cos\Omega = \cos\vartheta'$, $\hat{m} = \hat{z}$, $r'\cos\vartheta' = z'$, and

$$\frac{\partial u^{\text{inc}}}{\partial n} = \nabla u^{\text{inc}} \cdot \hat{n} = iku_0 e^{ikz'}(\hat{z} \cdot \hat{n}) \tag{1.51}$$

and also

$$\Phi_{\text{s}}^{\text{PO}}(\vartheta = 0) = \Phi_{\text{h}}^{\text{PO}}(\vartheta = 0) = -\frac{ik}{2\pi}\int_{S_{\text{il}}} (\hat{z} \cdot \hat{n})ds = \frac{ik}{2\pi}A, \tag{1.52}$$

where A is the area of the shadow cross-section (Fig. 1.4). Comparison of Equation (1.52) with Equation (1.45) shows that

$$\sigma^{\text{tot}} = \frac{4\pi}{k}\text{Im}\Phi \qquad (\vartheta = 0). \tag{1.53}$$

This equation is well known as the Optical Theorem (Born and Wolf, 1980).

1.3.4 Introducing the Notion of "Shadow Radiation"

This notion was introduced for electromagnetic waves by Ufimtsev (1968). It was additionally investigated in Ufimtsev (1990) and discussed in Ufimtsev (1996). A significant part of the results relating to electromagnetic shadow radiation was included in *Theory of Edge Diffraction in Electromagnetics* (Ufimtsev, 2003). In the present section, the notion of shadow radiation is introduced in conjunction with scalar waves. Consider again the reflection from an *infinite* perfectly reflecting plane (Fig. 1.6) located in an homogeneous medium. The scattered field in the region $z > 0$ is determined by the Helmholtz integral expression (Bakker and Copson, 1939)

$$u^{sc} = \frac{1}{4\pi} \int_S \left(u^{tot} \frac{\partial}{\partial n} \frac{e^{ikr}}{r} - \frac{\partial u^{tot}}{\partial n} \frac{e^{ikr}}{r} \right) ds, \tag{1.54}$$

where $u^{tot} = u^{inc} + u^{sc}$ is the total field, $ds = dxdy$ is a differential area of the infinite plane S ($z = 0$), and r is the distance between the integration and observation points.

Let the reflecting plane be acoustically soft. Then, on its surface,

$$u_S^{sc} = -u^{inc},$$

but

$$\frac{\partial u_S^{tot}}{\partial n} = 2\frac{\partial u^{inc}}{\partial n} = \frac{\partial u^{inc}}{\partial n} + \frac{\partial u^{inc}}{\partial n}. \tag{1.55}$$

Therefore, Equation (1.54) can be rewritten as

$$u_S^{tot,sc} = u_{S,1}^{sc} + u_{S,2}^{sc}, \tag{1.56}$$

where

$$u_{S,1}^{sc} = \frac{1}{4\pi} \int_S \left(u^{inc} \frac{\partial}{\partial n} \frac{e^{ikr}}{r} - \frac{\partial u^{inc}}{\partial n} \frac{e^{ikr}}{r} \right) ds, \tag{1.57}$$

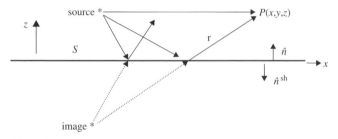

Figure 1.6 Reflection of waves from an infinite plane S in an homogeneous medium. The source is in the region $z > 0$. The total field in the shadow region ($z < 0$) equals zero.

and

$$u_{s,2}^{sc} = \frac{1}{4\pi} \int_{S} \left(-u^{inc} \frac{\partial}{\partial n} \frac{e^{ikr}}{r} - \frac{\partial u^{inc}}{\partial n} \frac{e^{ikr}}{r} \right) ds. \tag{1.58}$$

To evaluate the integral in Equation (1.57), we utilize the Helmholtz equivalency theorem (Bakker and Copson, 1939). According to this theorem, the field u created by the acoustic source at the point P in an homogeneous medium (Fig. 1.7) can be represented as the radiation generated by the equivalent sources u^{inc} and $\partial u^{inc}/\partial N$ distributed over the closed imaginary surface Σ of the volume V:

$$u(P) = \frac{1}{4\pi} \oint_{\Sigma} \left(u^{inc} \frac{\partial}{\partial N} \frac{e^{ikr}}{r} - \frac{\partial u^{inc}}{\partial N} \frac{e^{ikr}}{r} \right) ds = \begin{cases} 0, & \text{when } P \text{ inside } V \\ u^{inc}(P), & \text{when } P \text{ outside } V. \end{cases} \tag{1.59}$$

Here it is supposed that the source of the incident wave is located inside the volume V. One should emphasize the following wonderful property of this theorem. The field at the point P inside V or outside V does not depend on the shape of its surface Σ. One can deform this surface in any way, but the result of the integration in Equation (1.59) will be the same: $u(P) = 0$ if $P \in V$, and $u(P) = u^{inc}(P)$ if $P \notin V$.

In order to evaluate integral (1.57), we first apply the equivalency theorem (1.59) to the closed surface $\Sigma = S_R + H_R$ (Fig. 1.8). Here S_R is a circular plate with a radius R, which is a part of the infinite plane S shown in Figure 1.6, and H_R is a hemisphere with the same radius R. It is supposed that the source of the incident wave is inside the volume V.

According to the equivalency theorem (1.59),

$$\frac{1}{4\pi} \oint_{S_R+H_R} \left(u^{inc} \frac{\partial}{\partial N} \frac{e^{ikr}}{r} - \frac{\partial u^{inc}}{\partial N} \frac{e^{ikr}}{r} \right) ds = \begin{cases} 0, & \text{when } P \text{ inside } V \\ u^{inc}(P), & \text{when } P \text{ outside } V \end{cases} \tag{1.60}$$

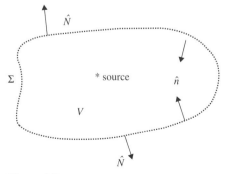

Figure 1.7 Illustration of the equivalency principle. Σ is an arbitrary imaginary surface covering a volume V of a free homogeneous medium, $\hat{n}$ and $\hat{N}$ are respectively the inward and outward unit vectors normal to Σ, and a source of the incident wave is inside V.

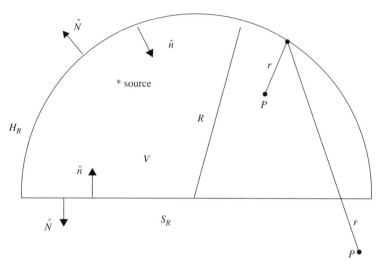

Figure 1.8   Surface of integration $\Sigma = S_R + H_R$ in Equation (1.60). A source of the incident wave is inside the volume V.

or after replacement of $\hat{N}$ by $(-\hat{n})$,

$$\frac{1}{4\pi} \oint_{S_R+H_R} \left(u^{\text{inc}} \frac{\partial}{\partial n} \frac{e^{ikr}}{r} - \frac{\partial u^{\text{inc}}}{\partial n} \frac{e^{ikr}}{r} \right) ds = \begin{cases} 0, & \text{when } P \text{ inside } V \\ -u^{\text{inc}}(P), & \text{when } P \text{ outside } V. \end{cases}$$

(1.61)

One can show that the field at the observation point P generated by the equivalent sources, distributed over H_R, vanishes when the radius R of H_R tends to infinity. Note also that with $R \to \infty$, the surface S_R is transformed into the infinite plane S. Taking into account these observations, we finally obtain the following values for the function (1.57):

$$u_{\text{s},1}^{\text{sc}} = \begin{cases} 0, & \text{in the region } z > 0 \\ -u^{\text{inc}}, & \text{in the region } z < 0. \end{cases}$$

(1.62)

The physical meaning of the field $u_{\text{s},1}^{\text{sc}}$ is clear. It cancels the incident wave in the region $z < 0$, creating the complete shadow there. That is why we call this field the *shadow radiation* and denote it by u^{sh}:

$$u^{\text{sh}} = \frac{1}{4\pi} \int_S \left(u^{\text{inc}} \frac{\partial}{\partial n} \frac{e^{ikr}}{r} - \frac{\partial u^{\text{inc}}}{\partial n} \frac{e^{ikr}}{r} \right) ds.$$

(1.63)

With respect to this equation (!), the surface S can be interpreted as *perfectly absorbing* (i.e., *black*), as it does not reflect the incident wave (Ufimtsev, 1968, 1990, 1996, 2003).

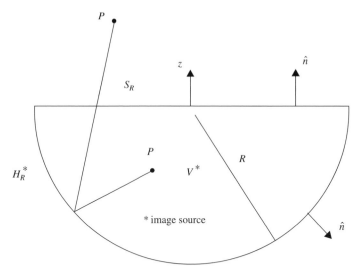

Figure 1.9   Illustration of the equivalency theorem applied to the reflected field. Here, as in Figure 1.8, S_R is a circular plate on the plane $z = 0$ with radius R, and H_R^* s a hemisphere with the same radius R.

To clarify the physical meaning of function $u_{S,2}^{sc}$, we introduce new denotations into Equation (1.58),

$$u_S^{refl} = -u^{inc}, \qquad \frac{\partial u_S^{refl}}{\partial n} = \frac{\partial u^{inc}}{\partial n}, \tag{1.64}$$

and apply the equivalency theorem to the surface $\Sigma = S_R + H_R^*$ shown in Figure 1.9. It is supposed that the image source, that is, the source of the reflected field, is inside the volume V^*.

According to the equivalency theorem,

$$u(P) = \frac{1}{4\pi} \oint_{S_R + H_R^*} \left(u^{refl} \frac{\partial}{\partial n} \frac{e^{ikr}}{r} - \frac{\partial u^{refl}}{\partial n} \frac{e^{ikr}}{r} \right) ds$$

$$= \begin{cases} 0, & \text{when } P \text{ inside } V^* \\ u^{refl}(P), & \text{when } P \text{ outside } V^*. \end{cases} \tag{1.65}$$

When $R \to \infty$, the integral over H_R^* tends to zero, S_R is transformed into the infinite plane S, and therefore

$$u_{S,2}^{sc} = u_S^{refl} = \frac{1}{4\pi} \int_S \left(u_S^{refl} \frac{\partial}{\partial n} \frac{e^{ikr}}{r} - \frac{\partial u_S^{refl}}{\partial n} \frac{e^{ikr}}{r} \right) ds$$

$$= \begin{cases} u_S^{refl}, & \text{in the region } z > 0 \\ 0, & \text{in the region } z < 0. \end{cases} \tag{1.66}$$

Thus, the function (1.58) represents the reflected field in the region $z > 0$.

The field (1.54) scattered by the hard infinite plane S also can be represented as the sum of the reflected field and the shadow radiation:

$$u_h^{sc} = u_h^{refl} + u^{sh}, \tag{1.67}$$

where

$$u_h^{refl} = \frac{1}{4\pi} \int_S \left(u_h^{refl} \frac{\partial}{\partial n} \frac{e^{ikr}}{r} - \frac{\partial u_h^{refl}}{\partial n} \frac{e^{ikr}}{r} \right) ds \tag{1.68}$$

and on the plane S

$$u_h^{refl} = u^{inc}, \qquad \frac{\partial u_h^{refl}}{\partial n} = -\frac{\partial u^{inc}}{\partial n}. \tag{1.69}$$

The shadow radiation does not depend on the boundary conditions and is the same both for the soft and hard planes. It is defined by Equation (1.63).

The above definitions of the reflected and shadow radiations are applicable to the PO field scattered by arbitrary soft and hard objects. In this case, however, the integration surface in (1.63), (1.66), and in (1.68) must be specified as the illuminated side (S_{il}) of the object (Fig. 1.4). Thus, in general,

$$u_{s,h}^{PO} = u_{s,h}^{refl} + u^{sh} \tag{1.70}$$

where

$$u_{s,h}^{refl} = \frac{1}{4\pi} \int_{S_{il}} \left(u_{s,h}^{refl} \frac{\partial}{\partial n} \frac{e^{ikr}}{r} - \frac{\partial u_{s,h}^{refl}}{\partial n} \frac{e^{ikr}}{r} \right) ds, \tag{1.71}$$

$$u^{sh} = \frac{1}{4\pi} \int_{S_{il}} \left(u^{inc} \frac{\partial}{\partial n} \frac{e^{ikr}}{r} - \frac{\partial u^{inc}}{\partial n} \frac{e^{ikr}}{r} \right) ds, \tag{1.72}$$

and $u_{s,h}^{refl}, \partial u_{s,h}^{refl}/\partial n$ are defined by Equations (1.64) and (1.69). Equation (1.72) can be interpreted as the generalization of the Kirchhoff definition for the *black bodies* suggested earlier by Ufimtsev (1968, 2003). It is clear that the PO formulation in the form of Equation (1.70) is valid for electromagnetic waves as well.

One should also notice another interesting relationship between the PO field and the shadow radiation. According to the definitions (1.31), (1.32) and (1.72) for these quantities, the shadow radiation can be represented in the form

$$u^{\text{sh}} = \frac{1}{2}(u_{\text{s}}^{\text{PO}} + u_{\text{h}}^{\text{PO}}). \tag{1.73}$$

If we now take into account Equation (1.37) and substitute it into (1.73), we immediately come to the fundamental conclusion that the shadow radiation exactly equals zero in the direction to the source of the incident wave. In other words, the *black bodies* (as they are defined above) do not generate the backscattering.

In contrast to the infinite plane problem, where the reflected field and shadow radiations exist in the separated half-spaces, the fields (1.71) and (1.72) caused by diffraction at the finite size objects exist in the whole surrounding space. However, their spatial distributions are different. The reflected field dominates in the ray region, and the shadow radiation concentrates at the shadow region and in its vicinity (Ufimtsev, 1968, 1990, 1996, 2003). Well-known manifestations of the shadow radiation are the phenomena Fresnel diffraction and forward scattering (Glaser, 1985; Ufimtsev, 1996; Willis, 1991).

Indeed, the diffraction bands bordering the geometrical optics shadow of opaque objects observed in Fresnel diffraction are nothing but the result of interference of an incident wave with shadow radiation. It is interesting that the history of diffraction as a science started in the seventeenth century with investigation of just this phenomenon (Grimaldi, Newton, Young, Fresnel). The forward scattering is the enhancement of the scattered field in the directions approaching the shadow boundary behind the object. It was extensively investigated both experimentally (Willis, 1991) and theoretically [Bowman et al. (1987)]. The numerical data for the field scattered by acoustically soft and hard objects, as well as perfectly conducting objects, presented in the work of Bowman et al. (1987), clearly illustrate the existence of this phenomenon. Our present analysis of the PO approximation reveals the nature of this phenomenon, which is inherent for scattering at any large opaque objects.

One should note that, due to transverse diffusion, the shadow radiation can penetrate far from the shadow region (Ufimtsev, 1990); see also Section 14.2 in the present book. Also, it gives origin to the edge waves, creeping waves, and surface diffracted rays (Ufimtsev, 1996, 2003).

1.3.5 Shadow Contour Theorem and the Total Scattering Cross-Section

Among the properties of shadow radiation, the most significant are the Shadow Contour Theorem and the Total Power of Shadow Radiation. They have already been established for electromagnetic waves (Ufimtsev, 1968, 1990, 1996, 2003) and will now be verified for acoustic waves.

Let us compare the shadow radiation generated by two scattering objects with different shapes, but with the same shadow contour (Fig. 1.10). Their illuminated sides are S_1 and S_2.

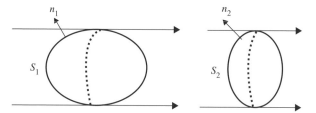

Figure 1.10   Two different objects with the same shadow contour (dotted line).

According to Equation (1.72),

$$u_1^{\text{sh}} = \frac{1}{4\pi} \int_{S_1} \left(u^{\text{inc}} \frac{\partial}{\partial n_1} \frac{e^{ikr}}{r} - \frac{\partial u^{\text{inc}}}{\partial n_1} \frac{e^{ikr}}{r} \right) ds$$

and

$$u_2^{\text{sh}} = \frac{1}{4\pi} \int_{S_2} \left(u^{\text{inc}} \frac{\partial}{\partial n_2} \frac{e^{ikr}}{r} - \frac{\partial u^{\text{inc}}}{\partial n_2} \frac{e^{ikr}}{r} \right) ds. \qquad (1.74)$$

The difference of these quantities can be written as

$$u_1^{\text{sh}} - u_2^{\text{sh}} = \frac{1}{4\pi} \int_{S_1+S_2} \left(u^{\text{inc}} \frac{\partial}{\partial n} \frac{e^{ikr}}{r} - \frac{\partial u^{\text{inc}}}{\partial n} \frac{e^{ikr}}{r} \right) ds, \qquad (1.75)$$

where $\hat{n} = \hat{n}_1$, $\hat{n} = -\hat{n}_2$ is the external normal to the surface $S_1 + S_2$ (Fig. 1.11). As a result of the Helmholtz equivalence principle (Bakker and Copson, 1939), the quantity (1.75) equals zero for observation points outside the volume enclosed by surface $S_1 + S_2$. Therefore

$$u_1^{\text{sh}} = u_2^{\text{sh}}. \qquad (1.76)$$

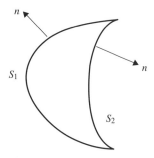

Figure 1.11   Surface $S_1 + S_2$ in an homogeneous medium. All sources and the observation points are outside the volume enclosed by this surface.

This equation represents the Shadow Contour Theorem:

> *The Shadow radiation does not depend on the whole shape of a scattering object and it is completely determined only by the size and geometry of the shadow contour.*

Now let us evaluate the total power of the reflected field and shadow radiation. The total power of the reflected field can be written as

$$P_{s,h}^{refl} = \frac{1}{2} \text{Re} \int_{S_{il}} (p_{s,h}^{refl})^* (\vec{v}_{s,h}^{refl} \cdot \hat{n}) ds \tag{1.77}$$

where, according to Equations (1.2), (1.41), (1.64), and (1.69),

$$
p_s^{refl} = -p^{inc} = -i\omega\rho u_0 e^{ikz}, \qquad \vec{v}_s^{refl} \cdot \hat{n} = \vec{v}^{inc} \cdot \hat{n} = iku_0 e^{ikz}(\hat{z} \cdot \hat{n}),
$$
$$
p_h^{refl} = p^{inc} = i\omega\rho u_0 e^{ikz}, \qquad \vec{v}_h^{refl} \cdot \hat{n} = -\vec{v}^{inc} \cdot \hat{n} = -iku_0 e^{ikz}(\hat{z} \cdot \hat{n}).
\tag{1.78}
$$

Substitution of these quantities into Equation (1.77) results in

$$P_s^{refl} = P_h^{refl} = \frac{1}{2} k^2 ZA |u_0|^2 = P^{inc} \cdot A \tag{1.79}$$

and

$$\sigma^{refl,tot} = A \tag{1.80}$$

where A is the area of the shadow region cross-section (Fig. 1.4). These equations show that the total power of the reflected field exactly equals the power of the intercepted incident rays. The scattering object only distributes them in the surrounding space. By changing the shape of the illuminated surface S_{il}, one can significantly decrease the backscattering by deflection of the reflected rays from the direction back to the source of the incident wave. This is the first basic idea used in stealth technology.

The total power of the shadow radiation is determined by

$$P^{sh,tot} = \frac{1}{2} \text{Re} \int_{S_{il}} (p^{inc})^* [\vec{v}^{inc} \cdot (-\hat{n})] ds, \tag{1.81}$$

The minus sign in front of $\vec{n}$ is chosen because on the surface of black bodies no radiation/reflection in the positive normal direction exists. According to Equations (1.41) and (1.78)

$$p^{inc} = i\omega\rho u_0 e^{ikz}, \quad \text{and} \quad (\vec{v}^{inc} \cdot \hat{n}) = iku_0 e^{ikz}(\hat{z} \cdot \hat{n}). \tag{1.82}$$

Substitution of Equation (1.82) into Equation (1.81) leads to

$$P^{\text{sh,tot}} = \frac{1}{2}k^2 ZA|u_0|^2 = P^{\text{inc}}A \qquad (1.83)$$

and

$$\sigma^{\text{sh,tot}} = A. \qquad (1.84)$$

Thus, the shadow radiation power equals the reflected power, and their sum exactly equals the total scattered power (1.42). This result illustrates the physics behind the fundamental diffraction law (1.45). It shows that objects with soft and hard boundary conditions reveal a dual nature. They can be interpreted as if they are simultaneously perfectly reflecting (with reflection coefficients $\mathcal{R}_s = -1$ and $\mathcal{R}_h = 1$) and perfectly absorbing (with $\mathcal{R} = 0$); that is, black. This law can now be written in the form

$$\sigma^{\text{tot}} = \sigma^{\text{refl,tot}} + \sigma^{\text{sh,tot}} = 2A, \qquad (1.85)$$

where A is the area of the shadow region cross-section.

From the equation $\sigma^{\text{refl,tot}} = A$, it is also clearly seen that the total power of the reflected waves does not depend on the object shaping if the area of shadow cross-section remains constant. However, this power can be decreased by absorbing coatings – this is the second basic idea of stealth technology. In contrast, the shadow radiation cannot be decreased by any absorbing coatings and it can be used for bistatic detection of large opaque objects with small backscattering cross-section (Ufimtsev, 1996).

1.3.6   Summary of Properties of Physical Optics Approximation

The PO approximation describes properly both all reflected rays away from the geometrical optics boundaries and the diffracted field near these boundaries, as well as near foci and caustics. Reflected rays are revealed by the asymptotic evaluation of the PO integrals. These integrals correctly predict the magnitude and position of main and near side lobes in the directivity pattern of the scattered field. However, these surface integrals are computer time consuming. Their transformation into line integrals reduces computer time, and is the subject of continuing research (Asvestas, 1985a,b, 1986, 1995; Gordon, 1994, 2003; Gordon and Bilow, 2002; Maggi, 1888; Meincke et al., 2003; Rubinowicz, 1917).

The *backscattered* field in the PO approximation possesses a special property. According to Equation (1.37), the fields scattered by the *soft* and *hard* objects (of the same shape and size) differ only in *sign*.

Separation of the PO field into the reflected field and the shadow radiation elucidates the scattering physics. In particular, it explains the fundamental law of diffraction theory, according to which the total scattering cross-section of large perfectly reflecting objects is double the area of their shadow cross-section. Well-known manifestations of shadow radiation are Fresnel diffraction and forward scattering.

The PO drawbacks are the following. It is not self-consistent. When the observation point approaches the scattering surface, the PO integrals do not reproduce the initial Geometrical Acoustics (GA) values for the surface field. Also, the PO field does not satisfy rigorously the boundary conditions and the reciprocity principle. The reason for these shortcomings is the Geometrical Optics (GO) approximation for the surface field, which does not include its diffraction components. The PO shortcomings are overcome in the PTD (Ufimtsev, 1962, 1991, 2003), which improves PO by taking into account the diffracted surface field.

1.4 NONUNIFORM COMPONENT OF INDUCED SURFACE FIELD

Surface fields u or $\partial u / \partial n$ induced by the incident wave on the scattering object can be considered as the sources of the scattered field. As noted in the Introduction, the central and original idea of PTD is the separation of these sources into *uniform* and *nonuniform* components:

$$j_{s,h} = j_{s,h}^{(0)} + j_{s,h}^{(1)}. \qquad (1.86)$$

The uniform component $j_{s,h}^{(0)}$ is defined by Equation (1.31) and represents the surface field induced on the infinite plane tangent to the object (Figs. 1.4 and 1.12). In the case of the incident plane wave, this field is *uniformly* distributed over the tangent plane. Its amplitude is constant and its phase is a linear function of the plane coordinates. That is why this component is called *uniform*. According to the definition (1.31), it can also be called the *geometrical optics* or *ray* component.

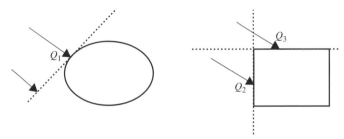

Figure 1.12   Uniform components of the field at points Q_1, Q_2, and Q_3 on the scattering objects are identical to those on the infinite tangential planes shown by the dotted lines.

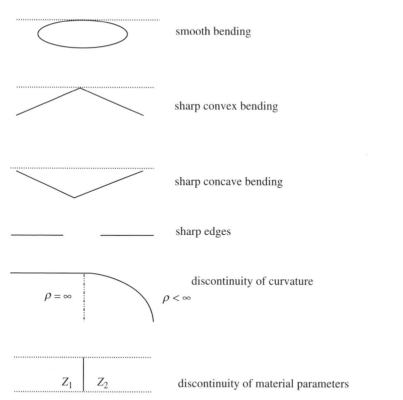

smooth bending

sharp convex bending

sharp concave bending

sharp edges

discontinuity of curvature

discontinuity of material parameters

Figure 1.13   Different shapes and structures where the incident wave generates the *nonuniform* scattering sources.

In contrast, the *nonuniform* component $j_{s,h}^{(1)}$ is the *diffraction part* of the surface field. It is caused by diffraction due to any deviation of the scattering surface from that of the tangential infinite plane. These deviations can be a smooth or sharp bending, sharp edges, discontinuity of curvature, discontinuity of material properties, apertures, small bumps and dips, and so on. (Fig. 1.13).

If the scattering object is convex and smooth and its dimensions and radii of curvature are large compared to the wavelength, then the induced nonuniform component concentrates near the boundary between the illuminated and shadowed surfaces (Fig. 1.4). This component is described asymptotically by the Fock functions (Fock, 1965). From the physical point of view it represents creeping waves that radiate surface diffracted rays. If the object possesses sharp edges, the nonuniform component concentrates in their vicinity (Fig. 1.14) and it is described asymptotically by the Sommerfeld functions (Sommerfeld, 1935) presented in Chapter 2. This form of the nonuniform components radiates the edge waves that are often called *fringe* waves. Similarly the nonuniform sources near the vertices radiate vertex waves.

Taking into account the diffraction/nonuniform components of the surface fields, PTD overcomes the PO shortcomings and provides more accurate asymptotic results for high-frequency scattered fields. The PTD separation of the surface fields into uniform and nonuniform components has proved to be very productive and is often used

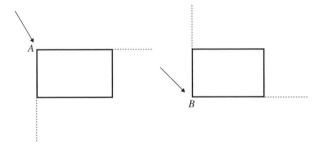

Figure 1.14   Nonuniform components of the surface field induced by the incident wave near edges A and B are asymptotically identical to those at the tangential wedges with infinite faces shown by dotted lines.

in diffraction theory. This concept is quite flexible. It can be extended for objects with other boundary conditions. It is also successfully used in hybrid techniques in combination with direct numerical methods. A proper choice of the uniform component depends on specific properties of the problem under investigation and can essentially facilitate its solution. See for example Ufimtsev (1998) and related references given in Ufimtsev (1996, 2003), as well as in the section "Additional references related to the PTD concept: Applications, modifications, and developments" shown at the end of this book.

1.5 ELECTROMAGNETIC WAVES

This section briefly presents the basic notions used in this book for the description of electromagnetic waves. This book studies the diffraction of electromagnetic waves at perfectly conducting bodies that are large compared to the wavelength. It is assumed that the waves and scattering objects are in free space (vacuum). The electric ($\vec{E}$) and magnetic vectors ($\vec{H}$) of the wave field are determined as

$$\vec{E} = \frac{i}{k} Z_0 \cdot [\nabla(\nabla \cdot \vec{A}^{\mathrm{e}}) + k^2 \vec{A}^{\mathrm{e}}] - \nabla \times \vec{A}^{\mathrm{m}} \tag{1.87}$$

and

$$\vec{H} = \frac{i}{kZ_0} [\nabla(\nabla \cdot \vec{A}^{\mathrm{m}}) + k^2 \vec{A}^{\mathrm{m}}] + \nabla \times \vec{A}^{\mathrm{e}}. \tag{1.88}$$

Here, $Z_0 = 1/Y_0 = \sqrt{\mu_0/\varepsilon_0} = 120\pi$ ohms is the impedance of free space, $k = \omega\sqrt{\varepsilon_0\mu_0} = 2\pi/\lambda$ is the wave number, and

$$\vec{A}^{\mathrm{e}} = \frac{1}{4\pi} \int \vec{j}^{\mathrm{e}} \frac{e^{ikr}}{r} dv \quad \text{and} \quad \vec{A}^{\mathrm{m}} = \frac{1}{4\pi} \int \vec{j}^{\mathrm{m}} \frac{e^{ikr}}{r} dv \tag{1.89}$$

are the electric and magnetic vector-potentials. They are the solutions of the equation

$$\Delta \vec{A}^{\mathrm{e,m}} + k^2 \vec{A}^{\mathrm{e,m}} = -\vec{j}^{\mathrm{e,m}}, \tag{1.90}$$

where $\vec{j}^{\mathrm{e}}(\vec{j}^{\mathrm{m}})$ is the electric (magnetic) current density of the field sources.

Notice that these Equations (1.87) to (1.90) are more convenient for calculation than those usually accepted in books on engineering electromagnetics. Indeed, Equations (1.89) and (1.90) have exactly the same form both for electric and magnetic potentials. The second terms (out of brackets) in Equations (1.87) and (1.88) do not contain the factors $1/\mu_0$ and $1/\varepsilon_0$, which eventually disappear in the integral field expressions.

In the far zone from a scattering object, one can use the following approximation (similar to Equations (1.16) and (1.17):

$$\vec{A}^{\mathrm{e,m}} = \frac{1}{4\pi} \frac{e^{ikR}}{R} \int \vec{j}^{\mathrm{e,m}} e^{-ikr' \cos\Omega} \, dv. \tag{1.91}$$

This leads to the field components

$$E_\vartheta = Z_0 H_\varphi = ik(Z_0 A_\vartheta^{\mathrm{e}} + A_\varphi^{\mathrm{m}}) \tag{1.92}$$

and

$$E_\varphi = -Z_0 H_\vartheta = ik(Z_0 A_\varphi^{\mathrm{e}} - A_\vartheta^{\mathrm{m}}). \tag{1.93}$$

The radial components E_R, H_R are of the order $1/R^2$ and are neglected here. The coordinate system is shown in Figure 1.2.

The scattering cross-section is determined by Equations (1.26) and (1.9) as

$$\sigma = 4\pi R^2 \left| \frac{\vec{E}^{\mathrm{sc}}}{\vec{E}^{\mathrm{inc}}} \right|^2. \tag{1.94}$$

Equation (1.45) for the total scattering cross-section is also valid for electromagnetic waves. In the case of an incident wave with a linear polarization, it can be represented in the form of Equation (1.53) as

$$\sigma^{\mathrm{tot}} = \frac{4\pi}{k} \mathrm{Im}[(E^{\mathrm{sc}}/E^{\mathrm{inc}}) \cdot \mathrm{Re}^{-ikR}], \tag{1.95}$$

where E^{sc} is the field scattered in the forward direction.

The PO approximation for the far field scattered by perfectly conducting objects is defined by

$$\vec{A}^{\mathrm{e}} = \frac{1}{4\pi} \frac{e^{ikR}}{R} \int_{S_{\mathrm{il}}} \vec{j}^{(0)} e^{-ikr' \cos\Omega} \, ds, \qquad \vec{A}^{\mathrm{m}} = 0, \tag{1.96}$$

where S_{il} is the illuminated side of the scattering surface and

$$\vec{j}^{(0)} = 2[\hat{n} \times \vec{H}^{\mathrm{inc}}] \tag{1.97}$$

is the *uniform component* of the surface electric current induced by the incident wave on the illuminated side of a scattering object. The paper by Ufimtsev (1999) describes in detail the properties of the PO approximation (see also Ruck et al. (1970)).

Equations (4.1.12) and (4.1.13) of Ufimtsev (2003) show that in the PO approximation, the field backscattered by *convex perfectly conducting* objects does not depend on the polarization of the incident wave.

Consider another important consequence of the PO approximations (4.1.12) and (4.1.13) of Ufimtsev (2003). These equations were derived under the following conditions:

- The incident wave is a plane wave propagating in the direction

$$\hat{k}^i = \hat{y}\sin\gamma + \hat{z}\cos\gamma.$$

- The observation point is in the backscattering direction $\hat{m} = -\hat{k}^i$ (in the plane yoz ($\varphi = -\pi/2$)).
- Equation (4.1.12) is valid for the incident wave with E-polarization, $E_x^{\text{inc}} = E_{0x}e^{ik(y\sin\gamma + z\cos\gamma)}$.
- Equation (4.1.13) is valid for the incident wave with H-polarization, $H_x^{\text{inc}} = H_{0x}e^{ik(y\sin\gamma + z\cos\gamma)}$.

In view of these comments, the PO approximations (4.1.12) and (4.1.13) in Ufimtsev (2003) can be written as

$$E_x^{(0)} = -\frac{ik}{2\pi}\frac{e^{ikR}}{R}\int_{S_{\text{il}}} E_x^{\text{inc}}e^{-ikr'\cos\Omega}(\hat{k}^i \cdot \hat{n})ds, \qquad (1.98)$$

and

$$H_x^{(0)} = \frac{ik}{2\pi}\frac{e^{ikR}}{R}\int_{S_{\text{il}}} H_x^{\text{inc}}e^{-ikr'\cos\Omega}(\hat{k}^i \cdot \hat{n})ds. \qquad (1.99)$$

Comparison of these equations with Equation (1.37) reveals the following fundamental relationships, which exist between the PO approximations for *backscattered* acoustic and electromagnetic waves:

$$E_x^{(0)} = u_s^{(0)} \quad \text{if} \quad E_x^{\text{inc}} = u^{\text{inc}} \qquad (1.100)$$

and

$$H_x^{(0)} = u_h^{(0)} \quad \text{if} \quad H_x^{\text{inc}} = u^{\text{inc}}. \qquad (1.101)$$

Utilizing the vector equivalency theorems (Ufimtsev, 2003) and the idea of Section 1.3.4, one can represent the PO field in a form similar to Equation (1.70):

$$\vec{E}^{\text{PO}} \equiv \vec{E}^{(0)} = \vec{E}^{\text{refl}} + \vec{E}^{\text{sh}}, \qquad \vec{H}^{\text{PO}} \equiv \vec{H}^{(0)} = \vec{H}^{\text{refl}} + \vec{H}^{\text{sh}}. \qquad (1.102)$$

Here, $\vec{E}^{\text{refl}}$, $\vec{H}^{\text{refl}}$ and $\vec{E}^{\text{sh}}$, $\vec{H}^{\text{sh}}$ are the reflected field and the shadow radiation, respectively. Their far-field approximations are

$$E_{\vartheta}^{\text{refl}} = \frac{ik}{4\pi} \frac{e^{ikR}}{R} \int_{S_{\text{il}}} \{Z_0[n \times \vec{H}^{\text{inc}}] \cdot \hat{\vartheta} + [\hat{n} \times \vec{E}^{\text{inc}}] \cdot \hat{\varphi}\} e^{-ikr' \cos \Omega} \, ds, \quad (1.103)$$

$$E_{\varphi}^{\text{refl}} = \frac{ik}{4\pi} \frac{e^{ikR}}{R} \int_{S_{\text{il}}} \{Z_0[n \times \vec{H}^{\text{inc}}] \cdot \hat{\varphi} - [\hat{n} \times \vec{E}^{\text{inc}}] \cdot \hat{\vartheta}\} e^{-ikr' \cos \Omega} \, ds, \quad (1.104)$$

$$E_{\vartheta}^{\text{sh}} = \frac{ik}{4\pi} \frac{e^{ikR}}{R} \int_{S_{\text{il}}} \{Z_0[n \times \vec{H}^{\text{inc}}] \cdot \hat{\vartheta} - [\hat{n} \times \vec{E}^{\text{inc}}] \cdot \hat{\varphi}\} e^{-ikr' \cos \Omega} \, ds, \quad (1.105)$$

$$E_{\varphi}^{\text{sh}} = \frac{ik}{4\pi} \frac{e^{ikR}}{R} \int_{S_{\text{il}}} \{Z_0[n \times \vec{H}^{\text{inc}}] \cdot \hat{\varphi} + [\hat{n} \times \vec{E}^{\text{inc}}] \cdot \hat{\vartheta}\} e^{-ikr' \cos \Omega} \, ds, \quad (1.106)$$

$$\vec{H}^{\text{refl}} = [\nabla R \times \vec{E}^{\text{refl}}]/Z_0,$$

$$\vec{H}^{\text{sh}} = [\nabla R \times \vec{E}^{\text{sh}}]/Z_0. \quad (1.107)$$

Here, $\hat{\vartheta}(\hat{\varphi})$ is the unit vector in the direction of the angle $\vartheta(\varphi)$ increase.

Suppose that the incident wave is given as

$$E_x^{\text{inc}} = Z_0 H_y^{\text{inc}} = E_{0x} e^{ikz}. \quad (1.108)$$

Then, one can derive the following relationships:

$$E_x^{\text{sh}} = \frac{ik}{2\pi} E_{0x} A \frac{e^{ikR}}{R}, \qquad E_x^{\text{refl}} = 0 \qquad (\vartheta = 0) \quad (1.109)$$

for the forward direction, and

$$E_x^{\text{sh}} = 0 \qquad (\vartheta = \pi) \quad (1.110)$$

for the backscattering direction. Here, A is the area of the shadow region cross-section (Fig. 1.4). Thus,

$$E_x^{\text{PO}} \equiv E_x^{\text{sh}} = E_{0x} \frac{ik}{2\pi} A \frac{e^{ikz}}{z} \qquad (\vartheta = 0) \quad (1.111)$$

for the forward direction, and

$$E_x^{\text{PO}} = -\frac{ik}{2\pi} E_{0x} \frac{e^{ikR}}{R} \int_{S_{\text{il}}} (\hat{n} \cdot \hat{z}) e^{i2kz'} \, ds \qquad (\vartheta = \pi) \quad (1.112)$$

for the backscattering direction.

The Shadow Contour Theorem established in Section 1.3.5 for acoustic waves is also valid for electromagnetic waves.

Shadow radiation can be interpreted as the field scattered by black bodies. Chapter 1 of Ufimtsev (2003) presents explicit expressions and the results of numerical calculation for the field scattered by arbitrary 2-D black cylinders and black bodies of revolution.

The definition of the nonuniform component of the surface sources introduced in Section 1.4 is also applicable for electric surface currents. Its modification is presented below in Sections 7.9.1 and 7.9.2.

PROBLEMS

1.1 The incident wave $u^{inc} = u_0 \exp[-ik(x \cos \varphi_0 + y \sin \varphi_0)]$ excites the scattering sources $j_s = 2\partial u^{inc}/\partial y$ on the illuminated side $(y = +0)$ of a soft infinite plane. $y = 0$ (Fig. P1.1). Start with the integral (1.10) and calculate the scattered field u_s^{sc} generated by these sources above and below the plane.

   (a) Express the integral over the variable ζ through the Hankel function (3.7). Use the integral representation (3.8) of this function and obtain the Fourier integral for a plane wave.

   (b) Consider the total field $u_s^t = u_s^{inc} + u_s^{sc}$ in the region $y < 0$ and realize the blocking role of the scattering sources j_s.

1.2 Solve the scattering problem similar to Problem 1.1, but for a hard reflecting plane.

1.3 The incident wave $E_z^{inc} = E_{0z} \exp[-ik(x \cos \varphi_0 + y \sin \varphi_0)]$ excites the surface current $\vec{j} = 2[\hat{y} \times \vec{H}^{inc}]$ on the illuminated side $(y = +0)$ of a perfectly conducting infinite plane (Fig. P1. 1). Start with Equations (1.87) and (1.89) and calculate the scattered field E_z^{sc} generated by these currents above and below the plane.

   (a) Express the integral over the variable ζ through the Hankel function (3.7). Use the integral representation (3.8) of this function and obtain the Fourier integral for a plane wave.

   (b) Consider the total field $E_z^t = E_z^{inc} + E_z^{sc}$ in the region $y < 0$ and realize the blocking role of the surface currents j_z.

1.4 Solve the problem analogous to Problem 1.3 but with the incident wave $H_z^{inc} = H_{0z} \exp[-ik(x \cos \varphi_0 + y \sin \varphi_0)]$. Start with Equation (1.88).

1.5 Suppose that the incident wave $u^{inc} = u_0 \exp[ik(x \cos \phi_0 + y \sin \phi_0)]$ hits a soft strip as shown in Figure 5.1. Use Equation (1.71) and calculate the *reflected part* of the PO field scattered by this strip.

   (a) Express the integral over the variable ζ through the Hankel function (3.7), apply its asymptotic approximation (2.29), and express the far field $(r \gg ka^2)$ in closed form.

   (b) Estimate the field in the directions $\phi = \phi_0, \phi = \pi - \phi_0, \phi = \pi + \phi_0$, and $\phi = -\phi_0$.

Figure P1.1   Excitation of an infinite plane by an incident wave.

    (c) Apply the optical theorem (5.16) to the field in the direction $\phi = \pi - \phi_0$ and provide the geometrical interpretation of the total reflecting cross-section.

    (d) Compute and plot the directivity pattern of the reflected field, setting $a = 2\lambda$, $\phi_0 = 45°$. What are the interesting properties of this field?

1.6 Solve the problem similar to Problem 1.5, but for a hard strip.

1.7 Suppose that the incident wave $u^{\text{inc}} = u_0 \exp[ik(x \cos \phi_0 + y \sin \phi_0)]$ hits a soft strip as shown in Figure 5.1. Use Equation (1.72) and calculate the *shadow radiation part* of the PO field scattered by this strip.

    (a) Express the integral over the variable ζ through the Hankel function (3.7), apply its asymptotic approximation (2.29), and calculate the far field ($r \gg ka^2$) in closed form.

    (b) Estimate the field in the directions $\phi = \phi_0$, $\phi = \pi - \phi_0$, and $\phi = \pi + \phi_0$.

    (c) Apply the optical theorem (5.16) to the field in the direction $\phi = \phi_0$ and give the geometrical interpretation of the total power of the shadow radiation.

    (d) Compute and plot the directivity pattern of the shadow radiation, setting $a = 2\lambda$, $\phi_0 = 45°$. What are the interesting properties of this field?

1.8 Is the difference between the reflected parts of the PO field scattered by soft and hard objects (of the same shape and size) illuminated by the same incident wave.

1.9 Is any difference between the shadow parts of the PO field scattered by soft and hard objects (of the same shape and size) illuminated by the same incident wave.

1.10 The incident wave $E_z^{\text{inc}} = E_{0z} \exp[ik(x \cos \phi_0 + y \sin \phi_0)]$ hits a perfectly conducting strip as shown in Figure 5.1. Calculate the *reflected part* of the PO scattered field.

    (a) Start with Equations (1.87), (1.88) and (1.89). Apply $\vec{j}^{\,\text{e,refl}} = \hat{n} \times \vec{H}^{\text{inc}}$, $\vec{j}^{\,\text{m,refl}} = \hat{n} \times \vec{E}^{\text{inc}}$. Prepare the integral expression for the reflected field.

    (b) Express the integral over the variable ζ through the Hankel function (3.7), apply its asymptotic approximation (2.29), and express the far field ($r \gg ka^2$) in closed form.

    (c) Estimate the field in the directions $\phi = \phi_0$, $\phi = \pi - \phi_0$, $\phi = \pi + \phi_0$, and $\phi = -\phi_0$.

    (d) Apply the optical theorem (5.16) to the field in the direction $\phi = \pi - \phi_0$ and provide the geometrical interpretation of the total reflecting cross-section.

    (e) Compute and plot the directivity pattern of the reflected field, setting $a = 2\lambda$, $\phi_0 = 45°$. What are the interesting properties of this field?

1.11 The incident wave $E_z^{\text{inc}} = E_{0z} \exp[ik(x \cos \phi_0 + y \sin \phi_0)]$ hits a perfectly conducting strip as shown in Figure 5.1. Calculate the *shadow radiation part* of the PO scattered field.

    (a) Start with Equations (1.87), (1.88) and (1.89). Apply

$$\vec{j}^{\,\text{e,sh}} = \hat{n} \times \vec{H}^{\text{inc}}, \quad \vec{j}^{\,\text{m,sh}} = -\hat{n} \times \vec{E}^{\text{inc}}.$$

    Prepare the integral expression for the shadow radiation.

    (b) Express the integral over the variable ζ through the Hankel function (3.7), apply its asymptotic approximation (2.29), and calculate the far field ($r \gg ka^2$) in closed form.

    (c) Estimate the field in the directions $\phi = \phi_0$, $\phi = \pi - \phi_0$, and $\phi = \pi + \phi_0$.

    (d) Apply the optical theorem (5.16) to the field in the direction $\phi = \phi_0$ and provide the geometrical interpretation of the total power of the shadow radiation.

    (e) Compute and plot the directivity pattern of the reflected field, setting $a = 2\lambda$, $\phi_0 = 45°$. What are the interesting properties of this field?

Chapter 2

Wedge Diffraction: Exact Solution and Asymptotics

The relationships $u_s = E_z$ and $u_h = H_z$ exist between the acoustic and electromagnetic fields in the 2-D wedge diffraction problem. Here, E_z and H_z are the components (of vectors $\vec{E}$ and $\vec{H}$) that are parallel to the edge of the wedge.

2.1 CLASSICAL SOLUTIONS

Diffraction at a wedge with a straight edge and infinite planar faces is an appropriate canonical problem to derive asymptotic expressions for the edge waves scattered from arbitrary curved edges. In the particular case of the wedge, which is a semi-infinite half-plane, the exact solution of this canonical problem was found by Sommerfeld (1896), who constructed so-called branched wave functions. Analysis of this work performed in Ufimtsev (1998) shows that Sommerfeld also developed almost everything that was necessary to obtain the solution for the wedge with an arbitrary angle between its faces. However, he missed the last step that led directly to the solution. This more general solution was found by Macdonald (1902) with the classical method of separation of variables in the wave equation. Later on, Sommerfeld also constructed the solution of the wedge diffraction problem by his method of branched wave functions and derived simple asymptotic expressions for the edge-diffracted waves (Sommerfeld, 1935).

Because the wedge diffraction problem is the basis for the construction of PTD, its solution is considered here in detail. First we derive this solution in the form of infinite series and then convert it to the Sommerfeld integrals convenient for asymptotic analysis. The material of this chapter, with the exception of Sections 2.6 and 2.7, is a scalar version of the theory developed by the author for electromagnetic waves (Ufimtsev, 1962).

Fundamentals of the Physical Theory of Diffraction. By Pyotr Ya. Ufimtsev
Copyright © 2007 John Wiley & Sons, Inc.

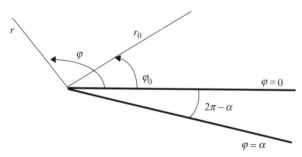

Figure 2.1   A wedge is excited by a filamentary source located at the line $r = r_0$, $\varphi = \varphi_0$.

The geometry of the problem is shown in Figure 2.1. A wedge with infinite planar faces $\varphi = 0$ and $\varphi = \alpha$ is located in a homogeneous medium. It is excited by a cylindrical wave. The source of this wave is a radiating filament with coordinates $r = r_0$, $\varphi = \varphi_0$. This is a two-dimensional problem where $\partial/\partial z \equiv 0$.

The field outside the wedge ($0 \le \varphi \le \alpha$) satisfies the wave equation

$$\Delta u + k^2 u = I_0 \delta(r - r_0, \varphi - \varphi_0) \tag{2.1}$$

and the boundary conditions

$$u_s = 0 \tag{2.2}$$

or

$$\partial u_h / \partial n = 0 \tag{2.3}$$

on the faces $\varphi = 0$ and $\varphi = \alpha$. In the case of electromagnetic waves, these boundary conditions are appropriate for the *perfectly conducting* wedge, and function u_s represents the z-component of electric field intensity $\vec{E}$, while function u_h is the z-component of magnetic field intensity $\vec{H}$. In the case of acoustic waves, condition (2.2) relates to the acoustically soft wedge, and (2.3) to the acoustically hard wedge.

Outside the immediate vicinity of the source, the field u satisfies the homogeneous wave equation

$$\Delta u + k^2 u = 0. \tag{2.4}$$

For the two-dimensional problem, the Laplacian operator is defined by

$$\Delta = \frac{\partial^2}{\partial r^2} + \frac{1}{r}\frac{\partial}{\partial r} + \frac{1}{r^2}\frac{\partial^2}{\partial \varphi^2}. \tag{2.5}$$

Using the classical method of separation of variables, we set in Equation (2.4)

$$u = R(r)\Phi(\varphi) \tag{2.6}$$

and substitute this u into Equation (2.4). After simple manipulations, the latter can be separated into two equations:

$$\frac{d^2R}{dx^2} + \frac{1}{x}\frac{dR}{dx} + \left(1 - \frac{v_l^2}{x^2}\right)R = 0, \qquad \text{with } x = kr, \tag{2.7}$$

and

$$\frac{d^2\Phi}{d\varphi^2} + v_l^2\Phi = 0. \tag{2.8}$$

The function Φ and the separation constants v_l are determined from the boundary conditions.

In the case of soft boundary conditions, according to Equations (2.2) and (2.8),

$$\Phi = \{\sin v_l\varphi\}, \qquad v_l = l\,\frac{\pi}{\alpha}, \qquad l = 1, 2, 3, \ldots, \tag{2.9}$$

and, in the case of hard boundary conditions, in accordance with Equations (2.3) and (2.8),

$$\Phi = \{\cos v_l\varphi\}, \qquad v_l = l\,\frac{\pi}{\alpha}, \qquad l = 0, 1, 2, 3, \ldots. \tag{2.10}$$

The Bessel and Hankel functions

$$R = \left\{\begin{matrix} J_{v_l}(kr) \\ H_{v_l}^{(1)}(kr) \end{matrix}\right\} \tag{2.11}$$

represent the solution of the radial equation (2.7). The Bessel functions $J_{v_l}(kr)$ can be used in the region $r \le r_0$, because they are finite at the edge $r = 0$, and the Hankel functions are appropriate in the region $r \ge r_0$, because they satisfy Sommerfeld's radiation condition at infinity:

$$\lim \sqrt{r}\left(\frac{du}{dr} - iku\right) = 0, \qquad \text{with } r \to \infty. \tag{2.12}$$

Hence, the solutions of Equation (2.4) can be written as

$$u_s = \begin{cases} \displaystyle\sum_{l=1}^{\infty} a_l J_{v_l}(kr) H_{v_l}^{(1)}(kr_0)\sin v_l\varphi_0 \sin v_l\varphi, & \text{with } r \le r_0 \\[4mm] \displaystyle\sum_{l=1}^{\infty} a_l J_{v_l}(kr_0) H_{v_l}^{(1)}(kr)\sin v_l\varphi_0 \sin v_l\varphi, & \text{with } r \ge r_0 \end{cases} \tag{2.13}$$

$$u_h = \begin{cases} \displaystyle\sum_{l=0}^{\infty} b_l J_{v_l}(kr) H_{v_l}^{(1)}(kr_0)\cos v_l\varphi_0 \cos v_l\varphi, & \text{with } r \le r_0 \\[4mm] \displaystyle\sum_{l=0}^{\infty} b_l J_{v_l}(kr_0) H_{v_l}^{(1)}(kr)\cos v_l\varphi_0 \cos v_l\varphi, & \text{with } r \ge r_0 \end{cases} \tag{2.14}$$

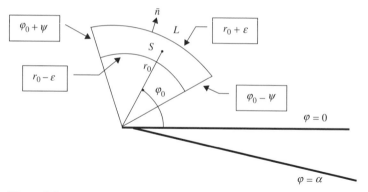

Figure 2.2   The integration contour L in the Green theorem (2.15).

These expressions satisfy the boundary conditions, as well as the reciprocity principles; that is, they do not change after interchanging r and r_0, φ and φ_0.

The unknown coefficients a_l and b_l can be found by applying the Green theorem

$$\oint_L \frac{\partial u}{\partial n} \, dl = \int_S \Delta u \, ds, \qquad ds = r \, dr \, d\varphi \qquad (2.15)$$

to the fields u_s and u_h in the region S bounded by the contour L shown in Figure 2.2. This contour consists of two arcs $r = r_0 - \varepsilon$, $r = r_0 + \varepsilon$ and two radial sides $\varphi = \varphi_0 - \psi$, $\varphi = \varphi_0 + \psi$.

Substitute $u_{s,h}$ into (2.15), take into account that, according to Equation (2.1),

$$\Delta u = -k^2 u + I_0 \delta(r - r_0, \varphi - \varphi_0), \qquad (2.16)$$

and let ε tend to zero. It is clear that

$$\int_S u_{s,h} \, ds \longrightarrow 0 \qquad \text{when } S \longrightarrow 0, \qquad (2.17)$$

as the Bessel and Hankel functions are finite at $r = r_0 > 0$.

As a result, the Green formula for $u_{s,h}$ transforms into

$$\int_{\varphi_0 - \psi}^{\varphi_0 + \psi} \left(\left. \frac{\partial u_{s,h}}{\partial r} \right|_{r=r_0+0} - \left. \frac{\partial u_{s,h}}{\partial r} \right|_{r=r_0-0} \right) r_0 \, d\varphi$$

$$= I_0 \int_{\varphi_0 - \psi}^{\varphi_0 + \psi} d\varphi \lim \int_{r_0 - \varepsilon}^{r_0 + \varepsilon} \delta(r - r_0, \varphi - \varphi_0) r \, dr, \qquad \text{with } \varepsilon \to 0. \qquad (2.18)$$

The two-dimensional Delta-function in polar coordinates equals

$$\delta(r - r_0, \varphi - \varphi_0) = \delta(r - r_0) \frac{1}{r} \delta(\varphi - \varphi_0), \qquad (2.19)$$

therefore

$$\int_{\varphi_0-\psi}^{\varphi_0+\psi} \left(\frac{\partial u_{s,h}}{\partial r}\bigg|_{r=r_0+0} - \frac{\partial u_{s,h}}{\partial r}\bigg|_{r=r_0-0} \right) r_0 \, d\varphi = I_0 \int_{\varphi_0-\psi}^{\varphi_0+\psi} \delta(\varphi - \varphi_0) d\varphi. \quad (2.20)$$

Equation (2.20) is valid for arbitrary limits of integration. This is possible if the integrands in the left and right sides are equal to each other:

$$\frac{\partial u_{s,h}}{\partial r}\bigg|_{r=r_0+0} - \frac{\partial u_{s,h}}{\partial r}\bigg|_{r=r_0-0} = \frac{1}{r_0} I_0 \delta(\varphi - \varphi_0). \quad (2.21)$$

This equation can be used to determine the unknown coefficients a_l and b_l in the expressions (2.13) and (2.14).

To do this, we substitute u_s of Equation (2.13) into Equation (2.21), multiply both sides by $\sin v_t \varphi$, where $v_t = t\pi/\alpha$, and integrate them over φ from 0 to α. Note that

$$\int_0^\alpha \sin v_l \varphi \sin v_t \varphi \, d\varphi = \begin{cases} \dfrac{1}{2}\alpha, & \text{with } l = t \\ 0, & \text{with } l \neq t \quad t = 1, 2, 3, \ldots, \end{cases} \quad (2.22)$$

and obtain

$$k r_0 \frac{\alpha}{2} a_l \left[J_{v_l}(kr_0) \frac{d}{dkr_0} H_{v_l}^{(1)}(kr_0) - H_{v_l}^{(1)}(kr_0) \frac{d}{dkr_0} J_{v_l}(kr_0) \right] = I_0. \quad (2.23)$$

The expression inside the brackets is the Wronskian

$$W[J_v(x), H_v^{(1)}(x)] = J_v(x) \frac{d}{dx} H_v^{(1)}(x) - H_v^{(1)}(x) \frac{d}{dx} J_v(x) = \frac{2i}{\pi x}. \quad (2.24)$$

From Equations (2.23) and (2.24) it follows that

$$a_l = \frac{\pi}{i\alpha} I_0. \quad (2.25)$$

In the case of the hard boundary conditions, we carry out similar manipulations. Substitute u_h of Equation (2.14) into Equation (2.21), multiply both sides by $\cos v_t \varphi$, and integrate over φ from 0 to α. As a result, we obtain

$$b_l = \varepsilon_l \frac{\pi}{i\alpha} I_0, \quad \text{with } \varepsilon_0 = \frac{1}{2}, \varepsilon_1 = \varepsilon_2 = \varepsilon_3 = \cdots = 1. \quad (2.26)$$

Thus, the coefficients a_l and b_l are found, and the functions $u_{s,h}$ are completely determined. Now we can write the final expressions for the total field excited by the external filamentary source.

In the case of soft boundary conditions, the field u_s is described by the following expressions

$$
u_s = \begin{cases}
\dfrac{\pi}{i\alpha} I_0 \displaystyle\sum_{l=1}^{\infty} J_{v_l}(kr) H_{v_l}^{(1)}(kr_0) \sin v_l \varphi_0 \sin v_l \varphi, & \text{with } r \leq r_0 \\[3mm]
\dfrac{\pi}{i\alpha} I_0 \displaystyle\sum_{l=1}^{\infty} J_{v_l}(kr_0) H_{v_l}^{(1)}(kr) \sin v_l \varphi_0 \sin v_l \varphi, & \text{with } r \geq r_0
\end{cases} \tag{2.27}
$$

In the case of hard boundary conditions, the field u_h is determined by

$$
u_h = \begin{cases}
\dfrac{\pi}{i\alpha} I_0 \displaystyle\sum_{l=0}^{\infty} \varepsilon_l J_{v_l}(kr) H_{v_l}^{(1)}(kr_0) \cos v_l \varphi_0 \cos v_l \varphi, & \text{with } r \leq r_0 \\[3mm]
\dfrac{\pi}{i\alpha} I_0 \displaystyle\sum_{l=0}^{\infty} \varepsilon_l J_{v_l}(kr_0) H_{v_l}^{(1)}(kr) \cos v_l \varphi_0 \cos v_l \varphi, & \text{with } r \geq r_0.
\end{cases} \tag{2.28}
$$

These expressions relate to the excitation of the field by a cylindrical wave, with a source term $I_0 \delta(r - r_0, \varphi - \varphi_0)$ around the wedge in the region $0 \leq \varphi \leq \alpha, 0 \leq r \leq \infty$. They can be modified for excitation by a plane wave.

2.2 TRANSITION TO THE PLANE WAVE EXCITATION

For the Hankel functions with large arguments ($kr_0 \to \infty$), one can use the asymptotic expression

$$
H_{v_l}^{(1)}(kr_0) \sim \sqrt{\frac{2}{\pi k r_0}} e^{i(kr_0 - \frac{\pi}{2}v_l - \frac{\pi}{4})} \approx H_0^{(1)}(kr_0) e^{-i\frac{\pi}{2}v_l}. \tag{2.29}
$$

The field Equations (2.27) and (2.28) can then be rewritten for the region $r < r_0$ as

$$
u_s = \frac{\pi}{i\alpha} I_0 H_0^{(1)}(kr_0) \sum_{l=1}^{\infty} e^{-i\frac{\pi}{2}v_l} J_{v_l}(kr) \sin v_l \varphi_0 \sin v_l \varphi, \tag{2.30}
$$

$$
u_h = \frac{\pi}{i\alpha} I_0 H_0^{(1)}(kr_0) \sum_{l=0}^{\infty} \varepsilon_l e^{-i\frac{\pi}{2}v_l} J_{v_l}(kr) \cos v_l \varphi_0 \cos v_l \varphi, \tag{2.31}
$$

or

$$u_s = \frac{\pi}{2i\alpha} I_0 H_0^{(1)}(kr_0) \sum_{l=0}^{\infty} e^{-i\frac{\pi}{2}v_l} J_{v_l}(kr)[\cos v_l(\varphi - \varphi_0) - \cos v_l(\varphi + \varphi_0)], \quad (2.32)$$

$$u_h = \frac{\pi}{2i\alpha} I_0 H_0^{(1)}(kr_0) \sum_{l=0}^{\infty} \varepsilon_l e^{-i\frac{\pi}{2}v_l} J_{v_l}(kr)[\cos v_l(\varphi - \varphi_0) + \cos v_l(\varphi + \varphi_0)].$$

$$(2.33)$$

To clarify these expressions, we consider the solution of the wave equation (2.1) in the free infinite homogeneous medium, without the scattering wedge. For this problem, it is convenient to use the new polar coordinates (ρ, ϕ) with the origin at the radiating source (r_0, φ_0). It is clear that due to the azimuthal symmetry of the problem, its solution is a function of only one variable, $u = u(\rho)$. In addition, as the wave equation is the differential equation of second order, its solution, in general, is the sum of two fundamental solutions of the related homogeneous equation:

$$u(\rho) = c_1 H_0^{(1)}(k\rho) + c_2 H_0^{(2)}(k\rho), \quad (2.34)$$

with constants c_1 and c_2. We retain here only the first term,

$$u(\rho) = c_1 H_0^{(1)}(k\rho), \quad (2.35)$$

because the second term, with $H_0^{(2)}(k\rho)$, does not satisfy the Sommerfeld radiation condition (2.12) and represents a nonphysical wave incoming from infinity. The constant c_1 is found again with the Green theorem (2.15) applied to the circular region S of a small radius ε and with the center at $\rho = 0$ (Fig. 2.3).

For small values $k\rho \ll 1$, the function $u(k\rho)$ and its normal derivative $du/dn = du/d\rho$ (at the boundary of the region S) are described by the asymptotic approximations

$$u(\rho) \approx c_1 \frac{i2}{\pi} \ln(k\rho), \qquad \frac{du(\rho)}{d\rho} \approx c_1 \frac{i2}{\pi\rho}, \qquad \text{with } k\rho \ll 1. \quad (2.36)$$

By substitution of these quantities into the Green theorem and taking the limit with $\varepsilon \to 0$, we find $c_1 = I_0/i4$ and

$$u(\rho) = \frac{1}{i4} I_0 H_0^{(1)}(k\rho). \quad (2.37)$$

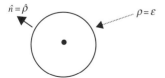

$\hat{n} = \hat{\rho}$

$\rho = \varepsilon$

Figure 2.3 A circular region $(0 \le \rho \le \varepsilon, 0 \le \phi \le 2\pi)$ with the radiating source at the center.

This solution explains the physical meaning of the factor in front of the series in Equations (2.32) and (2.33). It actually represents the field of the incident wave on the edge of the wedge

$$u_0 = \frac{1}{4i} I_0 H_0^{(1)}(kr_0).$$ (2.38)

When $r_0 \rightarrow \infty$ and $I_0 \rightarrow \infty$, this field can be interpreted as the plane wave traveling to the wedge from the direction $\varphi = \varphi_0$:

$$u^{inc} = u_0 e^{-ikr\cos(\varphi-\varphi_0)}.$$ (2.39)

As a result, one can rewrite Equations (2.32) and (2.33) in the classical Sommerfeld form (Sommerfeld, 1935):

$$u_s = u_0 \cdot [u(kr, \varphi - \varphi_0) - u(kr, \varphi + \varphi_0)]$$ (2.40)

$$\text{for } u^{inc} = u_0 e^{-ikr\cos(\varphi-\varphi_0)}$$

$$u_h = u_0 \cdot [u(kr, \varphi - \varphi_0) + u(kr, \varphi + \varphi_0)]$$ (2.41)

where

$$u(kr, \psi) = \frac{2\pi}{\alpha} \sum_{l=0}^{\infty} \varepsilon_l e^{-i(\pi/2)v_l} J_{v_l}(kr) \cos v_l \psi.$$ (2.42)

Equations (2.40) to (2.42) finally determine the total field generated by the *incident plane wave* in the presence of the perfectly reflecting wedge.

2.3 CONVERSION OF THE SERIES SOLUTION TO THE SOMMERFELD INTEGRALS

In his work Sommerfeld (1935) presented the solution of the wedge diffraction problem in integral form and then transformed it into infinite series. Here, we will perform a reverse procedure, and convert the infinite series (2.40) and (2.41) into integrals. To do this, we use the following expression of the Bessel function (Sommerfeld, 1935):

$$J_{v_l}^{(kr)} = \frac{1}{2\pi} \int_I^{III} e^{i[kr\cos\beta + v_l(\beta - \pi/2)]} \, d\beta,$$ (2.43)

where the integration contour is shown in Figure 2.4. Then the function $u(kr, \psi)$ can be represented as

$$u(kr, \psi) = \frac{1}{2\alpha} \int_I^{III} e^{ikr\cos\beta} \left[1 + \sum_{l=1}^{\infty} e^{iv_l(\beta - \pi + \psi)} + \sum_{l=1}^{\infty} e^{iv_l(\beta - \pi - \psi)} \right] d\beta,$$ (2.44)

with $v_l = l\pi/\alpha$. Here, the series are geometrical progressions. Their summation leads to

$$u(kr, \psi) = \frac{1}{2\alpha} \int_I^{III} e^{ikr\cos\beta} \left[\frac{1}{1 - e^{i\frac{\pi}{\alpha}(\beta-\pi+\psi)}} - \frac{1}{1 - e^{-i\frac{\pi}{\alpha}(\beta-\pi-\psi)}} \right] d\beta. \quad (2.45)$$

With a new variable $\beta' = \beta - \pi$, this becomes

$$u(kr, \psi) = \frac{1}{2\alpha} \int_{I'}^{III'} e^{-ikr\cos\beta'} \left[\frac{1}{1 - e^{i\frac{\pi}{\alpha}(\beta'+\psi)}} - \frac{1}{1 - e^{-i\frac{\pi}{\alpha}(\beta'-\psi)}} \right] d\beta'. \quad (2.46)$$

Here, the integration contour is shifted by $-\pi$ compared to the contour shown in Figure 2.4. According to the difference inside the brackets, the function $u(kr, \psi)$ can be represented as the sum of two integrals. In the integral related to the first term inside the brackets, we replace β' by β. In the integral related to the second term inside the brackets, we change β' by $-\beta$. As a result, we arrive at the Sommerfeld integral (Sommerfeld, 1935)

$$u(kr, \psi) = \frac{1}{2\alpha} \int_C \frac{e^{-ikr\cos\beta}}{1 - e^{i\frac{\pi}{\alpha}(\beta+\psi)}} d\beta. \quad (2.47)$$

The integration contour C, consisting of two branches, is shown in Figure 2.5.

The integrand of $u(kr, \psi)$ possesses the first-order poles

$$\beta_m = 2\alpha m - \psi, \qquad \text{with } m = 0, \pm 1, \pm 2, \pm 3, \ldots. \quad (2.48)$$

Application of the Cauchy theorem to the integral over the closed contour C–D (Fig. 2.5) results in

$$u(kr, \psi) = v(kr, \psi) + e^{-ikr\cos\psi}, \qquad \text{with } -\pi < \psi < \pi,$$

$$u(kr, \psi) = v(kr, \psi), \qquad \text{with } \pi < \psi < 2\alpha - \pi, \qquad (2.49)$$

$$u(kr, \psi) = v(kr, \psi) + e^{-ikr\cos(2\alpha-\psi)}, \qquad \text{with } 2\alpha - \pi < \psi < 2\alpha + \pi,$$

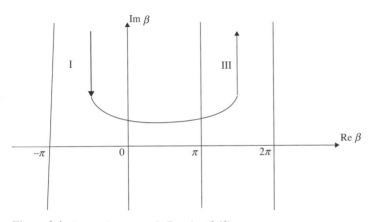

Figure 2.4   Integration contour in Equation (2.43).

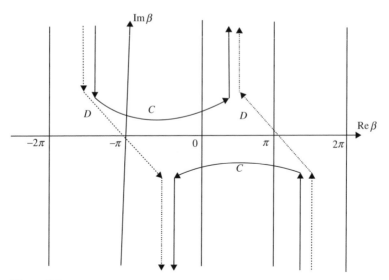

Figure 2.5 Integration contours in Equations (2.47) and (2.50).

where

$$v(kr, \psi) = \frac{1}{2\alpha} \int_D \frac{e^{-ikr\cos\beta}}{1 - e^{i\frac{\pi}{\alpha}(\beta+\psi)}} \, d\beta. \tag{2.50}$$

The integration contour D consists of two branches (Fig. 2.5). In the integral over the left branch, we replace the variable β by $\zeta - \pi$, and in the integral over the right branch we put $\beta = \zeta + \pi$. Then the function $v(kr, \psi)$ transforms into the integral over the contour D_0 (Fig. 2.6)

$$v(kr, \psi) = i \frac{\sin\frac{\pi}{n}}{2\pi n} \int_{D_0} \frac{e^{ikr\cos\zeta}}{\cos\frac{\pi}{n} - \cos\frac{\zeta+\psi}{n}} \, d\zeta \tag{2.51}$$

where $n = \alpha/\pi$.

The physical interpretation of Equation (2.49) is the following. The function $v(kr, \psi)$ describes the diffracted part of the field, and the residues relate to the geometrical optics. This interpretation becomes clear if we consider functions $u(kr, \varphi - \varphi_0)$ and $u(kr, \varphi + \varphi_0)$.

In the case $0 < \varphi_0 < \alpha - \pi$, when only one face ($\varphi = 0$) of the wedge is illuminated (Fig. 2.7), these functions are determined by

$$\left.\begin{array}{l} u(kr, \varphi - \varphi_0) = v(kr, \varphi - \varphi_0) + e^{-ikr\cos(\varphi-\varphi_0)} \\ u(kr, \varphi + \varphi_0) = v(kr, \varphi + \varphi_0) + e^{-ikr\cos(\varphi+\varphi_0)} \end{array}\right\} \quad \text{with } 0 < \varphi < \pi - \varphi_0 \tag{2.52}$$

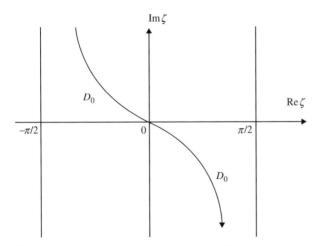

Figure 2.6 Integration contour in Equation (2.51).

$$
\left.\begin{aligned}
u(kr, \varphi - \varphi_0) &= v(kr, \varphi - \varphi_0) + e^{-ikr\cos(\varphi - \varphi_0)} \\
u(kr, \varphi + \varphi_0) &= v(kr, \varphi + \varphi_0)
\end{aligned}\right\} \quad \text{with } \pi - \varphi_0 < \varphi < \pi + \varphi_0
$$

$$(2.53)$$

$$
\left.\begin{aligned}
u(kr, \varphi - \varphi_0) &= v(kr, \varphi - \varphi_0) \\
u(kr, \varphi + \varphi_0) &= v(kr, \varphi + \varphi_0)
\end{aligned}\right\} \quad \text{with } \pi + \varphi_0 < \varphi < \alpha. \qquad (2.54)
$$

In Equations (2.52) and (2.53), the term $e^{-ikr\cos(\varphi - \varphi_0)}$ determines the incident plane wave, which exists only in the illuminated region, $0 < \varphi < \pi + \varphi_0$, and the term $e^{-ikr\cos(\varphi + \varphi_0)}$ relates to the reflected plane wave existing in the region $0 < \varphi < \pi - \varphi_0$. In agreement with the geometrical optics, Equation (2.54) does not contain either the incident or reflected plane waves, because the region $\pi + \varphi_0 < \varphi < \alpha$ is shadowed by the wedge. The boundaries of the incident and reflected plane waves are shown in Figure 2.7.

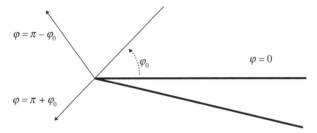

Figure 2.7 The incident plane wave propagates from the direction $\varphi = \varphi_0$. The line $\varphi = \pi - \varphi_0$ is the boundary of the reflected wave, and the line $\varphi = \pi + \varphi_0$ is the shadow boundary.

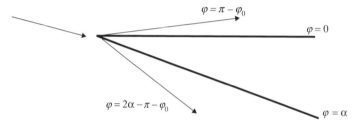

Figure 2.8   The incident plane wave illuminates both faces of the wedge. The line $\varphi = \pi - \varphi_0$ is the boundary of the wave reflected from the face $\varphi = 0$, and the line $\varphi = 2\alpha - \pi - \varphi_0$ is the boundary of the wave reflected from the face $\varphi = \alpha$.

Figure 2.8 illustrates the situation when both faces of the wedge are illuminated. In this case $\alpha - \pi < \varphi_0 < \pi$, and functions $u(kr, \varphi - \varphi_0)$, $u(kr, \varphi + \varphi_0)$ are determined by

$$\left. \begin{aligned} u(kr, \varphi - \varphi_0) &= v(kr, \varphi - \varphi_0) + e^{-ikr\cos(\varphi - \varphi_0)} \\ u(kr, \varphi + \varphi_0) &= v(kr, \varphi + \varphi_0) + e^{-ikr\cos(\varphi + \varphi_0)} \end{aligned} \right\} \quad \text{with } 0 < \varphi < \pi - \varphi_0$$

$$(2.55)$$

$$\left. \begin{aligned} u(kr, \varphi - \varphi_0) &= v(kr, \varphi - \varphi_0) + e^{-ikr\cos(\varphi - \varphi_0)} \\ u(kr, \varphi + \varphi_0) &= v(kr, \varphi + \varphi_0) \end{aligned} \right\} \quad \text{with } \pi - \varphi_0 < \varphi < 2\alpha - \pi - \varphi_0$$

$$(2.56)$$

and

$$\left. \begin{aligned} u(kr, \varphi - \varphi_0) &= v(kr, \varphi - \varphi_0) + e^{-ikr\cos(\varphi - \varphi_0)} \\ u(kr, \varphi + \varphi_0) &= v(kr, \varphi + \varphi_0) + e^{-ikr\cos(2\alpha - \varphi - \varphi_0)} \end{aligned} \right\} \quad \text{with } 2\alpha - \pi - \varphi_0 < \varphi < \alpha$$

$$(2.57)$$

The term $e^{-ikr\cos(2\alpha - \varphi - \varphi_0)}$ describes the plane wave reflected from the face $\varphi = \alpha$.

2.4 THE SOMMERFELD RAY ASYMPTOTICS

> The relationships $u_s = E_z$ and $u_h = H_z$ exist between the acoustic and electromagnetic edge-diffracted rays.

A simple asymptotic expression for the function $v(kr, \psi)$ with $kr \gg 1$ can be found by the steepest descent method (Copson, 1965; Murray, 1984). With this purpose we replace the integration variable in Equation (2.51) by

$$s = \sqrt{2}e^{i\pi/4} \sin \frac{\zeta}{2}. \tag{2.58}$$

Then $s^2 = i(1 - \cos \zeta)$ and

$$v(kr, \psi) = \frac{\sin \dfrac{\pi}{n}}{n\pi \sqrt{2}} e^{i(kr+\pi/4)} \int_{-\infty}^{\infty} \frac{e^{-krs^2}\, ds}{\left(\cos \dfrac{\pi}{n} - \cos \dfrac{\psi + \zeta}{n}\right)\cos \dfrac{\zeta}{2}} \qquad (2.59)$$

where $n = \alpha/\pi$.

Here, $s = 0$ is the saddle point. Indeed, when the point s moves from the saddle point along the imaginary axis, the function $\exp(-krs^2)$ increases most rapidly. In contrast, this function decreases most rapidly when the point s moves away from the saddle point along the real axis. Because of that, the vicinity of the saddle point provides the major contribution to the integral when $kr \gg 1$. According to the steepest descent method, the slowly varying factor of the integrand is expanded into the Taylor power series near the saddle point, and then it is integrated term by term. If the integrand expansion is convergent only in the vicinity of the saddle point, the series obtained after integration will be semiconvergent, that is, asymptotic. Retaining the first term in this series for the function $v(kr, \psi)$, we obtain

$$v(kr, \psi) \sim \frac{\sin \dfrac{\pi}{n}}{n\pi \sqrt{2}} \frac{e^{i(kr+\pi/4)}}{\cos \dfrac{\pi}{n} - \cos \dfrac{\psi}{n}} \int_{-\infty}^{\infty} e^{-krs^2}\, ds = \frac{\dfrac{1}{n}\sin \dfrac{\pi}{n}}{\cos \dfrac{\pi}{n} - \cos \dfrac{\psi}{n}} \frac{e^{i(kr+\pi/4)}}{\sqrt{2\pi kr}}.$$

$$(2.60)$$

The next terms of the asymptotic series for function $v(kr, \psi)$ are small quantities of order $(kr)^{-3/2}$ and higher. The asymptotic expression (2.60) is valid under the condition $\sqrt{kr}|\cos(\psi/2)| \gg 1$ and describes cylindrical waves diverging from the edge, that is, the edge waves.

According to Equations (2.40), (2.49), and (2.60), the wave diffracted at the edge of the acoustically soft wedge is determined as

$$u_s^d = u_0[v(kr, \varphi - \varphi_0) - v(kr, \varphi + \varphi_0)] \sim u_0 f(\varphi, \varphi_0, \alpha) \frac{e^{i(kr+\pi/4)}}{\sqrt{2\pi kr}}, \qquad (2.61)$$

where

$$f(\varphi, \varphi_0, \alpha) = \frac{\sin \dfrac{\pi}{n}}{n} \left(\frac{1}{\cos \dfrac{\pi}{n} - \cos \dfrac{\varphi - \varphi_0}{n}} - \frac{1}{\cos \dfrac{\pi}{n} - \cos \dfrac{\varphi + \varphi_0}{n}} \right). \qquad (2.62)$$

Equations (2.41), (2.49), and (2.60) determine the wave arising at the edge of the acoustically hard wedge

$$
u_{\mathrm{h}}^{\mathrm{d}} = u_0[v(kr, \varphi - \varphi_0) + v(kr, \varphi + \varphi_0)] \sim u_0 g(\varphi, \varphi_0, \alpha) \frac{e^{i(kr + \pi/4)}}{\sqrt{2\pi kr}}, \qquad (2.63)
$$

where

$$
g(\varphi, \varphi_0, \alpha) = \frac{\sin \dfrac{\pi}{n}}{n} \left(\frac{1}{\cos \dfrac{\pi}{n} - \cos \dfrac{\varphi - \varphi_0}{n}} + \frac{1}{\cos \dfrac{\pi}{n} - \cos \dfrac{\varphi + \varphi_0}{n}} \right). \qquad (2.64)
$$

Functions f and g describe the directivity patterns of the edge waves. The asymptotic expressions (2.60) to (2.64) were introduced by Sommerfeld (1935) and are well known. It is easy to verify that they satisfy the boundary conditions (2.2) and (2.3).

One should mention that the two-dimensional edge waves (2.61) and (2.63) can be interpreted as continuous sets of edge diffracted rays. They arise due to diffraction, but propagate from the edge in the first asymptotic approximation as ordinary rays in accordance with geometrical optics laws for 2-D fields. As shown in the work of Pelosi et al. (1998), the edge diffracted rays had already been visually observed already by Newton, although he did not use such a terminology. The term "diffracted ray" was introduced by Kalashnikov (1912), who was also the first to present an objective experimental proof of the existence of edge diffracted rays by recording them on a photographic plate. Theoretically, their existence was established first by Rubinowicz (1924) and later on by many other researchers. Keller (1962) formulated the concept of diffracted rays in a general form.

Here it is also pertinent to remind one about Sommerfeld's warning against the too formal ray interpretation of diffraction phenomena. He wrote that shining diffraction points on edges do not exist in reality and they are just optical illusions: "Das ist naturlich eine optische Tauschung" (Sommerfeld, 1896, p. 369). He explained that such seemingly shining edge points are the result of our perception, or in Sommerfeld's words, the result of "analytical continuation" of diffracted rays by our eyes.

Because of the ray structure of edge waves, the asymptotic expressions (2.61) to (2.64) derived above can be called the ray asymptotics, as emphasized in the title of the present section. These asymptotics have an essential drawback. They are not valid near the shadow boundary ($\varphi \approx \pi + \varphi_0$) and near the boundaries of reflected plane waves ($\varphi \approx \pi - \varphi_0, \varphi \approx 2\alpha - \pi - \varphi_0$). The mathematical reason for this drawback is given in the following. Two poles

$$
s_1 = \sqrt{2} e^{i\pi/4} \sin \frac{\pi - \psi}{2}, \qquad s_2 = \sqrt{2} e^{i\pi/4} \sin\left(\alpha - \frac{\pi + \psi}{2}\right) \qquad (2.65)
$$

of the integrand in Equation (2.59) approach the saddle point $s = 0$ when $\psi = \varphi \pm \varphi_0 \to \pi$ and $\psi = \varphi + \varphi_0 \to 2\alpha - \pi$. In this case, the Taylor expansion for the integrand becomes meaningless because its terms tend to infinity. There is a physical background behind this mathematics. The vicinity of boundaries of incident and reflected waves is the region of the effective transverse diffusion where the field cannot be described in terms of diffracted rays and has a more complex structure. This phenomenon is considered in detail in Section 5.5 of Ufimtsev (2003).

2.5 THE PAULI ASYMPTOTICS

In 1938, Pauli suggested the asymptotic expansion for the function $v(kr, \psi)$ that is valid at the geometrical optics boundaries $\varphi = \pi \pm \varphi_0$ and transforms to the Sommerfeld asymptotics away from these boundaries (Pauli, 1938). In this section we provide the derivation for the first term of the Pauli expansion. Usually, for engineering analysis, only the first terms in asymptotic expansions are of practical value. Higher-order terms commonly are not utilized, because they are smaller in magnitude, and are quite complicated to evaluate. Besides, the high-order terms can occur beyond the frames of validity of idealized mathematical models used for description of real physical phenomena. That is why we focus here on the first asymptotic term.

According to Equation (2.59),

$$v(kr, \psi) = \frac{\sin \dfrac{\pi}{n}}{n\pi \sqrt{2}} e^{i(kr+\pi/4)} \int_{-\infty}^{\infty} \frac{e^{-krs^2} \, ds}{\left(\cos \dfrac{\pi}{n} - \cos \dfrac{\psi + \zeta}{n}\right) \cos \dfrac{\zeta}{2}}, \qquad (2.66)$$

where

$$s = \sqrt{2} e^{i\pi/4} \sin \frac{\zeta}{2}, \qquad s^2 = i(1 - \cos \zeta), \qquad n = \frac{\alpha}{\pi}. \qquad (2.67)$$

Let us multiply and divide the integrand by

$$\cos \psi + \cos \zeta = i(s^2 - s_0^2) \qquad \text{with } s_0^2 = 2i \cos^2 \frac{\psi}{2}. \qquad (2.68)$$

Then

$$v(kr, \psi) = \frac{\sin \dfrac{\pi}{n}}{n\pi \sqrt{2}} e^{i(kr-\pi/4)} \int_{-\infty}^{\infty} f(s, \psi) \frac{e^{-krs^2}}{s^2 - s_0^2} \, ds, \qquad (2.69)$$

where the poles $s = \pm s_0 = \pm \sqrt{2} e^{i\frac{\pi}{4}} \cos \frac{\psi}{2}$ are outside the integration contour and approach it at the saddle point $s = 0$ when $\psi \to \pi$. The function

$$f(s, \psi) = \frac{\cos \psi + \cos \zeta}{\left(\cos \dfrac{\pi}{n} - \cos \dfrac{\psi + \zeta}{n}\right) \cos \dfrac{\zeta}{2}} \qquad (2.70)$$

does not have a pole at the saddle point $s = 0$ ($\zeta = 0$) when $\psi = \varphi \pm \varphi_0 \to \pi$. Therefore it can be expanded into a regular Taylor series. By integrating this series term by term, Pauli obtained the asymptotic expansion of function $v(kr, \psi)$ for large argument kr. The first term of this expansion is determined by

$$v(kr, \psi) = \frac{\sqrt{2} \sin \dfrac{\pi}{n}}{\pi} \cdot \frac{1 + \cos \psi}{\cos \dfrac{\pi}{n} - \cos \dfrac{\psi}{n}} e^{i(kr - \pi/4)} \int_0^\infty \frac{e^{-krs^2}}{s^2 - s_0^2} ds. \qquad (2.71)$$

Here the integral can be represented as

$$\int_0^\infty \frac{e^{-krs^2}}{s^2 - s_0^2} ds = e^{-krs_0^2} \int_0^\infty ds \int_{kr}^\infty e^{-(s^2 - s_0^2)t} dt. \qquad (2.72)$$

By changing the order of integration we obtain

$$\int_0^\infty \frac{e^{-krs^2}}{s^2 - s_0^2} ds = e^{-krs_0^2} \int_{kr}^\infty e^{s_0^2 t} \frac{dt}{\sqrt{t}} \int_0^\infty e^{-x^2} dx = \frac{\sqrt{\pi}}{2} e^{-krs_0^2} \int_{kr}^\infty e^{i|s_0|^2 t} \frac{dt}{\sqrt{t}}$$

$$= \frac{\sqrt{\pi}}{|s_0|} e^{-ikr|s_0|^2} \int_{\sqrt{kr}|s_0|}^\infty e^{iq^2} dq. \qquad (2.73)$$

As a result,

$$v(kr, \psi) = \frac{2}{n} \frac{\sin \dfrac{\pi}{n} \left| \cos \dfrac{\psi}{2} \right|}{\cos \dfrac{\pi}{n} - \cos \dfrac{\psi}{n}} e^{-ikr \cos \psi} \frac{e^{-i\frac{\pi}{4}}}{\sqrt{\pi}} \int_{\sqrt{2kr}\left|\cos \frac{\psi}{2}\right|}^\infty e^{iq^2} dq \qquad (2.74)$$

or

$$v(kr, \psi) = \frac{2}{n} \frac{\sin \dfrac{\pi}{n} \cos \dfrac{\psi}{2}}{\cos \dfrac{\pi}{n} - \cos \dfrac{\psi}{n}} e^{-ikr \cos \psi} \frac{e^{-i\frac{\pi}{4}}}{\sqrt{\pi}} \cdot \int_{\sqrt{2kr} \cos \frac{\psi}{2}}^{\infty \cos \frac{\psi}{2}} e^{iq^2} dq. \qquad (2.75)$$

This expression represents the slightly modified first term in the Pauli asymptotic expansion. The next term is of order $(kr)^{-1/2}$ near the boundaries $\varphi = \pi \pm \varphi_0$ and it is of order $(kr)^{-3/2}$ away from them.

The upper limit of the Fresnel integral in Equation (2.75) should be read as $\mathrm{sgn}(\cos \psi/2)\infty$. It always equals infinity but changes its sign when the observation point intersects the geometrical optics boundaries ($\psi = \varphi \pm \varphi_0 = \pi$). Here, the function $v(kr, \psi)$ undergoes the discontinuity and in this way it ensures the continuity of the function $u(kr, \psi)$ and therefore the continuity of the total field. Indeed, by using the formula

$$\int_0^\infty e^{iq^2} dq = \frac{\sqrt{\pi}}{2} e^{i\frac{\pi}{4}} \qquad (2.76)$$

one can show that

$$v(kr, \pi + 0) = \frac{1}{2}e^{ikr}, \qquad v(kr, \pi - 0) = -\frac{1}{2}e^{ikr} \tag{2.77}$$

and

$$u(kr, \pi \pm 0) = \frac{1}{2}e^{ikr}. \tag{2.78}$$

Also, with the help of the asymptotic approximations,

$$\int_{\infty}^{p} e^{iq^2}\,dq \sim \frac{e^{ip^2}}{2ip}, \qquad \int_{-\infty}^{-p} e^{iq^2}\,dq \sim -\frac{e^{ip^2}}{2ip}, \qquad \text{with } p \gg 1, \tag{2.79}$$

it is easy to verify that the Pauli expression (2.75) converts to the Sommerfeld asymptotics (2.60) under the condition $\sqrt{kr}|\cos \psi/2| \gg 1$.

As shown in Section 5.5 of Ufimtsev (2003), the Pauli asymptotics (2.75) can be considered as a "stenographic form" of the more physically meaningful expression

$$v(kr, \psi) = V\left[\sqrt{\frac{kr}{2}}(\psi - \pi)\right]e^{ikr} + \left(\frac{\frac{1}{n}\sin\frac{\pi}{n}}{\cos\frac{\pi}{n} - \cos\frac{\psi}{n}} - \frac{1}{\psi - \pi}\right)\frac{e^{i\left(kr+\frac{\pi}{4}\right)}}{\sqrt{2\pi kr}} \tag{2.80}$$

with

$$V(\tau) = e^{-i\tau^2}\frac{e^{-i\frac{\pi}{4}}}{\sqrt{\pi}}\int_{\tau}^{\infty \operatorname{sgn} \tau} e^{iq^2}\,dq, \tag{2.81}$$

which follows from the solution of the parabolic equation. Here, the first term $V[\sqrt{kr/2}(\psi - \pi)]e^{ikr}$ describes the transverse diffusion of the wave field in the vicinity of the geometrical optics boundaries and does not depend on the reflective properties of the wedge faces. The second term in Equation (2.80) can be interpreted as the diffraction background.

It is of interest that in the particular case when the angle $\alpha = 2\pi$ and the wedge transforms into the half-plane, the Pauli asymptotics (2.75) transforms to the function

$$v(kr, \psi) = e^{-ikr\cos\psi}\frac{e^{-i\pi/4}}{\sqrt{\pi}}\int_{\infty\cos\frac{\psi}{2}}^{\sqrt{2kr}\cos\frac{\psi}{2}} e^{iq^2}\,dq \tag{2.82}$$

and provides the exact (!) solution to the half-plane diffraction problem. Indeed, in this case, $n = 2$ and Equation (2.51) becomes

$$v(kr, \psi) = -\frac{i}{4\pi}\int_{D_0} \frac{e^{ikr\cos\zeta}}{\cos\dfrac{\psi + \zeta}{2}}\,d\zeta, \tag{2.83}$$

which can be converted to the Fresnel integral. To do this, let us separate the contour D_0 (see Fig. 2.6) into two parts at the point $\zeta = 0$. Summation of the integrals over these parts of the integration contour leads to the expression

$$v(kr, \psi) = -\frac{i}{4\pi} \int_0^{\frac{\pi}{2}-i\infty} e^{ikr\cos\zeta} \left(\frac{1}{\cos\dfrac{\psi+\zeta}{2}} + \frac{1}{\cos\dfrac{\psi-\zeta}{2}} \right) d\zeta$$

$$= -\frac{i}{\pi} \cos\frac{\psi}{2} \int_0^{\frac{\pi}{2}-i\infty} e^{ikr\cos\zeta} \left(\frac{\cos\dfrac{\zeta}{2}}{\cos\psi + \cos\zeta} \right) d\zeta. \qquad (2.84)$$

We then introduce the integration variable $s = \sqrt{2}e^{i\frac{\pi}{4}} \sin\frac{\zeta}{2}$ and apply the procedure outlined in Equations (2.68) to (2.71). As a result we obtain

$$v(kr, \psi) = -\frac{\sqrt{2}}{\pi} \cos\frac{\psi}{2} e^{i(kr-\frac{\pi}{4})} \int_0^\infty \frac{e^{-krs^2}}{s^2 - s_0^2} ds. \qquad (2.85)$$

With the help of Equation (2.73) this expression transforms to Equation (2.82). The latter, together with Equations (2.39) to (2.41) and (2.52) to (2.54) provides the exact solution to the half-plane diffraction problem.

Thus, the Pauli asymptotics (2.75) possesses valuable properties. It is simple. It provides the exact solution to the half-plane diffraction problem. It describes both the transverse diffusion of the wave field near the geometrical optics boundaries and the diffracted rays away from these boundaries. However, it is not free from certain drawbacks. These drawbacks are as follows:

- The total field $u_{s,h}$ determined with the Pauli asymptotics (2.75) exactly satisfies the boundary conditions (2.2) and (2.3) on the face $\varphi = 0$. However, on face $\varphi = \alpha$, these boundary conditions are satisfied only asymptotically, when $\sqrt{kr}|\cos\psi/2| \gg 1$ and the Pauli asymptotics converts to the Sommerfeld expression (2.60).

- The Pauli asymptotics (2.75) provides correct values for the wave field in the direction of the shadow boundary ($\varphi = \pi + \varphi_0$) and in the direction $\varphi = \pi - \varphi_0$ of the plane wave reflected from the face $\varphi = 0$. However, it fails at the direction $\varphi = 2\alpha - \pi - \varphi_0$ of the plane wave reflected from the face $\varphi = \alpha$ (Fig. 2.8). It predicts a wrong infinite value for the field in this direction.

One can suggest the following remedy to diminish these drawbacks, to some extent. The asymptotics (2.75) should be used only in the region $0 \le \varphi \le \alpha/2$. In order to calculate the field in the rest of the region $\alpha/2 \le \varphi \le \alpha$, it is necessary to introduce new polar coordinates with the angle φ' measured from the face $\varphi = \alpha$ and then to apply the expression (2.75) in the region $0 < \varphi' \le \alpha/2$. In this way, one can obtain correct values for the field at the boundary of the plane wave reflected from the face $\varphi = \alpha$ and satisfy the boundary conditions on this face, but at the expense of the field discontinuity in the direction $\varphi = \alpha/2$.

The mentioned discontinuity of the field at $\varphi = \alpha/2$ is manifestation of the fact that asymptotics (2.75) does not satisfy the fundamental physical principle. It is not invariant with respect to choice of the coordinate system. Indeed, if we choose the polar coordinates φ' and φ_0' measured from the face $\varphi = \alpha$, the Pauli asymptotics leads to the relationships

$$v^{\text{Pauli}}(kr, \varphi' - \varphi_0') = v^{\text{Pauli}}[kr, \alpha - \varphi - (\alpha - \varphi_0)] = v^{\text{Pauli}}(kr, \varphi - \varphi_0) \quad (2.86)$$

and

$$v^{\text{Pauli}}(kr, \varphi' + \varphi_0') = v^{\text{Pauli}}(kr, 2\alpha - \varphi - \varphi_0) \neq v^{\text{Pauli}}(kr, \varphi + \varphi_0). \quad (2.87)$$

The last inequality indicates that, strictly speaking, the Pauli asymptotics does not satisfy the invariance principle. However, it satisfies this principle approximately, when it transforms into the Sommerfeld ray asymptotics.

In the next section, we derive new asymptotics applicable in all scattering directions ($0 < \varphi < \alpha$).

2.6 UNIFORM ASYMPTOTICS: EXTENSION OF THE PAULI TECHNIQUE

Here we derive asymptotic expressions under the condition that the incident wave does not undergo double and higher-order multiple reflections at faces of the wedge. This condition is always realized for convex wedges ($\pi < \alpha \leq 2\pi$) and also for the concave wedges/horns ($\pi/2 < \alpha < \pi$), but only for certain directions of the incident wave. However, the theory developed below can be easily extended for any narrow horns ($0 < \alpha < \pi/2$) with multiple reflections.

Now we return to Equation (2.66) and we observe that only two poles,

$$s_1 = \sqrt{2}e^{i\pi/4}\cos\frac{\psi}{2} \quad \text{and} \quad s_2 = -\sqrt{2}e^{i\pi/4}\cos\left(\alpha - \frac{\psi}{2}\right), \quad (2.88)$$

can approach the saddle point $s = 0$ when $\psi = \varphi \pm \varphi_0 \to \pi$ or $\psi = \varphi + \varphi_0 \to 2\alpha - \pi$. The pole s_1 approaches the saddle point when the direction of observation φ tends to the shadow boundary $\varphi = \pi + \varphi_0$ or to the boundary $\varphi = \pi - \varphi_0$ of the wave reflected from the face $\varphi = 0$ (Fig. 2.7). The pole s_2 approaches the saddle point when the direction φ tends to the boundary $\varphi = 2\alpha - \pi - \varphi_0$ of the wave reflected from the face $\varphi = \alpha$ (Fig. 2.8). All other poles in (2.66) can be ignored as they are aside the integration contour and never reach the saddle point in the absence of multiple reflections.

Taking these observations into account we multiply and divide the integrand in Equation (2.66) by the factor

$$(\cos\zeta + \cos\psi)[\cos\zeta + \cos(2\alpha - \psi)] = -(s^2 - s_1^2)(s^2 - s_2^2) \quad (2.89)$$

and obtain

$$v(kr, \psi) = -\frac{\sin \dfrac{\pi}{n}}{n\pi \sqrt{2}} e^{i(kr+\pi/4)} \int_{-\infty}^{\infty} f(s, \psi) \frac{e^{-krs^2}}{(s^2 - s_1^2)(s^2 - s_2^2)} ds, \qquad (2.90)$$

where

$$f(s, \psi) = \frac{(\cos \zeta + \cos \psi)[\cos \zeta + \cos(2\alpha - \psi)]}{\left(\cos \dfrac{\pi}{n} - \cos \dfrac{\psi + \zeta}{n}\right) \cos \dfrac{\zeta}{2}}. \qquad (2.91)$$

This function is finite and continuous in the vicinity of the saddle point $s = 0$ and therefore it can be expanded into the Taylor series. The integration of this series in Equation (2.90) leads to the uniform asymptotic expansion for $v(kr, \psi)$ valid for any angles φ and φ_0. One should note that only the even terms with factors s^{2m} ($m = 1, 2, 3, \ldots$) give nonzero contributions to the integral in Equation (2.90). We retain and calculate here the two first terms of this asymptotic expansion. The second term is retained because for $\alpha \to 2\pi$ it partially contains the quantity of the same order as the first term. The related asymptotic expression for the function $v(kr, \psi)$ is determined by

$$v(kr, \psi) = v_1(kr, \psi) + v_2(kr, \psi) \qquad (2.92)$$

where

$$v_1(kr, \psi) = -\frac{\sin \dfrac{\pi}{n}}{n\pi \sqrt{2}} e^{i(kr+\pi/4)} f(0, \psi) \int_{-\infty}^{\infty} \frac{e^{-krs^2} ds}{(s^2 - s_1^2)(s^2 - s_2^2)}, \qquad (2.93)$$

$$v_2(kr, \psi) = -\frac{\sin \dfrac{\pi}{n}}{n\pi 2\sqrt{2}} e^{i(kr+\pi/4)} \frac{d^2 f(0, \psi)}{ds^2} \int_{-\infty}^{\infty} \frac{e^{-krs^2} s^2 ds}{(s^2 - s_1^2)(s^2 - s_2^2)}, \qquad (2.94)$$

and

$$\int_{-\infty}^{\infty} \frac{e^{-krs^2}}{(s^2 - s_1^2)(s^2 - s_2^2)} ds = \frac{2}{(s_1^2 - s_2^2)} \left[\int_0^{\infty} \frac{e^{-krs^2} ds}{(s^2 - s_1^2)} - \int_0^{\infty} \frac{e^{-krs^2} ds}{(s^2 - s_2^2)} \right], \qquad (2.95)$$

$$\int_{-\infty}^{\infty} \frac{e^{-krs^2} s^2}{(s^2 - s_1^2)(s^2 - s_2^2)} ds = \frac{2}{(s_1^2 - s_2^2)} \left[\int_0^{\infty} \frac{e^{-krs^2} s^2 ds}{(s^2 - s_1^2)} - \int_0^{\infty} \frac{e^{-krs^2} s^2 ds}{(s^2 - s_2^2)} \right], \qquad (2.96)$$

$$\int_0^{\infty} \frac{e^{-krs^2} s^2}{s^2 - s_{1,2}^2} ds = \int_0^{\infty} e^{-krs^2} ds + s_{1,2}^2 \int_0^{\infty} \frac{e^{-krs^2}}{s^2 - s_{1,2}^2} ds. \qquad (2.97)$$

According to Equation (2.73), these integrals are reduced to the Fresnel integrals. As a result we obtain the following asymptotic expression

$$
v(kr, \psi) = \frac{\dfrac{1}{n} \sin \dfrac{\pi}{n}}{\cos \dfrac{\pi}{n} - \cos \dfrac{\psi}{n}} \cdot \frac{1}{\sin \alpha \sin(\alpha - \psi)} \cdot \frac{e^{-i\frac{\pi}{4}}}{\sqrt{\pi}}
$$

$$
\left[P(\alpha, \psi) \cos^3 \left(\frac{\psi}{2} \right) e^{-ikr \cos \psi} \int_{\sqrt{2kr} \cos \frac{\psi}{2}}^{\infty \cos \frac{\psi}{2}} e^{iq^2} dq \right.
$$

$$
\left. - Q(\alpha, \psi) \cos^3 \left(\alpha - \frac{\psi}{2} \right) e^{-ikr \cos(2\alpha - \psi)} \int_{\sqrt{2kr} \cos(\alpha - \frac{\psi}{2})}^{\infty \cos(\alpha - \frac{\psi}{2})} e^{iq^2} dq \right]
$$

$$(2.98)$$

where

$$
P(\alpha, \psi) = 2 - \left[\cos \left(\alpha - \frac{\psi}{2} \right) \right]^2 \left[1 + \frac{4}{n^2} \left(\frac{1 + \sin^2 \dfrac{\psi}{n} - \cos \dfrac{\pi}{n} \cos \dfrac{\psi}{n}}{\left(\cos \dfrac{\pi}{n} - \cos \dfrac{\psi}{n} \right)^2} \right) \right]
$$

$$(2.99)$$

and

$$
Q(\alpha, \psi) = 2 - \left(\cos \frac{\psi}{2} \right)^2 \left[1 + \frac{4}{n^2} \frac{1 + \sin^2 \dfrac{\psi}{n} - \cos \dfrac{\pi}{n} \cos \dfrac{\psi}{n}}{\left(\cos \dfrac{\pi}{n} - \cos \dfrac{\psi}{n} \right)^2} \right].
$$

$$(2.100)$$

One can show that away from the boundaries of reflected plane waves, where $\sqrt{kr} \left| \cos \frac{\psi}{2} \right| \gg 1$ and $\sqrt{kr} \left| \cos(\alpha - \frac{\psi}{2}) \right| \gg 1$, the asymptotics (2.98) transforms into the ray-type asymptotics

$$
v(kr, \psi) \sim \frac{\dfrac{1}{n} \sin \dfrac{\pi}{n}}{\cos \dfrac{\pi}{n} - \cos \dfrac{\psi}{n}} \frac{e^{i(kr + \pi/4)}}{\sqrt{2\pi kr}} + \frac{\dfrac{1}{n} \sin \dfrac{\pi}{n}}{\cos \dfrac{\pi}{n} - \cos \dfrac{\psi}{n}}
$$

$$
\times \left[1 + \frac{4}{n^2} \left(\frac{1 + \sin^2 \dfrac{\psi}{n} - \cos \dfrac{\pi}{n} \cos \dfrac{\psi}{n}}{\left(\cos \dfrac{\pi}{n} - \cos \dfrac{\psi}{n} \right)^2} \right) \right] \frac{e^{i(kr - \pi/4)}}{4\sqrt{\pi} (2kr)^{3/2}}.
$$

$$(2.101)$$

We emphasize that only the first term here can be interpreted in terms of diffracted rays, and the second term, as well as all high-order asymptotic terms not presented in Equation (2.101), have a pronounced wave nature.

At the boundaries of the reflected waves, the function (2.98) is discontinuous,

$$v(kr, \pi \pm 0) = v(kr, 2\alpha - \pi \mp 0)$$

$$= \pm \frac{1}{2} e^{ikr} - \frac{1}{n} \cot \frac{\pi}{n} \sin \alpha e^{ikr \cos 2\alpha} \frac{e^{-i\pi/4}}{\sqrt{\pi}} \int_{\sqrt{2kr} \sin \alpha}^{\infty \sin \alpha} e^{iq^2} dq, \quad (2.102)$$

and compensates the discontinuities in the geometrical optics part of the total field. Under the condition $\sqrt{2kr} |\sin \alpha| \gg 1$, it follows from Equation (2.102) that

$$v(kr, \pi \pm 0) = v(kr, 2\alpha - \pi \mp 0) = \pm \frac{1}{2} e^{ikr} - \frac{1}{2n} \cot \frac{\pi}{n} \frac{e^{i(kr+\pi/4)}}{\sqrt{2\pi kr}}. \quad (2.103)$$

It is easy to check that the asymptotics (2.98) provides the correct result $v(kr, \psi) = 0$ in the limiting case $\alpha = \pi$, when the wedge transforms into the infinite plane and the diffracted field vanishes. Taking into account that $P(2\pi, \psi) = Q(2\pi, \psi) = 0$ and applying L'Hospital's rule, one can show that, in the other limiting case when $\alpha \to 2\pi$, the asymptotics (2.98) converts into the function (2.82) related to the exact solution of the half-plane diffraction problem.

Now let us estimate the accuracy of asymptotics (2.98). This expression represents the sum of the two first terms in the asymptotic series resulting from the term-by-term integration of the Taylor series for the integrand in Equation (2.90). Therefore, the error of Equation (2.98) is the magnitude of the order of the neglected third asymptotic term. It is determined by the integral

$$\int_{-\infty}^{\infty} \frac{e^{-krs^2} s^4}{s^2 - s_{1,2}^2} ds = \begin{cases} O[(kr)^{-3/2}] & \text{with } s_{1,2} = 0 \\ O[(kr)^{-5/2}] & \text{with } \sqrt{kr} |s_{1,2}| \gg 1, \end{cases} \quad (2.104)$$

where the values $s_{1,2} = 0$ relate to the geometrical optics boundaries of the incident and reflected waves. Note also that the asymptotic expressions for the total field based on Equation (2.98) satisfy the boundary conditions (2.2) and (2.3) not rigorously, but only asymptotically under the condition $\sqrt{kr} |s_{1,2}| \gg 1$.

In Equation (2.104) we have used the symbol $O[(kr)^m]$ to show the behavior of the integral under the condition $kr \gg 1$. This is an ordinary definition accepted in the asymptotic theory: The expression $f(x) = O(x^m)$ means that $\lim[f(x)/x^m] = \text{const}$ with $x \to \infty$. This common asymptotic terminology is used throughout the book.

As was mentioned in the previous section, the Pauli asymptotics (2.75) is not invariant with respect to the choice of the coordinate system. A similar situation happens with asymptotic expressions (2.98):

$$v[kr, \alpha - \varphi - (\alpha - \varphi_0)] \neq v(kr, \varphi - \varphi_0) \quad (2.105)$$

and

$$v[kr, \alpha - \varphi + (\alpha - \varphi_0)] = v(kr, \varphi + \varphi_0). \quad (2.106)$$

In addition, the asymptotics (2.98) does not satisfy the reciprocity principle:

$$v(kr, \varphi - \varphi_0) \neq v(kr, \varphi_0 - \varphi). \tag{2.107}$$

However, these shortcomings are admissible for asymptotic expressions, which satisfy the rigorous laws only approximately. Deviations from these laws are asymptotically small and they are beyond the accuracy of these asymptotics. For example, asymptotics (2.102) and (2.103) for the field at the geometrical optics boundaries are indeed invariant with respect to the choice of polar coordinates and according to estimations (2.104) their error is a small value of the order of $(kr)^{-3/2}$. In addition, the ray-type asymptotics (2.101) is invariant in the same sense and it rigorously satisfies the reciprocity principle.

The found asymptotics (2.98) is convenient for field analysis. In the particular case when $\alpha = 2\pi$ and the wedge transforms into the half-plane, it provides the exact solution. This asymptotics is simple and applicable for all observation angles $0 \leq \varphi \leq \alpha$. However, because of its asymptotic origination, it satisfies the boundary conditions, reciprocity, and invariance principles not rigorously, but only approximately.

The asymptotics (2.98) represents the combination of the two first terms in the total asymptotic series for the function (2.90). Calculation of higher-order terms in this asymptotic series is straightforward, but we did not derive them because of their small practical value, as mentioned in the beginning of Section 2.5.

2.7 COMMENTS ON ALTERNATIVE ASYMPTOTICS

Oberhettinger and Tuzhilin (Bowmam et al., 1987) derived alternative asymptotic solutions for the wedge diffraction problem. These authors used another technique to treat the singularities in the exact solution. They included into the rigorous integrals two special identical terms with opposite signs. One of these terms completely cancels the singularity of the original integrand. Therefore, a regular technique can be applied to derive the asymptotic expansion for this part of the field. The second additional term after integration generates the Fresnel integral that describes the basic features of the field in the vicinity of the geometrical optics boundaries. These asymptotics have a structure similar to that of Equation (2.80). The only difference is that in Equation (2.80) the Fresnel integral and the term $1/(\psi - \pi)$ relate to diffraction at a black half-plane, and in the Oberhettinger and Tuzhilin expressions the similar quantities relate to diffraction at a perfectly reflecting half-plane.

The Oberhettinger asymptotics (Bowman et al., 1987) is the alternative to the Pauli expansion. Just like the Pauli expansion, it also fails in the direction of the plane wave reflected from the face $\varphi = \alpha$ and does not satisfy rigorously the invariance principle. The asymptotic solution by Tuzhilin (Bowman et al., 1987) is the alternative to the asymptotics (2.98). With an appropriate choice of different numbers n in Equations (6.21) and (6.25) of (Bowman et al., 1987), the Tuzhilin asymptotics can properly treat different singularities but do not satisfy rigorously the boundary conditions on the face $\varphi = 0$.

Notice also that the Pauli technique was extended by Clemmow (1950) to the case when several poles of the integrand can approach the saddle point. However, he did not apply his theory to the wedge diffraction problem. Instead he utilized it for the investigation of scattering at a black half-plane that does not represent a boundary value problem.

This section studies the edge waves scattered from a wedge with semi-infinite *planar* faces. A review on edge waves arising in two-dimensional structures with *curved* concave and convex faces is given in the paper by Molinet (2005).

PROBLEMS

2.1   The incident wave $u^{\text{inc}} = u_0 \exp[-ikr\cos(\varphi - \varphi_0)]$ generates the field inside a 2-D soft corner (a wedge with the angle $\alpha < \pi$ between its faces). Derive an exact expression for the total field inside the corner with the angle $\alpha = \pi/2$. Does this field contain the edge wave or not. Explain why?

2.2   Solve the problem similar to Problem 2.1, but for the hard corner with $\alpha = \pi/2$.

2.3   The incident wave $E_z^{\text{inc}} = E_{0z} \exp[-ikr\cos(\varphi - \varphi_0)]$ generates the field inside a 2-D perfectly conducting corner (a wedge with the angle $\alpha < \pi$ between its faces). Derive an exact expression for the total field inside the corner with the angle $\alpha = \pi/2$. Does this field contain the edge wave or not. Explain why?

2.4   Solve the problem similar to Problem 2.3, but with the incident wave

$$H_z^{\text{inc}} = H_{0z} \exp[-ikr\cos(\varphi - \varphi_0)].$$

2.5   Use the exact solution (2.40) for a soft wedge ($\alpha > \pi$) and derive asymptotic approximations for the scattering sources $j_s = \partial u/\partial n$ induced on both faces of the wedge, close to the edge ($kr \ll 1$).

(a) Apply the asymptotics of the Bessel functions:

$$J_0(x) \approx 1 - \left(\frac{x}{2}\right)^2, \qquad J_v(x) \approx \frac{1}{\Gamma(v+1)} \left(\frac{x}{2}\right)^v$$

(b) Derive the asymptotic expression for the source function $j_s(kr, \alpha)$. Realize the different behavior of this function for the angle $\alpha > \pi$ and $\alpha < \pi$.

2.6   Analyze the problem similar to Problem 2.5, but for a hard wedge.

2.7   Explore the electromagnetic version of Problem 2.5 for the incident wave $E_z^{\text{inc}} = E_{0z} \exp[-ikr\cos(\varphi - \varphi_0)]$ exciting a perfectly conducting wedge. Analyze the surface currents close to the edge ($kr \ll 1$).

2.8   Explore the problem similar to Problem 2.7, but for the incident wave $H_z^{\text{inc}} = H_{0z} \exp[-ikr\cos(\varphi - \varphi_0)]$ exciting a perfectly conducting wedge. Analyze the surface currents.

2.9   Use the Sommerfeld ray asymptotics (2.61) and (2.63). Calculate the surface sources j_s and j_h far from the edge ($kr \gg 1$). They are the waves running from the edge. At which wedge (soft or hard) do these waves decrease faster? Explain the origination of the phase factor $\exp(i\pi/4)$.

2.10 Use the Sommerfeld ray asymptotics (2.61) and (2.63) for electromagnetic waves with
E- and H-polarization, respectively. Calculate the surface currents far from the edge
($kr \gg 1$). They are the waves running from the edge. Which waves (with component j_z
or j_r) decrease faster? Explain the origination of the phase factor $\exp(i\pi/4)$.

2.11 Find the first two terms of the asymptotc expansion (with $p \to \infty$) for the integrals

$$\int_p^\infty e^{ix^2} x \, dx, \qquad \int_p^\infty e^{ix^2} \frac{dx}{x^n} \quad \text{with } n \geq 1.$$

2.12 An acoustic wave hits a wedge barrier (Fig. 2.7). Behind which barrier (soft or hard)
is the intensity of the diffracted acoustic wave higher? Explore this problem utilizing the Sommerfeld ray asymptotics (2.61) and (2.63). Compute and plot the ratio
$f(\varphi, \varphi_0, \alpha)/g(\varphi, \varphi_0, \alpha)$. Explore these two examples:
(a) $\alpha = 315°$, $\varphi_0 = 45°$, $225° \leq \varphi \leq \alpha$,
(b) $\alpha = 355°$, $\varphi_0 = 45°$, $225° \leq \varphi \leq \alpha$.
Prove analytically that this ratio equals unity at the shadow boundary $\varphi = 180° + \varphi_0$.

2.13 An electromagnetic wave hits a perfectly conducting wedge (Fig. 2.7). Compare the
intensity of edge diffracted waves with E_z- and H_z-polarization. Which of them is more
intensive in the shadow region? Use the Sommerfeld ray asymptotics (2.61) and (2.63).
Compute and plot the ratio $f(\varphi, \varphi_0, \alpha)/g(\varphi, \varphi_0, \alpha)$. Explore these two examples:
(a) $\alpha = 300°$, $\varphi_0 = 45°$, $225° \leq \varphi \leq \alpha$,
(b) $\alpha = 360°$, $\varphi_0 = 45°$, $225° \leq \varphi \leq \alpha$.
Prove analytically that this ratio equals unity at the shadow boundary $\varphi = 180° + \varphi_0$.

2.14 Suppose that the incident wave $E_z^{\text{inc}} = E_{0z} \exp(ikx)$ undergoes grazing diffraction at a
perfectly conducting wedge. This is a special case of diffraction illustrated in Figure 2.8
when $\varphi_0 \to \pi$. The problem is to calculate the surface currents on the face $\varphi = 0$. (The
acoustic version of this problem is the grazing diffraction of a plane wave at a soft
wedge.) In this case, the geometrical optics part of the field above the face $\varphi = 0$ equals
zero. According to the exact solution (2.40), the total field in the vicinity of this face
equals

$$E_z = E_{0z}[v(kr, \varphi - \pi) - v(kr, \varphi + \pi)]$$

with function $v(kr, \psi)$ defined by the integral (2.51). To calculate the surface current,
complete the following steps:

(a) Write the integral expression for the current

$$j_z = -H_r = Y_0 \frac{i}{kr} \frac{dE_z}{d\varphi}, \qquad \text{with } \varphi = 0 \quad \text{and} \quad Y_0 = 1/Z_0,$$

(b) In the integrand, replace the differential operator $d/d\varphi$, by $d/d\zeta$, where ζ is the
integration variable,

(c) Integrate by parts,

(d) Apply the stationary phase technique and show that

$$j_z \approx 2E_{0z} Y_0 \frac{e^{ikr + i\pi/4}}{\sqrt{2\pi kr}} \qquad \text{with } kr \gg 1.$$

2.15 This problem is analogous to Problem 2.14. Analyze the grazing diffraction of the wave $H_z^{inc} = H_{0z} \exp(ikx)$ at a perfectly conducting wedge. Calculate the surface current induced on the face $\varphi = 0$.

(a) Use the exact solution (2.41), (2.51),

(b) Apply the stationary phase technique and show that

$$j_x \approx H_{0z} \left[1 - \frac{1}{n} \cot \frac{\pi}{n} \frac{e^{ikr + i\pi/4}}{\sqrt{2\pi kr}} \right], \qquad \text{with } kr \gg 1.$$

Chapter **3**

Wedge Diffraction: The Physical Optics Field

> The relationships $u_s^{(0)} = E_z^{(0)}$ and $u_h^{(0)} = H_z^{(0)}$ exist between the PO acoustic and electromagnetic fields. An exception is for an oblique incidence (see Section 4.3).

The exact expressions for the scattered field were derived in the previous chapter. In the present chapter, we calculate the Physical Optics (PO) part of the scattered field, that is, the field generated by the *uniform* component of the induced surface sources. It will be used in the next chapter to examine the field radiated by the *nonuniform* sources as the difference between the exact and PO fields.

3.1 ORIGINAL PO INTEGRALS

The geometry of the problem is shown in Figure 3.1. A perfectly reflecting wedge located in a homogeneous medium is excited by a plane wave

$$u^{\text{inc}} = u_0 e^{-ikr\cos(\varphi - \varphi_0)} = u_0 e^{-ik(x\cos\varphi_0 + y\sin\varphi_0)}, \tag{3.1}$$

where we assume that $0 \le \varphi_0 \le \pi$. In the PO approximation, the surface sources (induced on the face $\varphi = 0$) are determined according to Equation (1.31):

$$j_s^{(0)} = -u_0 2ik\sin\varphi_0 e^{-ikx\cos\varphi_0}, \qquad j_h^{(0)} = u_0 2e^{-ikx\cos\varphi_0}. \tag{3.2}$$

Equation (1.32) determines the field radiated by these sources:

$$u_s^{(0)} = u_0 \frac{ik\sin\varphi_0}{2\pi} \int_0^\infty e^{-ik\xi\cos\varphi_0} \, d\xi \int_{-\infty}^\infty \frac{e^{ik\sqrt{(x-\xi)^2 + y^2 + \zeta^2}}}{\sqrt{(x-\xi)^2 + y^2 + \zeta^2}} \, d\zeta \tag{3.3}$$

Fundamentals of the Physical Theory of Diffraction. By Pyotr Ya. Ufimtsev
Copyright © 2007 John Wiley & Sons, Inc.

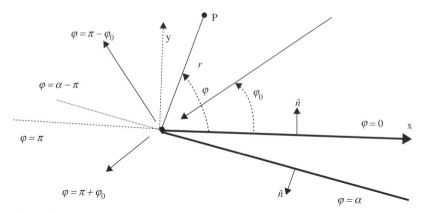

Figure 3.1 A wedge and related coordinates.

and

$$u_{\mathrm{h}}^{(0)} = u_0 \frac{1}{2\pi} \int_0^\infty e^{-ik\xi \cos\varphi_0} \, d\xi \int_{-\infty}^\infty \frac{\partial}{\partial n} \frac{e^{ik\sqrt{(x-\xi)^2+y^2+\zeta^2}}}{\sqrt{(x-\xi)^2 + y^2 + \zeta^2}} \, d\zeta. \qquad (3.4)$$

Here, we use the denotations

$$\frac{\partial}{\partial n} f(r) = \nabla' f(r) \cdot \hat{n} = -\nabla f(r) \cdot \hat{n} = -\frac{\partial}{\partial y} f(r) \qquad (3.5)$$

where ∇' and ∇ are the gradient operators applied to coordinates of the integration and observation points, respectively. In view of Equation (3.5), the field $u_{\mathrm{h}}^{(0)}$ can be written as

$$u_{\mathrm{h}}^{(0)} = -u_0 \frac{1}{2\pi} \frac{\partial}{\partial y} \int_0^\infty e^{-ik\xi \cos\varphi_0} \, d\xi \int_{-\infty}^\infty \frac{e^{ik\sqrt{(x-\xi)^2+y^2+\zeta^2}}}{\sqrt{(x-\xi)^2 + y^2 + \zeta^2}} \, d\zeta. \qquad (3.6)$$

In Equations (3.3) and (3.6), the integral over the variable ζ can be expressed through the Hankel function. We utilize two integral forms for this function. The first form

$$H_0^{(1)}(kd) = \frac{1}{i\pi} \int_{-\infty}^\infty \frac{e^{ik\sqrt{d^2+\zeta^2}}}{\sqrt{d^2 + \zeta^2}} \, d\zeta \qquad (3.7)$$

follows from the table formula 8.421.11 of Gradshteyn and Ryzhik (1994). The second form

$$H_0^{(1)}\left(k\sqrt{d^2 + z^2}\right) = \frac{1}{\pi} \int_{-\infty}^\infty \frac{e^{i(vd-wz)}}{v} \, dw \qquad (3.8)$$

(where $v = \sqrt{k^2 - w^2}$, Im $v > 0$, and $d > 0$) can be verified by its conversion to the Sommerfeld formula (Sommerfeld, 1935)

$$H_0^{(1)}(\rho) = \frac{1}{\pi} \int_{-\delta+i\infty}^{\delta-i\infty} e^{i\rho \cos \beta} \, d\beta, \qquad 0 \le \delta \le \pi, \tag{3.9}$$

by setting $w = k \sin t$, $v = k \cos t$, and $k\sqrt{d^2 + z^2} = \rho$.

Application of Equation (3.7) leads to

$$u_s^{(0)} = -u_0 \frac{k \sin \varphi_0}{2} \int_0^\infty e^{-ik\xi \cos \varphi_0} H_0^{(1)} \left(k\sqrt{(x-\xi)^2 + y^2} \right) d\xi \tag{3.10}$$

and

$$u_h^{(0)} = -u_0 \frac{i}{2} \frac{\partial}{\partial y} \int_0^\infty e^{-ik\xi \cos \varphi_0} H_0^{(1)} \left(k\sqrt{(x-\xi)^2 + y^2} \right) d\xi. \tag{3.11}$$

Then we use Equation (3.8) and find

$$u_s^{(0)} = -u_0 \frac{k \sin \varphi_0}{2\pi} \int_{-\infty}^\infty \frac{e^{iv|y|}}{v} e^{-iwx} \, dw \int_0^\infty e^{i(w - k \cos \varphi_0)\xi} \, d\xi \tag{3.12}$$

and

$$u_h^{(0)} = \text{sgn}(y) u_0 \frac{1}{2\pi} \int_{-\infty}^\infty e^{i(v|y|-wx)} \, dw \int_0^\infty e^{i(w - k \cos \varphi_0)\xi} \, d\xi. \tag{3.13}$$

To ensure the convergence of the internal integrals, we impose the condition Im$(w - k \cos \varphi_0) > 0$ and obtain

$$u_s^{(0)} = u_0 \frac{i \sin \varphi_0}{2\pi} I_s, \qquad u_h^{(0)} = \text{sgn}(y) u_0 \frac{1}{i2\pi} I_h \tag{3.14}$$

where

$$I_s = k \int_{-\infty}^\infty \frac{e^{i(v|y|-wx)}}{v \cdot (k \cos \varphi_0 - w)} \, dw, \qquad I_h = \int_{-\infty}^\infty \frac{e^{i(v|y|-wx)}}{k \cos \varphi_0 - w} \, dw \tag{3.15}$$

and the integration contour skirts above the pole $w = k \cos \varphi_0$.

3.2 CONVERSION OF THE PO INTEGRALS TO THE CANONICAL FORM

In integrals I_s and I_h, we introduce the polar coordinates by the relationships

$$x = r \cos \varphi, \qquad |y| = \begin{cases} r \sin \varphi, & \text{with } \varphi < \pi \\ -r \sin \varphi, & \text{with } \varphi > \pi \end{cases} \tag{3.16}$$

and change the integration variable w by ξ, setting

$$w = -k \cos \xi, \qquad v = k \sin \xi, \qquad \mathrm{Im}\, v > 0. \tag{3.17}$$

Then,

$$v|y| - wx = \begin{cases} kr \cos(\xi - \varphi), & \text{with } \varphi < \pi \\ kr \cos(\xi + \varphi), & \text{with } \varphi > \pi. \end{cases} \tag{3.18}$$

The equations

$$\begin{aligned} w &= -k \cos \xi' \cosh \xi'' + ik \sin \xi' \sinh \xi'', \\ v &= k \sin \xi' \cosh \xi'' + ik \cos \xi' \sinh \xi'' \end{aligned} \tag{3.19}$$

and the condition

$$\mathrm{Im}\, v = k \cos \xi' \sinh \xi'' > 0 \tag{3.20}$$

determine the integration path F in the complex plane $\xi = \xi' + i\xi''$ as shown in Figure 3.2.

After these manipulations we obtain the following expressions:

$$I_s = \int_F \frac{e^{ikr\cos(\xi - \varphi)}}{\cos \xi + \cos \varphi_0}\, d\xi, \qquad \text{with } \varphi < \pi, \tag{3.21}$$

$$I_s = \int_F \frac{e^{ikr\cos(\xi + \varphi)}}{\cos \xi + \cos \varphi_0}\, d\xi, \qquad \text{with } \varphi > \pi, \tag{3.22}$$

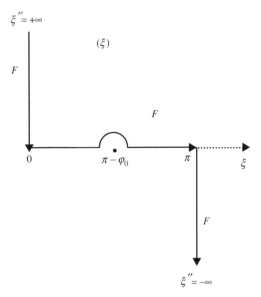

Figure 3.2   Integration contour F in the complex plane $\xi = \xi' + i\xi''$.

and

$$I_h = \int_F \frac{e^{ikr\cos(\xi-\varphi)}\sin\xi}{\cos\xi + \cos\varphi_0} d\xi, \qquad \text{with } \varphi < \pi, \tag{3.23}$$

$$I_h = \int_F \frac{e^{ikr\cos(\xi+\varphi)}\sin\xi}{\cos\xi + \cos\varphi_0} d\xi, \qquad \text{with } \varphi > \pi, \tag{3.24}$$

where, as is shown in Figure 3.2, the integration contour F skirts above the pole $\xi = \pi - \varphi_0$.

In Figures 3.3 and 3.4, we introduce two additional contours G_1 and G_2, related to the cases $\varphi < \pi$ and $\varphi > \pi$, respectively. In the dashed regions of these figures, the relations $\text{Im}\cos(\xi - \varphi) > 0$ and $\text{Im}\cos(\xi + \varphi) > 0$ are valid, which guarantees the convergence of integrals I_s and I_h. We then apply the Cauchy residue theorem

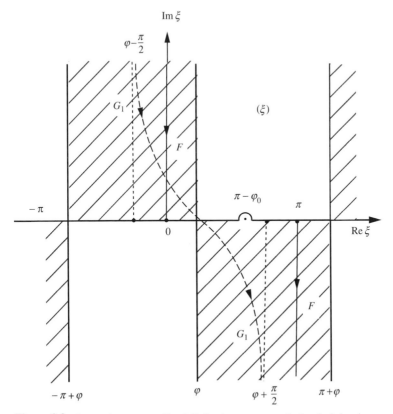

Figure 3.3   Integration contours F and G_1 for the case $\varphi < \pi$. In the shaded regions, $\text{Im}\cos(\xi - \varphi) > 0$.

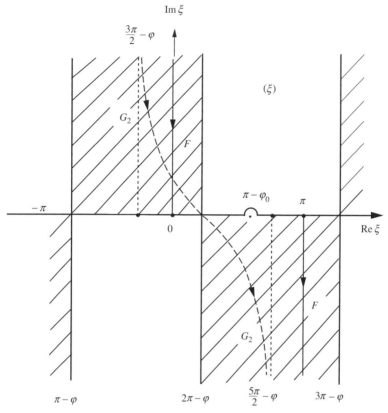

Figure 3.4 Integration contours F and G_2 for the case $\varphi > \pi$. In the shaded regions, $\operatorname{Im} \cos(\xi + \varphi) > 0$.

to the integrals over the closed contours $F - G_{1,2}$ (they are closed at the infinity, $\xi'' = \pm\infty$) and obtain

$$I_s = \int_{G_1} \frac{e^{ikr\cos(\xi-\varphi)}}{\cos\xi + \cos\varphi_0}\, d\xi + \frac{2\pi i}{\sin\varphi_0} e^{-ikr\cos(\varphi+\varphi_0)}, \qquad \text{with } 0 \le \varphi \le \pi - \varphi_0,$$

(3.25)

$$I_s = \int_{G_1} \frac{e^{ikr\cos(\xi-\varphi)}}{\cos\xi + \cos\varphi_0}\, d\xi, \qquad \text{with } \pi - \varphi_0 < \varphi < \pi,$$

(3.26)

$$I_s = \int_{G_2} \frac{e^{ikr\cos(\xi+\varphi)}}{\cos\xi + \cos\varphi_0}\, d\xi, \qquad \text{with } \pi < \varphi < \pi + \varphi_0,$$

(3.27)

and

$$I_s = \int_{G_2} \frac{e^{ikr\cos(\xi+\varphi)}}{\cos\xi + \cos\varphi_0}\, d\xi + \frac{2\pi i}{\sin\varphi_0} e^{-ikr\cos(\varphi-\varphi_0)}, \qquad \text{with } \pi + \varphi_0 \le \varphi \le 2\pi.$$

(3.28)

In these equations, we change with integration variable ξ by $\zeta + \varphi$ in the region $0 \leq \varphi < \pi$, and by $\zeta + 2\pi - \varphi$ in the region $\pi < \varphi < 2\pi$. As a result, the integrals in Equations (3.25) to (3.28) are transformed into the integrals over the contour D_0 shown in Figure 2.6:

$$\int_{G_1} \frac{e^{ikr\cos(\xi-\varphi)}}{\cos\xi + \cos\varphi_0}\, d\xi = \int_{D_0} \frac{e^{ikr\cos\zeta}\,d\zeta}{\cos\varphi_0 + \cos(\zeta+\varphi)}, \qquad \text{with } 0 \leq \varphi < \pi \qquad (3.29)$$

and

$$\int_{G_2} \frac{e^{ikr\cos(\xi+\varphi)}}{\cos\xi + \cos\varphi_0}\, d\xi = \int_{D_0} \frac{e^{ikr\cos\zeta}\,d\zeta}{\cos\varphi_0 + \cos(\zeta-\varphi)}, \qquad \text{with } \pi \leq \varphi < 2\pi. \qquad (3.30)$$

We omit similar transformations for integral I_h and present only their results:

$$I_h = \int_{D_0} \frac{e^{ikr\cos\zeta}\sin(\zeta+\varphi)}{\cos\varphi_0 + \cos(\zeta+\varphi)}\, d\zeta + 2\pi i e^{-ikr\cos(\varphi+\varphi_0)}, \qquad \text{with } 0 \leq \varphi \leq \pi - \varphi_0,$$
$$(3.31)$$

$$I_h = \int_{D_0} \frac{e^{ikr\cos\zeta}\sin(\zeta+\varphi)}{\cos\varphi_0 + \cos(\zeta+\varphi)}\, d\zeta, \qquad \text{with } \pi - \varphi_0 \leq \varphi < \pi, \qquad (3.32)$$

$$I_h = \int_{D_0} \frac{e^{ikr\cos\zeta}\sin(\zeta-\varphi)}{\cos\varphi_0 + \cos(\zeta-\varphi)}\, d\zeta, \qquad \text{with } \pi \leq \varphi \leq \pi + \varphi_0, \qquad (3.33)$$

$$I_h = \int_{D_0} \frac{e^{ikr\cos\zeta}\sin(\zeta-\varphi)}{\cos\varphi_0 + \cos(\zeta-\varphi)}\, d\zeta + 2\pi i e^{-ikr\cos(\varphi-\varphi_0)}, \qquad \text{with } \pi + \varphi_0 \leq \varphi \leq 2\pi.$$
$$(3.34)$$

Finally one can represent the scattered field in the following form:

$$u_s^{(0)} = u_0 v_s^{+}(kr,\varphi,\varphi_0) + \begin{cases} -u_0 e^{-ikr\cos(\varphi+\varphi_0)}, & \text{with } 0 \leq \varphi \leq \pi - \varphi_0 \\ 0, & \text{with } \pi - \varphi_0 < \varphi < \pi, \end{cases} \qquad (3.35)$$

$$u_s^{(0)} = u_0 v_s^{-}(kr,\varphi,\varphi_0) + \begin{cases} 0, & \text{with } \pi \leq \varphi \leq \pi + \varphi_0 \\ -u_0 e^{-ikr\cos(\varphi-\varphi_0)}, & \text{with } \pi + \varphi_0 < \varphi < 2\pi, \end{cases} \qquad (3.36)$$

and

$$u_h^{(0)} = u_0 v_h^{+}(kr,\varphi,\varphi_0) + \begin{cases} u_0 e^{-ikr\cos(\varphi+\varphi_0)}, & \text{with } 0 \leq \varphi \leq \pi - \varphi_0 \\ 0, & \text{with } \pi - \varphi_0 < \varphi < \pi, \end{cases} \qquad (3.37)$$

$$u_h^{(0)} = u_0 v_h^{-}(kr,\varphi,\varphi_0) + \begin{cases} 0, & \text{with } \pi \leq \varphi \leq \pi + \varphi_0 \\ -u_0 e^{-ikr\cos(\varphi-\varphi_0)}, & \text{with } \pi + \varphi_0 < \varphi < 2\pi, \end{cases} \qquad (3.38)$$

where

$$v_s^\pm(kr, \varphi, \varphi_0) = \frac{i \sin \varphi_0}{2\pi} \int_{D_0} \frac{e^{ikr \cos \zeta}}{\cos \varphi_0 + \cos(\zeta \pm \varphi)} \, d\zeta, \tag{3.39}$$

and

$$v_h^\pm(kr, \varphi, \varphi_0) = \pm \frac{1}{i2\pi} \int_{D_0} \frac{e^{ikr \cos \zeta} \sin(\zeta \pm \varphi)}{\cos \varphi_0 + \cos(\zeta \pm \varphi)} \, d\zeta. \tag{3.40}$$

The above expressions for $u_s^{(0)}$ and $u_h^{(0)}$ determine the PO field scattered by the face $\varphi = 0$ of the wedge. They are valid under the condition $0 \leq \varphi_0 < \alpha - \pi$.

However, in the case $\alpha - \pi < \varphi_0 < \pi$, the other face ($\varphi = \alpha$) is also illuminated and generates the additional scattered field. This field is also determined by Equations (3.35) to (3.38), where one should replace φ by $\alpha - \varphi$ and φ_0 by $\alpha - \varphi_0$. Thus, the PO field $u_s^{(0)}$ scattered in this case by both faces of the wedge is described by the following equations:

$$u_s^{(0)} = u_0[v_s^+(kr, \varphi, \varphi_0) + v_s^-(kr, \alpha - \varphi, \alpha - \varphi_0)]$$
$$+ \begin{cases} -u_0 e^{-ikr \cos(\varphi + \varphi_0)}, & \text{with } 0 \leq \varphi \leq \pi - \varphi_0 \\ 0, & \text{with } \pi - \varphi_0 < \varphi < \alpha - \pi. \end{cases} \tag{3.41}$$

$$u_s^{(0)} = u_0[v_s^+(kr, \varphi, \varphi_0) + v_s^+(kr, \alpha - \varphi, \alpha - \varphi_0)], \quad \text{with } \alpha - \pi < \varphi < \pi, \tag{3.42}$$

$$u_s^{(0)} = u_0[v_s^-(kr, \varphi, \varphi_0) + v_s^+(kr, \alpha - \varphi, \alpha - \varphi_0)], \quad \text{with } \pi < \varphi < 2\alpha - \pi - \varphi_0, \tag{3.43}$$

and

$$u_s^{(0)} = u_0[v_s^-(kr, \varphi, \varphi_0) + v_s^+(kr, \alpha - \varphi, \alpha - \varphi_0)]$$
$$- u_0 e^{-ikr \cos(2\alpha - \varphi - \varphi_0)}, \quad \text{with } 2\alpha - \pi - \varphi_0 < \varphi < \alpha. \tag{3.44}$$

Similar equations are valid for the field $u_h^{(0)}$ in the case $\alpha - \pi < \varphi_0 < \pi$:

$$u_h^{(0)} = u_0[v_h^+(kr, \varphi, \varphi_0) + v_h^-(kr, \alpha - \varphi, \alpha - \varphi_0)]$$
$$+ \begin{cases} u_0 e^{-ikr \cos(\varphi + \varphi_0)}, & \text{with } 0 \leq \varphi \leq \pi - \varphi_0 \\ 0, & \text{with } \pi - \varphi_0 < \varphi < \alpha - \pi, \end{cases} \tag{3.45}$$

$$u_{\mathrm{h}}^{(0)} = u_0[v_{\mathrm{h}}^+(kr, \varphi, \varphi_0) + v_{\mathrm{h}}^+(kr, \alpha - \varphi, \alpha - \varphi_0)], \qquad \text{with } \alpha - \pi < \varphi < \pi,$$

$$(3.46)$$

$$u_{\mathrm{h}}^{(0)} = u_0[v_{\mathrm{h}}^-(kr, \varphi, \varphi_0) + v_{\mathrm{h}}^+(kr, \alpha - \varphi, \alpha - \varphi_0)], \qquad \text{with } \pi < \varphi < 2\alpha - \pi - \varphi_0,$$

$$(3.47)$$

$$u_{\mathrm{h}}^{(0)} = u_0[v_{\mathrm{h}}^-(kr, \varphi, \varphi_0) + v_{\mathrm{h}}^+(kr, \alpha - \varphi, \alpha - \varphi_0)]$$

$$+ u_0 e^{-ikr\cos(2\alpha - \varphi - \varphi_0)}, \qquad \text{with } 2\alpha - \pi - \varphi_0 < \varphi < \alpha.$$

$$(3.48)$$

As is seen in Equations (3.41) to (3.48), the PO scattered field consists of the reflected plane waves and the diffracted part described by functions $v_{\mathrm{s}}^{\pm}$ and $v_{\mathrm{h}}^{\pm}$. Simple asymptotic expressions for the functions $v_{\mathrm{s,h}}^{\pm}$ and for the diffracted field are derived in the next section.

3.3 RAY ASYMPTOTICS FOR THE PO DIFFRACTED FIELD

The relationships $u_{\mathrm{s}} = E_z$ and $u_{\mathrm{h}} = H_z$ exist between the acoustic and electromagnetic fields studied in this section.

Again, as in Section 2.4, we introduce in the integrals (3.39) and (3.40) the new variable $s = \sqrt{2}e^{i\pi/4}\sin\frac{\zeta}{2}$ and transform functions $v_{\mathrm{s,h}}^{\pm}$ as

$$v_{\mathrm{s}}^{\pm}(kr, \varphi, \varphi_0) = \frac{\sin\varphi_0}{\sqrt{2\pi}} e^{i(kr + \pi/4)} \int_{-\infty}^{\infty} \frac{e^{-krs^2}\, ds}{[\cos\varphi_0 + \cos(\zeta \pm \varphi)]\cos\frac{\zeta}{2}}, \qquad (3.49)$$

$$v_{\mathrm{h}}^{\pm}(kr, \varphi, \varphi_0) = \mp \frac{1}{\sqrt{2\pi}} e^{i(kr + \pi/4)} \int_{-\infty}^{\infty} \frac{\sin(\zeta \pm \varphi)e^{-krs^2}\, ds}{[\cos\varphi_0 + \cos(\zeta \pm \varphi)]\cos\frac{\zeta}{2}}. \qquad (3.50)$$

Then the standard saddle point technique leads to the asymptotic expressions

$$v_{\mathrm{s}}^{\pm}(kr, \varphi, \varphi_0) \sim \frac{\sin\varphi_0}{\cos\varphi + \cos\varphi_0} \frac{e^{i(kr + \pi/4)}}{\sqrt{2\pi kr}}, \qquad (3.51)$$

and

$$v_{\mathrm{h}}^{\pm}(kr, \varphi, \varphi_0) \sim -\frac{\sin\varphi}{\cos\varphi + \cos\varphi_0} \frac{e^{i(kr + \pi/4)}}{\sqrt{2\pi kr}}. \qquad (3.52)$$

These asymptotics are valid under the conditions $\sqrt{kr}|\cos(\varphi \pm \varphi_0)/2| \gg 1$, that is, away from the geometrical optics boundaries, where the diffracted field has a ray structure and can be interpreted in terms of edge diffracted rays.

In view of Equations (3.51) and (3.52), the ray asymptotics for the diffracted part of the fields (3.41) to (3.44) and (3.45) to (3.48) can be written in the following form:

$$u_s^{(0)d} \sim u_0 f^{(0)}(\varphi, \varphi_0) \frac{e^{i(kr+\pi/4)}}{\sqrt{2\pi kr}} \qquad (3.53)$$

and

$$u_h^{(0)d} \sim u_0 g^{(0)}(\varphi, \varphi_0) \frac{e^{i(kr+\pi/4)}}{\sqrt{2\pi kr}}. \qquad (3.54)$$

The directivity patterns of these diffracted rays are different for the situations with one or two illuminated faces. In the case $0 < \varphi_0 < \alpha - \pi$, when only the face $\varphi = 0$ is illuminated by the incident wave,

$$f^{(0)}(\varphi, \varphi_0) = \frac{\sin \varphi_0}{\cos \varphi + \cos \varphi_0}, \qquad g^{(0)}(\varphi, \varphi_0) = -\frac{\sin \varphi}{\cos \varphi + \cos \varphi_0}. \qquad (3.55)$$

However, if both faces are illuminated ($\alpha - \pi < \varphi_0 < \pi$),

$$f^{(0)}(\varphi, \varphi_0) = \frac{\sin \varphi_0}{\cos \varphi + \cos \varphi_0} + \frac{\sin(\alpha - \varphi_0)}{\cos(\alpha - \varphi) + \cos(\alpha - \varphi_0)} \qquad (3.56)$$

and

$$g^{(0)}(\varphi, \varphi_0) = -\frac{\sin \varphi}{\cos \varphi + \cos \varphi_0} - \frac{\sin(\alpha - \varphi)}{\cos(\alpha - \varphi) + \cos(\alpha - \varphi_0)}. \qquad (3.57)$$

It is seen that these functions are singular at the boundaries of reflected plane waves, where $\varphi = \pi \pm \varphi_0$ or $\varphi = 2\alpha - \pi - \varphi_0$. The asymptotics of the PO diffracted field, which are uniformly valid around the wedge, can be easily derived by the application of the Pauli technique demonstrated in Sections 2.5 and 2.6. We do not consider them here, because our main purpose in the canonical wedge diffraction problem is the calculation of the field generated by the *nonuniform* component of the surface sources.

PROBLEMS

3.1   Analyze the accuracy of the PO approximation. Compute and plot the relative error

$$\delta f^{(0)}(\varphi, \varphi_0, \alpha) = |F(\varphi, \varphi_0, \alpha) - 1| \cdot 100\%,$$

where

$$F(\varphi, \varphi_0, \alpha) = f^{(0)}(\varphi, \varphi_0)/f(\varphi, \varphi_0, \alpha).$$

Consider these two examples:

(a) $\alpha = 270°$, $\varphi_0 = 45°$, $0 \le \varphi \le \alpha$,

(b) $\alpha = 360°$, $\varphi_0 = 45°$, $0 \le \varphi \le \alpha$.

Prove analytically that $F(\pi \pm \varphi_0, \varphi_0, \alpha) = 1$. Formulate your conclusion regarding $\delta f^{(0)}$.

3.2 Analyze the accuracy of the PO approximation. Compute and plot the relative error

$$\delta g^{(0)}(\varphi, \varphi_0, \alpha) = |G(\varphi, \varphi_0, \alpha) - 1| \cdot 100\%,$$

where

$$G(\varphi, \varphi_0, \alpha) = g^{(0)}(\varphi, \varphi_0)/g(\varphi, \varphi_0, \alpha).$$

Consider these two examples:

(a) $\alpha = 270°$, $\varphi_0 = 45°$, $0 \le \varphi \le \alpha$,

(b) $\alpha = 360°$, $\varphi_0 = 45°$, $0 \le \varphi \le \alpha$.

Prove analytically that $G(\pi \pm \varphi_0, \varphi_0, \alpha) = 1$. Formulate your conclusion regarding $\delta g^{(0)}$.

3.3 The Sommerfeld function $f(\varphi, \varphi_0, \alpha)$ satisfies the reciprocity principle, but its PO approximation $f^{(0)}(\varphi, \varphi_0)$ does not. Analyze the PO deviations from the reciprocity principle. Compute the relative level of these deviations,

$$d_f(\varphi, \varphi_0) = |F(\varphi, \varphi_0, \alpha) - F(\varphi_0, \varphi, \alpha)| \cdot 100\%,$$

where

$$F(\varphi, \varphi_0, \alpha) = f^{(0)}(\varphi, \varphi_0)/f(\varphi, \varphi_0, \alpha).$$

Investigate the case with $\alpha = 350°$, set $\Delta\varphi = \Delta\varphi_0 = 70°$. Prepare a (6×6) square table with numerical data for the deviations d_f. Formulate your conclusion.

3.4 The Sommerfeld function $g(\varphi, \varphi_0, \alpha)$ satisfies the reciprocity principle, but its PO approximation $g^{(0)}(\varphi, \varphi_0)$ does not. Analyze the PO deviations from the reciprocity principle. Compute the relative level of these deviations,

$$d_g(\varphi, \varphi_0) = |G(\varphi, \varphi_0, \alpha) - G(\varphi_0, \varphi, \alpha)| \cdot 100\%,$$

where

$$G(\varphi, \varphi_0, \alpha) = g^{(0)}(\varphi, \varphi_0, \alpha)/g(\varphi, \varphi_0, \alpha).$$

Investigate the case with $\alpha = 350°$, set $\Delta\varphi = \Delta\varphi_0 = 70°$. Prepare a (6×6) square table with numerical data for the deviations d_g. Formulate your conclusion.

Chapter 4

Wedge Diffraction: Radiation by the Nonuniform Component of Surface Sources

The relationships $u_s^{(1)} = E_z^{(1)}$ and $u_h^{(1)} = H_z^{(1)}$ exist between the acoustic and electromagnetic fields generated by the nonuniform sources $j^{(1)}$. An exception exists for an oblique incidence (see Section 4.3).

We have now reached the moment when we can construct the integral and asymptotic representations for the field radiated by the *nonuniform* component of the surface sources, which are induced at the wedge by the incident wave. The exact expressions for the total field generated around the wedge have been derived in Chapter 2. The Physical Optics (PO) part of this field (which is generated by the *uniform* component of the surface sources) has been studied in Chapter 3. The contribution to the diffracted field by the *nonuniform* component is the difference between the exact total field and its PO part. This contribution is investigated in this chapter.

4.1 INTEGRALS AND ASYMPTOTICS

According to the exact solution (see Equations (2.40), (2.41), and (2.52) to (2.57)), the total field around the wedge consists of the diffracted and geometrical optics parts

$$u_{s,h}^t = u_{s,h}^d + u_{s,h}^{go}, \tag{4.1}$$

where $u_{s,h}^d$ is described by the functions $v(kr, \psi)$, and $u_{s,h}^{go}$ is the sum of the incident and reflected plane waves. Equations (3.35) and (3.36), (3.37) and (3.38), (3.41) to (3.44), and (3.45) to (3.48) represent the scattered field in the PO approximation.

Fundamentals of the Physical Theory of Diffraction. By Pyotr Ya. Ufimtsev
Copyright © 2007 John Wiley & Sons, Inc.

By summation with the incident wave (3.1), these equations determine the PO part of the total field:

$$u_{s,h}^{(0)t} = u_{s,h}^{(0)d} + u_{s,h}^{go}, \tag{4.2}$$

where $u_{s,h}^{(0)d}$ is the diffracted part of the field, which is described by functions $v_{s,h}^{\pm}$. The geometrical optics part $u_{s,h}^{go}$ of the PO field is the same quantity as that in Equation (4.1).

The field (4.1) is generated by total surface source $j_{s,h} = j_{s,h}^{(0)} + j_{s,h}^{(1)}$, consisting of the uniform and nonuniform components, and the PO field (4.2) is radiated only by the uniform component $j_{s,h}^{(0)}$. Therefore, the field created by the nonuniform component is the difference

$$u_{s,h}^{(1)} = u_{s,h}^{t} - u_{s,h}^{(0)t} = u_{s,h}^{d} - u_{s,h}^{(0)d}. \tag{4.3}$$

In the case $0 < \varphi_0 < \alpha - \pi$, when only one face ($\varphi = 0$) is illuminated, this field is determined by

$$u_s^{(1)}/u_0 = v(kr, \varphi - \varphi_0) - v(kr, \varphi + \varphi_0) - \begin{cases} v_s^+(kr, \varphi, \varphi_0), & \text{with } 0 \le \varphi < \pi \\ v_s^-(kr, \varphi, \varphi_0), & \text{with } \pi < \varphi \le \alpha, \end{cases} \tag{4.4}$$

and

$$u_h^{(1)}/u_0 = v(kr, \varphi - \varphi_0) + v(kr, \varphi + \varphi_0) - \begin{cases} v_h^+(kr, \varphi, \varphi_0), & \text{with } 0 \le \varphi < \pi \\ v_h^-(kr, \varphi, \varphi_0), & \text{with } \pi < \varphi \le \alpha. \end{cases} \tag{4.5}$$

In the case $\alpha - \pi < \varphi_0 < \pi$, when both faces are illuminated, the field $u_{s,h}^{(1)}$ is determined as

$$u_s^{(1)}/u_0 = v(kr, \varphi - \varphi_0) - v(kr, \varphi + \varphi_0)$$

$$- \begin{cases} v_s^+(kr, \varphi, \varphi_0) + v_s^-(kr, \alpha - \varphi, \alpha - \varphi_0), & \text{with } 0 \le \varphi < \alpha - \pi \\ v_s^+(kr, \varphi, \varphi_0) + v_s^+(kr, \alpha - \varphi, \alpha - \varphi_0), & \text{with } \alpha - \pi < \varphi < \pi \\ v_s^-(kr, \varphi, \varphi_0) + v_s^+(kr, \alpha - \varphi, \alpha - \varphi_0), & \text{with } \pi < \varphi \le \alpha, \end{cases} \tag{4.6}$$

$$u_h^{(1)}/u_0 = v(kr, \varphi - \varphi_0) + v(kr, \varphi + \varphi_0)$$

$$- \begin{cases} v_h^+(kr, \varphi, \varphi_0) + v_h^-(kr, \alpha - \varphi, \alpha - \varphi_0), & \text{with } 0 \le \varphi < \alpha - \pi \\ v_h^+(kr, \varphi, \varphi_0) + v_h^+(kr, \alpha - \varphi, \alpha - \varphi_0), & \text{with } \alpha - \pi < \varphi < \pi \\ v_h^-(kr, \varphi, \varphi_0) + v_h^+(kr, \alpha - \varphi, \alpha - \varphi_0), & \text{with } \pi < \varphi \le \alpha. \end{cases} \tag{4.7}$$

The function $v(kr, \psi)$ is defined in Equation (2.51) and the functions $v_{s,h}^{\pm}(kr, \varphi, \varphi_0)$ in Equations (3.39) and (3.40) by the integrals in the complex plane over the same contour D_0 shown in Figure 2.6. Therefore, the field $u_{s,h}^{(1)}$ can be represented as the integral over the contour D_0 with the integrand consisting of a linear combination of the integrands related to functions v and $v_{s,h}^{\pm}$:

$$u_{s,h}^{(1)} = \int_{D_0} U_{s,h}(\alpha, \varphi, \varphi_0, \zeta) e^{ikr \cos \zeta} d\zeta. \tag{4.8}$$

For the observation points far away from the edge ($kr \gg 1$), this integral can be asymptotically evaluated by the saddle point method (Copson, 1965; Murray, 1984). To do this, we first replace the integration variable ζ by $s = \sqrt{2} e^{i\pi/4} \sin \frac{\zeta}{2}$ and transform the integral to the form

$$u_{s,h}^{(1)} = \sqrt{2} e^{-i\frac{\pi}{4}} e^{ikr} \int_{-\infty}^{\infty} \frac{U_{s,h}[\alpha, \varphi, \varphi_0, \zeta(s)]}{\cos[\zeta(s)/2]} e^{-krs^2} ds. \tag{4.9}$$

We then expand the integrand $U_{s,h}/\cos(\zeta/2)$ into the Taylor series in the vicinity of the saddle point $s = 0$, retain only the first term in this series, and obtain the asymptotic expression

$$u_{s,h}^{(1)} = \sqrt{2} e^{-i\frac{\pi}{4}} e^{ikr} U_{s,h}(\alpha, \varphi, \varphi_0, 0) \int_{-\infty}^{\infty} e^{-krs^2} ds + O\left[\int_{-\infty}^{\infty} e^{-krs^2} s^2 ds \right] \tag{4.10}$$

or

$$u_{s,h}^{(1)} = \sqrt{2\pi} e^{-i\frac{\pi}{4}} \frac{e^{ikr}}{\sqrt{kr}} U_{s,h}(\alpha, \varphi, \varphi_0, 0) + O\left[(kr)^{-\frac{3}{2}} \right], \tag{4.11}$$

which holds under the condition $kr \gg 1$.

Finally, one can rewrite the above expression (4.11) in terms of the functions f, $g, f^{(0)}, g^{(0)}$ introduced in Sections 2.4 and 3.3:

$$u_s^{(1)} \sim u_0 f^{(1)}(\varphi, \varphi_0, \alpha) \frac{e^{i(kr+\pi/4)}}{\sqrt{2\pi kr}}, \tag{4.12}$$

$$u_h^{(1)} \sim u_0 g^{(1)}(\varphi, \varphi_0, \alpha) \frac{e^{i(kr+\pi/4)}}{\sqrt{2\pi kr}}, \tag{4.13}$$

where

$$f^{(1)}(\varphi, \varphi_0, \alpha) = f(\varphi, \varphi_0, \alpha) - f^{(0)}(\varphi, \varphi_0), \tag{4.14}$$

and

$$g^{(1)}(\varphi, \varphi_0, \alpha) = g(\varphi, \varphi_0, \alpha) - g^{(0)}(\varphi, \varphi_0). \tag{4.15}$$

Thus, the field generated by the nonuniform component $j_{s,h}^{(1)}$ represents by itself the cylindrical wave diverging from the edge of the wedge. This form of the field is a consequence of the fact that the nonuniform component $j_{s,h}^{(1)}$ concentrates near the edge. Just for this reason the quantity $j_{s,h}^{(1)}$ is sometimes called the *fringe* component. Approximations (4.12) and (4.13) reveal a ray structure of this part of the diffracted field and because of that they can be termed ray asymptotics.

The directivity patterns of the field (4.12) and (4.13) possess a wonderful property. In contrast to the functions $f, g, f^{(0)}, g^{(0)}$, which are *singular* at the geometrical optics boundaries, the functions $f^{(1)}$ and $g^{(1)}$ are *finite* there. It turns out that the singularities of functions f and g are totally cancelled by the singularities of functions $f^{(0)}$ and $g^{(0)}$, respectively. The following equations determine the finite values of functions $f^{(1)}$ and $g^{(1)}$ for these special directions.

For the direction $\varphi = \pi - \varphi_0$ (which is the boundary of the plane wave reflected from the face $\varphi = 0$, Fig. 2.7), the functions $f^{(1)}$ and $g^{(1)}$ have the values

$$\begin{Bmatrix} f^{(1)} \\ g^{(1)} \end{Bmatrix} = \frac{\dfrac{1}{n}\sin\dfrac{\pi}{n}}{\cos\dfrac{\pi}{n} - \cos\dfrac{\pi - 2\varphi_0}{n}} + \frac{1}{2}\cot\varphi_0 \pm \frac{1}{2n}\cot\frac{\pi}{n} \qquad (4.16)$$

when $0 < \varphi_0 < \alpha - \pi$, and

$$\begin{Bmatrix} f^{(1)} \\ g^{(1)} \end{Bmatrix} = \frac{\dfrac{1}{n}\sin\dfrac{\pi}{n}}{\cos\dfrac{\pi}{n} - \cos\dfrac{\pi - 2\varphi_0}{n}} + \frac{1}{2}\cot\varphi_0 \pm \frac{1}{2n}\cot\frac{\pi}{n}$$

$$- \begin{Bmatrix} \dfrac{\sin(\alpha - \varphi_0)}{\cos(\alpha - \varphi_0) - \cos(\alpha + \varphi_0)} \\ \dfrac{\sin(\alpha + \varphi_0)}{\cos(\alpha - \varphi_0) - \cos(\alpha + \varphi_0)} \end{Bmatrix} \qquad (4.17)$$

when $\alpha - \pi < \varphi_0 < \pi$.

For the direction $\varphi = \pi + \varphi_0$ (which is the shadow boundary of the incident wave, Fig. 2.7), the values of functions $f^{(1)}$ and $g^{(1)}$ are determined by

$$\begin{Bmatrix} f^{(1)} \\ g^{(1)} \end{Bmatrix} = \mp\frac{\dfrac{1}{n}\sin\dfrac{\pi}{n}}{\cos\dfrac{\pi}{n} - \cos\dfrac{\pi + 2\varphi_0}{n}} \pm \frac{1}{2}\cot\varphi_0 - \frac{1}{2n}\cot\frac{\pi}{n} \qquad (4.18)$$

when $0 < \varphi_0 < \pi$. The direction $\varphi = \pi + \varphi_0$ under the condition $\alpha - \pi < \varphi_0 < \pi$ is inside the wedge and is not of interest.

In the case $\alpha - \pi < \varphi_0 < \pi$, when both faces of the wedge are illuminated (Fig. 2.8), the functions $f^{(1)}$ and $g^{(1)}$ have the following values at the direction $\varphi = 2\alpha - \pi - \varphi_0$ (which is the boundary of the plane waves reflected from the face $\varphi = \alpha$):

$$
\begin{Bmatrix} f^{(1)} \\ g^{(1)} \end{Bmatrix} = \frac{\dfrac{1}{n}\sin\dfrac{\pi}{n}}{\cos\dfrac{\pi}{n} - \cos\dfrac{\varphi - \varphi_0}{n}} + \frac{1}{2}\cot(\alpha - \varphi_0) \pm \frac{1}{2n}\cot\frac{\pi}{n}
$$

$$
+ \begin{Bmatrix} -\dfrac{\sin\varphi_0}{\cos\varphi + \cos\varphi_0} \\[2ex] \dfrac{\sin\varphi}{\cos\varphi + \cos\varphi_0} \end{Bmatrix}. \tag{4.19}
$$

One should note that the functions $f^{(1)}$ and $g^{(1)}$ are singular in the two special directions

$$
\varphi = 0, \qquad \text{when } \varphi_0 = \pi \tag{4.20}
$$

and

$$
\varphi = \alpha, \qquad \text{when } \varphi_0 = \alpha - \pi, \tag{4.21}
$$

which relate to the grazing reflections from the faces under the grazing incidence. This is a special case when the integrand in (4.9) cannot be expanded into the Taylor

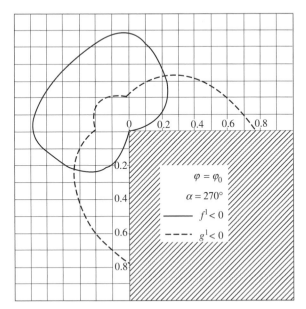

Figure 4.1   Directivity patterns of edge waves radiated by nonuniform components of the surface sources. The function $f^{(1)}(g^{(1)})$ corresponds to the case of the acoustically soft (hard) wedge; they also describe the $E_z(H_z)$ component of the electromagnetic wave scattered at the perfectly conducting wedge. Reprinted from Ufimtsev (1957) with permission of Zhurnal Tekhnicheskoi Fiziki.

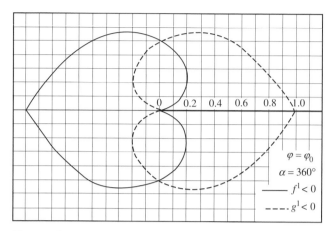

Figure 4.2   Directivity patterns of edge waves radiated by nonuniform components of the surface sources. The function $f^{(1)}(g^{(1)})$ corresponds to the case of the acoustically soft (hard) half-plane; they also describe the $E_z(H_z)$ component of the electromagnetic wave scattered at the perfectly conducting half-plane. Reprinted from Ufimtsev (1957) with permission of Zhurnal Tekhnicheskoi Fiziki.

series because its terms become infinite. Section 7.9 develops a special version of PTD that is free from the grazing singularity.

Figures 4.1 and 4.2 illustrate the behavior and beauty of functions $f^{(1)}$ and $g^{(1)}$.

4.2 INTEGRAL FORM OF FUNCTIONS $f^{(1)}$ AND $g^{(1)}$

It is well known in antenna and scattering theories that the directivity pattern of the far field can be considered as a conformal Fourier transform of the radiating/scattering sources distributed over antennas/scatterers. This is clearly seen in Equation (1.19). In this section we will establish this type of relationship between the directivity patterns $f^{(1)}$, $g^{(1)}$ and their sources $j_s^{(1)}$, $j_h^{(1)}$ at the wedge.

The geometry of the problem is shown in Figure 3.1 and the incident wave is given by Equation (3.1). The nonuniform components of the surface sources

$$j_s^{(1)} = u_0 J_s^{(1)}, \qquad j_h^{(1)} = u_0 J_h^{(1)} \tag{4.22}$$

radiate the field defined by Equation (1.10). We consider first the radiation from the wedge face $\varphi = 0$:

$$u_s^{(1)} = -\frac{u_0}{4\pi} \int_0^\infty J_s^{(1)}(k\xi, \varphi_0)\mathrm{d}\xi \int_{-\infty}^\infty \frac{e^{ik\sqrt{(x-\xi)^2+y^2+\zeta^2}}}{\sqrt{(x-\xi)^2+y^2+\zeta^2}}\,\mathrm{d}\zeta, \tag{4.23}$$

$$u_h^{(1)} = -\frac{u_0}{4\pi}\frac{\partial}{\partial y}\int_0^\infty J_h^{(1)}(k\xi,\varphi_0)d\xi \int_{-\infty}^\infty \frac{e^{ik\sqrt{(x-\xi)^2+y^2+\zeta^2}}}{\sqrt{(x-\xi)^2+y^2+\zeta^2}}d\zeta. \qquad (4.24)$$

In view of Equation (3.7),

$$u_s^{(1)} = -u_0\frac{i}{4}\int_0^\infty J_s^{(1)}(k\xi,\varphi_0)H_0^{(1)}\left(k\sqrt{(x-\xi)^2+y^2}\right)d\zeta \qquad (4.25)$$

and

$$u_h^{(1)} = -u_0\frac{i}{4}\frac{\partial}{\partial y}\int_0^\infty J_h^{(1)}(k\xi,\varphi_0)H_0^{(1)}\left(k\sqrt{(x-\xi)^2+y^2}\right)d\zeta. \qquad (4.26)$$

According to Equations (2.61) and (2.63), the functions $J_s^{(1)}$ and $J_h^{(1)}$ decrease as $(k\xi)^{-3/2}$ and $(k\xi)^{-1/2}$, respectively, with increasing distance ξ from the edge. At a certain distance $\xi = \xi_{\text{effective}}$, these functions are sufficiently small and can be approximated by zero for $\xi \geq \xi_{\text{eff}}$. In the far zone, where $r \gg k\xi_{\text{eff}}^2$, the Hankel function in Equations (4.25) and (4.26) can be replaced by its asymptotics (2.29). This leads to the asymptotic expressions

$$u_s^{(1)} \sim -u_0\frac{e^{i(kr+\pi/4)}}{2\sqrt{2\pi kr}}\int_0^{\xi_{\text{eff}}} J_s^{(1)}(k\xi,\varphi_0)e^{-ik\xi\cos\varphi}\,d\xi \qquad (4.27)$$

and

$$u_h^{(1)} \sim -u_0 ik\sin\varphi\frac{e^{i(kr+\pi/4)}}{2\sqrt{2\pi kr}}\int_0^{\xi_{\text{eff}}} J_h^{(1)}(k\xi,\varphi_0)e^{-ik\xi\cos\varphi}\,d\xi. \qquad (4.28)$$

These expressions describe the field radiated from the face $\varphi = 0$. By the replacement of φ by $\alpha - \varphi$ and φ_0 by $\alpha - \varphi_0$, one can find the field radiated from the face $\varphi = \alpha$. The total field created by both faces must be equal to that of Equations (4.12) and (4.13). By equating them we obtain the useful relationships

$$f^{(1)}(\varphi,\varphi_0,\alpha) = -\frac{1}{2}\left[\int_0^{\xi_{\text{eff}}} J_s^{(1)}(k\xi,\varphi_0)e^{-ik\xi\cos\varphi}\,d\xi\right.$$

$$\left. + \int_0^{\xi_{\text{eff}}} J_s^{(1)}(k\xi,\alpha-\varphi_0)e^{-ik\xi\cos(\alpha-\varphi)}\,d\xi\right] \qquad (4.29)$$

and

$$g^{(1)}(\varphi,\varphi_0,\alpha) = -\frac{ik}{2}\left[\sin\varphi\int_0^{\xi_{\text{eff}}} J_h^{(1)}(k\xi,\varphi_0)e^{-ik\xi\cos\varphi}\,d\xi\right.$$

$$\left. + \sin(\alpha-\varphi)\int_0^{\xi_{\text{eff}}} J_h^{(1)}(k\xi,\alpha-\varphi_0)e^{-ik\xi\cos(\alpha-\varphi)}\,d\xi\right]. \qquad (4.30)$$

These expressions show that the functions $f^{(1)}$ and $g^{(1)}$ can also be interpreted as the directivity patterns of elementary diffracted waves generated (in the plane normal to the edge) by the sources distributed at the wedge along the lines normal to the edge.

4.3 OBLIQUE INCIDENCE OF A PLANE WAVE AT A WEDGE

For an oblique incidence, the relationship $u_s^{(t)} = E_z^{(t)}$ exists for the diffracted rays generated by the *total surface currents*, $j^{(t)} = j^{(0)} + j^{(1)}$. However, $u_s^{(0,1)}$ is not equal to $E_z^{(0,1)}$, because of the polarization coupling in the electromagnetic PO field.

The complete equivalence $u_h^{(0,1,t)} = H_z^{(0,1,t)}$ exists between the acoustic and electromagnetic diffracted rays.

Here, we extend the results of Section 4.1 to the general case when the incident wave propagates under the oblique direction to the edge (Fig. 4.3). It is given by the equation

$$u^{\text{inc}} = u_0 e^{ik(x\cos\tilde{\alpha}+y\cos\beta+z\cos\gamma)} \tag{4.31}$$

with $0 < \gamma \le \pi/2$. We use the "tilde-hat" for the angle $\tilde{\alpha}$ to avoid possible confusion with the external angle α of the wedge.

The boundary conditions on the wedge faces are shown in Equations (2.2) and (2.3). To satisfy these conditions, the diffracted field must have the same dependence on the coordinate z as the incident wave (4.31):

$$u^{\text{d}} = u(r,\varphi)e^{ikz\cos\gamma}. \tag{4.32}$$

The substitution of this function u^{d} into the wave equation (2.4) leads to the equation for the function $u(r,\varphi)$,

$$\Delta u(r,\varphi) + k_1^2 u(r,\varphi) = 0, \qquad \text{with } k_1 = k\sin\gamma, \tag{4.33}$$

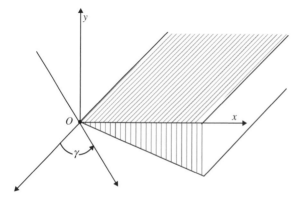

Figure 4.3 Oblique incidence of a plane wave at a wedge.

where the Laplacian operator Δ is defined by Equation (2.5). It is also expedient to represent the incident wave (4.31) in the form of Equation (3.1):

$$u^{\text{inc}} = u_0 e^{ikz \cos \gamma} e^{-ik_1(x \cos \varphi_0 + y \sin \varphi_0)}, \tag{4.34}$$

where

$$\sin \gamma \cos \varphi_0 = -\cos \tilde{\alpha}, \qquad \sin \gamma \sin \varphi_0 = -\cos \beta \tag{4.35}$$

and

$$\tan \varphi_0 = \frac{\cos \beta}{\cos \tilde{\alpha}}, \qquad \text{with } 0 \leq \varphi_0 < \pi. \tag{4.36}$$

Thus, we reduced the three-dimensional (3-D) diffraction problem for the oblique incidence to the 2-D problem for the normal incidence ($\gamma = \pi/2$) considered in Chapter 2. The solution for the oblique incidence can be automatically found by simple replacements in the solution for the normal incidence. Namely, the quantity u_0 should be replaced by $u_0 e^{ikz \cos \gamma}$, the wave number k by $k_1 = k \sin \gamma$, and the angle φ_0 by $\varphi_0 = \arctan(\cos \beta / \cos \tilde{\alpha})$.

This rule has been established here for the exact solution of the wedge diffraction problem and for its asymptotics. One can show that it is also valid for the PO part of the field. First, by the substitution of the incident wave (4.34) into Equation (1.31) we find the PO surface sources

$$j_s^{(0)} = -u_0 e^{ikz \cos \gamma} 2ik_1 \sin \varphi_0 e^{-ik_1 x \cos \varphi_0}, \tag{4.37}$$

$$j_h^{(0)} = u_0 e^{ikz \cos \gamma} 2e^{ik_1 x \cos \varphi_0}. \tag{4.38}$$

Comparison with Equation (3.2) shows that the sources (4.37) and (4.38) satisfy the above rule for the transition from normal to oblique incidence. If the sources of the field satisfy this rule, one can expect the generated field does too. To verify that, we substitute Equations (4.37) and (4.38) into the original expressions (1.32) for the PO field:

$$u_s^{(0)} = u_0 \frac{ik_1 \sin \varphi_0}{2\pi} \int_0^\infty e^{-ik_1 \xi \cos \varphi_0} d\xi \int_{-\infty}^\infty e^{ik\zeta \cos \gamma} \frac{e^{ik\sqrt{(x-\xi)^2 + y^2 + (z-\zeta)^2}}}{\sqrt{(x-\xi)^2 + y^2 + (z-\zeta)^2}} d\zeta, \tag{4.39}$$

$$u_h^{(0)} = -u_0 \frac{1}{2\pi} \frac{\partial}{\partial y} \int_0^\infty e^{-ik_1 \xi \cos \varphi_0} d\xi \int_{-\infty}^\infty e^{ik\zeta \cos \gamma} \frac{e^{ik\sqrt{(x-\xi)^2 + y^2 + (z-\zeta)^2}}}{\sqrt{(x-\xi)^2 + y^2 + (z-\zeta)^2}} d\zeta, \tag{4.40}$$

or

$$u_{\mathrm{s}}^{(0)} = u_0 e^{ikz \cos \gamma} \frac{ik_1 \sin \varphi_0}{2\pi} \int_0^\infty e^{-ik_1\xi \cos \varphi_0} d\xi \int_{-\infty}^\infty e^{iks \cos \gamma} \frac{e^{ik\sqrt{(x-\xi)^2+y^2+s^2}}}{\sqrt{(x-\xi)^2 + y^2 + s^2}} ds, \tag{4.41}$$

$$u_{\mathrm{h}}^{(0)} = -u_0 e^{ikz \cos \gamma} \frac{1}{2\pi} \frac{\partial}{\partial y} \int_0^\infty e^{-ik_1\xi \cos \varphi_0} d\xi \int_{-\infty}^\infty e^{iks \cos \gamma} \frac{e^{ik\sqrt{(x-\xi)^2+y^2+s^2}}}{\sqrt{(x-\xi)^2 + y^2 + s^2}} ds. \tag{4.42}$$

The integrals over the variable s still contain the wave number k for the normal incidence and need further investigation. Let us rewrite them as

$$\int_{-\infty}^\infty e^{iks \cos \gamma} \frac{e^{ik\sqrt{D^2+s^2}}}{\sqrt{D^2 + s^2}} ds, \quad \text{with } D = \sqrt{(x-\xi)^2 + y^2}. \tag{4.43}$$

We then return to the integral (3.8) for the Hankel function and make the following changes: $w = s$, $z = t$, $d = -ip$, $k = -iD$, $q = \sqrt{p^2 - t^2}$ with $D > 0$ and $\mathrm{Im}\, q > 0$. After these manipulations it follows from (3.8) that

$$H_0^{(1)}(qD) = \frac{1}{i\pi} \int_{-\infty}^\infty e^{-ist} \frac{e^{ip\sqrt{D^2+s^2}}}{\sqrt{D^2 + s^2}} ds = \frac{1}{i\pi} \int_{-\infty}^\infty e^{ist} \frac{e^{ip\sqrt{D^2+s^2}}}{\sqrt{D^2 + s^2}} ds. \tag{4.44}$$

By setting here $t = k \cos \gamma$, $p = k$, $q = \sqrt{p^2 - t^2} = k \sin \gamma = k_1$, we find

$$\int_{-\infty}^\infty e^{iks \cos \gamma} \frac{e^{ik\sqrt{D^2+s^2}}}{\sqrt{D^2 + s^2}} ds = i\pi H_0^{(1)}(k_1 D). \tag{4.45}$$

This relationship allows one to rewrite the PO fields (4.41) and (4.42) in the form

$$u_{\mathrm{s}}^{(0)} = -u_0 e^{ikz \cos \gamma} \frac{k_1 \sin \varphi_0}{2} \int_0^\infty e^{-ik_1\xi \cos \varphi_0} H_0^{(1)}\left(k_1 \sqrt{(x-\xi)^2 + y^2}\right) d\xi, \tag{4.46}$$

$$u_{\mathrm{s}}^{(0)} = -u_0 e^{ikz \cos \gamma} \frac{i}{2} \frac{\partial}{\partial y} \int_0^\infty e^{-ik_1\xi \cos \varphi_0} H_0^{(1)}\left(k_1 \sqrt{(x-\xi)^2 + y^2}\right) d\xi. \tag{4.47}$$

Comparison with expressions (3.10) and (3.11) finally confirms that the PO fields really satisfy the above rule for the transition to oblique incidence.

Thus, it has been proved that this rule is applicable both to the exact solution and to the PO approximation. Hence, this rule is also applicable to their difference, which is the field $u_{\mathrm{s,h}}^{(1)}$ generated by the nonuniform component $j_{\mathrm{s,h}}^{(1)}$ of the surface sources of the diffracted field. By the application of this rule to Equations (4.12) and (4.13), one can easily find the field $u_{\mathrm{s,h}}^{(1)}$ generated under the oblique incidence:

$$u_{\mathrm{s}}^{(1)} \sim u_0 f^{(1)}(\varphi, \varphi_0, \alpha) \frac{e^{i(k_1 r + \pi/4)}}{\sqrt{2\pi k_1 r}} e^{ikz \cos \gamma}, \tag{4.48}$$

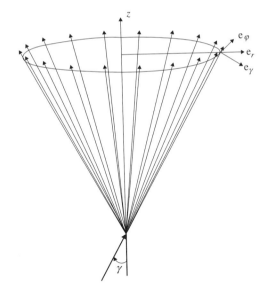

Figure 4.4   Cone of diffracted rays.

$$u_{\mathrm{h}}^{(1)} \sim u_0 \, g^{(1)}(\varphi, \varphi_0, \alpha) \frac{e^{i(k_1 r + \pi/4)}}{\sqrt{2\pi k_1 r}} e^{ikz \cos \gamma}, \tag{4.49}$$

where the functions $f^{(1)}$ and $g^{(1)}$ are defined in Section 4.1 and the angle φ_0 is determined by Equation (4.36).

The waves (4.48) and (4.49) have the form of conic waves and can be interpreted in terms of diffracted rays. Indeed, their eikonal

$$S = z \cos \gamma + r \sin \gamma \tag{4.50}$$

describes the phase fronts in the form of conic surfaces where $S = \mathrm{const}$. The gradient of the eikonal

$$\nabla S = \hat{z} \cos \gamma + \hat{r} \sin \gamma \tag{4.51}$$

indicates the directions of the edge diffracted rays. They are distributed over a cone surface shown in Figure 4.4. The axis of this cone is directed along the edge. All rays form the same angle γ with the edge as the incident ray.

In the case of electromagnetic waves

$$E_z^{\mathrm{inc}} = E_{0z} e^{ikz \cos \gamma} e^{-ik_1 r \cos(\varphi - \varphi_0)} \tag{4.52}$$

and

$$H_z^{\mathrm{inc}} = H_{0z} e^{ikz \cos \gamma} e^{-ik_1 r \cos(\varphi - \varphi_0)} \tag{4.53}$$

incident at a perfectly conducting wedge, the following aymptotics describe the diffracted far field (Ufimtsev, 2003):

$$E_z^{(0)} = [E_{0z}f^{(0)}(\varphi, \varphi_0) - H_{0z}\cos\gamma]\frac{e^{ik_1 r + i\pi/4}}{\sqrt{2\pi k_1 r}}e^{ikz\cos\gamma}, \tag{4.54}$$

$$H_z^{(0)} = H_{0z}g^{(0)}(\varphi, \varphi_0)\frac{e^{ik_1 r + i\pi/4}}{\sqrt{2\pi k_1 r}}e^{ikz\cos\gamma}, \tag{4.55}$$

$$E_z^{(1)} = [E_{0z}f^{(1)}(\varphi, \varphi_0, \alpha) + H_{0z}\cos\gamma]\frac{e^{ik_1 r + i\pi/4}}{\sqrt{2\pi k_1 r}}e^{ikz\cos\gamma}, \tag{4.56}$$

$$H_z^{(1)} = H_{0z}g^{(1)}(\varphi, \varphi_0, \alpha)\frac{e^{ik_1 r + i\pi/4}}{\sqrt{2\pi k_1 r}}e^{ikz\cos\gamma}. \tag{4.57}$$

The second term inside the brackets in Equation (4.54) shows the polarization coupling in the PO field under the oblique incidence. It also reveals itself in the second term of Equation (4.56). However, the field generated by the total surface current $\vec{j}^{(t)} = \vec{j}^{(0)} + \vec{j}^{(1)}$ is free from this polarization coupling,

$$E_z^{(t)} = E_{0z}f(\varphi, \varphi_0, \alpha)\frac{e^{ik_1 r + i\pi/4}}{\sqrt{2\pi k_1 r}}e^{ikz\cos\gamma}, \tag{4.58}$$

$$H_z^{(t)} = H_{0z}g(\varphi, \varphi_0, \alpha)\frac{e^{ik_1 r + i\pi/4}}{\sqrt{2\pi k_1 r}}e^{ikz\cos\gamma}. \tag{4.59}$$

One should mention that the diffraction cone was first discovered theoretically for arbitrary curved edges by Rubinowicz (1924). In the case of curved edges, the axis of the diffraction cone is directed along the tangent to the edge at the diffraction point. He made this discovery by the asymptotic evaluation of the Kirchhoff diffraction integral (Rubinowicz, 1924). He also established that the edge diffracted rays satisfy Fermat's principle. Later on, this concept of edge diffracted rays was included into the Geometrical Theory of Diffraction (Keller, 1962). The *ray interpretation of the edge diffracted waves* (4.48) and (4.49) was also suggested in the PTD (Ufimtsev, 1962). Senior and Uslenghi (1972) proved the existence of these rays in experiments with the diffraction of laser radiation.

PROBLEMS

Functions $f(\varphi, \varphi_0, \alpha)$, $g(\varphi, \varphi_0, \alpha)$ as well as functions $f^{(0)}(\varphi, \varphi_0)$, $g^{(0)}(\varphi, \varphi_0)$ are singular at the geometrical optics boundaries of the incident and reflected rays. Verify that their differences $f^{(1)}(\varphi, \varphi_0, \alpha)$ and $g^{(1)}(\varphi, \varphi_0, \alpha)$ are finite there.

4.1   Prove Equation (4.16).

4.2   Prove Equation (4.17).

4.3   Prove Equation (4.18).

4.4   Prove Equation (4.19).

Chapter 5

First-Order Diffraction at Strips and Polygonal Cylinders

> The relationship $u_s = E_z$ and $u_h = H_z$ exist between the acoustic and electromagnetic fields for these problems.

In Chapters 3 and 4, we have built a foundation for the solution of 2-D diffraction problems. General asymptotic expressions have been derived for the first-order edge diffracted waves generated both by the uniform and nonuniform components of the surface sources. In the present chapter, this general theory is applied to high-frequency diffraction at strips and cylinders with triangular cross-sections. These specific diffraction problems have been comprehensively studied in the existing literature. In particular, the uniform asymptotic expressions (with arbitrary high asymptotic precision) for the directivity pattern and for the surface field at the strips have been derived in Ufimtsev (1969, 1970, 2003). In these publications one can also find many other references related to the strip diffraction problem. Among them one should note the first and classical solution by Schwarzschild (1902). High-frequency diffraction at polygonal cylinders was investigated in Morse (1964) and Borovikov (1966). We now consider these problems again, to demonstrate the first applications of PTD.

5.1 DIFFRACTION AT A STRIP

The geometry of the problem is shown in Figure 5.1. The soft (1.5) or hard (1.6) boundary conditions are imposed at the strip. The incident wave is given by

$$u^{inc} = u_0 e^{ik(x \cos \phi_0 + y \sin \phi_0)}, \qquad \text{with} - \pi/2 < \phi_0 < \pi/2. \qquad (5.1)$$

Fundamentals of the Physical Theory of Diffraction. By Pyotr Ya. Ufimtsev
Copyright © 2007 John Wiley & Sons, Inc.

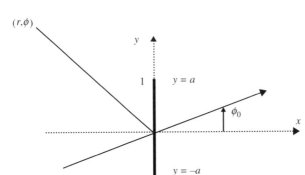

Figure 5.1   Cross-section of the strip by the plane $z = \text{const}$.

The diffracted field is investigated around the strip in the directions $-\pi/2 < \phi < 3\pi/2$. For description of edge waves we utilize the local coordinates r_1, φ_1, φ_{01} and r_2, φ_2, φ_{02}, which are measured from the illuminated side of the strip (Fig. 5.2).

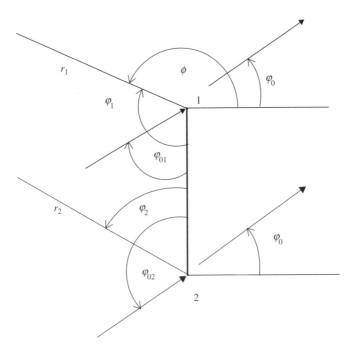

Figure 5.2   Local coordinates for edge waves.

5.1.1 Physical Optics Part of the Scattered Field

The Physical Optics (PO) part of the scattered field generated by the uniform components (1.31) of the surface sources is determined by the integrals (1.32). However, this integration process can be avoided. It turns out that the far field can be immediately calculated as the sum of two edge waves described in general form by Equations (3.53) and (3.54):

$$u_s^{(0)} = u_0 \left[f^{(0)}(\varphi_1, \varphi_{01}) \frac{e^{i(kr_1+\pi/4)}}{\sqrt{2\pi kr_1}} + f^{(0)}(\varphi_2, \varphi_{02}) \frac{e^{i(kr_2+\pi/4)}}{\sqrt{2\pi kr_2}} \right], \qquad (5.2)$$

$$u_h^{(0)} = u_0 \left[g^{(0)}(\varphi_1, \varphi_{01}) \frac{e^{i(kr_1+\pi/4)}}{\sqrt{2\pi kr_1}} + g^{(0)}(\varphi_2, \varphi_{02}) \frac{e^{i(kr_2+\pi/4)}}{\sqrt{2\pi kr_2}} \right]. \qquad (5.3)$$

For the far zone ($r \gg ka^2$), these expressions can be simplified:

$$u_s^{(0)} = u_0 \left[f^{(0)}(1) e^{ika(\sin\phi_0 - \sin\phi)} + f^{(0)}(2) e^{-ika(\sin\phi_0 - \sin\phi)} \right] \frac{e^{i(kr+\pi/4)}}{\sqrt{2\pi kr}}, \qquad (5.4)$$

$$u_h^{(0)} = u_0 \left[g^{(0)}(1) e^{ika(\sin\phi_0 - \sin\phi)} + g^{(0)}(2) e^{-ika(\sin\phi_0 - \sin\phi)} \right] \frac{e^{i(kr+\pi/4)}}{\sqrt{2\pi kr}}, \qquad (5.5)$$

where

$$f^{(0)}(1) \equiv f^{(0)}(\varphi_1, \varphi_{01}), \qquad f^{(0)}(2) \equiv f^{(0)}(\varphi_2, \varphi_{02}), \qquad (5.6)$$

and

$$g^{(0)}(1) \equiv g^{(0)}(\varphi_1, \varphi_{01}), \qquad g^{(0)}(2) \equiv g^{(0)}(\varphi_2, \varphi_{02}). \qquad (5.7)$$

In accordance with Equation (3.55), functions $f^{(0)}$ and $g^{(0)}$ are defined in terms of the basic coordinates ϕ and ϕ_0 as

$$f^{(0)}(1) = -f^{(0)}(2) = \frac{\cos\phi_0}{\sin\phi_0 - \sin\phi} \qquad (5.8)$$

and

$$g^{(0)}(1) = -g^{(0)}(2) = \frac{\cos\phi}{\sin\phi_0 - \sin\phi}, \qquad (5.9)$$

with $-\pi/2 < \phi < 3\pi/2$ and $-\pi/2 < \phi_0 < \pi/2$.

The field expressions (5.4) and (5.5) possesses a wonderful property. Although all functions (5.8) and (5.9) are singular for the directions $\phi = \phi_0$ and $\phi = \pi - \phi_0$, their combinations in Equations (5.4) and (5.5) are always finite due to the relationships $f^{(0)}(1) = -f^{(0)}(2)$ and $g^{(0)}(1) = -g^{(0)}(2)$. This property of expressions

(5.4) and (5.5) becomes obvious when they are written in the explicit form

$$u_s^{(0)} = u_0 \Phi_s^{(0)}(\phi, \phi_0) \frac{e^{i(kr+\pi/4)}}{\sqrt{2\pi kr}}, \tag{5.10}$$

$$u_h^{(0)} = u_0 \Phi_h^{(0)}(\phi, \phi_0) \frac{e^{i(kr+\pi/4)}}{\sqrt{2\pi kr}}, \tag{5.11}$$

where

$$\Phi_s^{(0)}(\phi, \phi_0) = i2 \cos \phi_0 \frac{\sin[ka(\sin \phi - \sin \phi_0)]}{\sin \phi - \sin \phi_0}, \tag{5.12}$$

and

$$\Phi_h^{(0)}(\phi, \phi_0) = i2 \cos \phi \frac{\sin[ka(\sin \phi - \sin \phi_0)]}{\sin \phi - \sin \phi_0}. \tag{5.13}$$

Now it is clear that

$$\Phi_s^{(0)}(\phi_0, \phi_0) = \Phi_h^{(0)}(\phi_0, \phi_0) = i2ka \cos \phi_0 \tag{5.14}$$

for the forward direction $\phi = \phi_0$, and

$$\Phi_s^{(0)}(\pi - \phi_0, \phi_0) = -\Phi_h^{(0)}(\pi - \phi_0, \phi_0) = i2ka \cos \phi_0 \tag{5.15}$$

for the specular direction $\phi = \pi - \phi_0$.

In accordance with the 2-D form of the optical theorem (Ufimtsev, 2003), the total scattering cross-section is defined as

$$\sigma^{tot} = \frac{2}{k} \text{Im} \Phi(\phi_0, \phi_0), \tag{5.16}$$

which, following Equation (5.14), equals

$$\sigma_s^{(0)tot} = \sigma_h^{(0)tot} = 2A, \tag{5.17}$$

where $A = 2a \cos \phi_0$ is the "width" of the incident wave part intercepted by the strip (Fig. 5.3). Equation (5.17) can be also interpreted as the total scattered power per unit length of the strip in the z-axis direction.

It is expedient to introduce the 2-D bistatic scattering cross-section σ by a relationship similar to Equation (1.24):

$$P_{av}^{sc} = \frac{\sigma \cdot P_{av}^{inc}}{2\pi r} \tag{5.18}$$

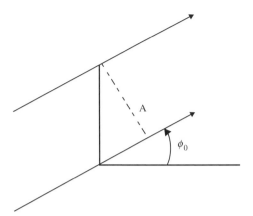

Figure 5.3   Cross-section A of the incident wave intercepted by the strip.

where

$$P^{\text{inc}}_{\text{av}} = \frac{1}{2}k^2 Z |u_0|^2, \qquad P^{\text{sc}}_{\text{av}} = \frac{1}{2}k^2 Z |u^{\text{sc}}|^2. \tag{5.19}$$

Therefore in view of Equations (5.10) and (5.11), the bistatic scattering cross-section is defined by

$$\sigma^{(0)}_{\text{s,h}} = \frac{1}{k} \left| \Phi^{(0)}_{\text{s,h}}(\phi, \phi_0) \right|^2. \tag{5.20}$$

This is a general definition of the bistatic scattering cross-section σ for any 2-D scattered fields represented in the form of Equations (5.10) and (5.11).

In the case of backscattering when $\phi = \pi + \phi_0$, the directivity patterns are given by the simple expression

$$\Phi^{(0)}_{\text{s}} = -\Phi^{(0)}_{\text{h}} = i \cot \phi_0 \sin(2ka \sin \phi_0). \tag{5.21}$$

Finally one should notice another peculiarity of the directivity patterns (5.12) and (5.13) for the field generated by the uniform components of the surface sources. They have exact zeros in the directions where

$$ka(\sin \phi_0 - \sin \phi) = \pm n\pi, \qquad \text{with } n = 1, 2, 3, \ldots. \tag{5.22}$$

This property is the consequence of the relationships $f^{(0)}(1) = -f^{(0)}(2)$ and $g^{(0)}(1) = -g^{(0)}(2)$.

5.1.2 Total Scattered Field

The nonuniform components of scattering sources concentrate near the edges of the strip and generate the two edge waves described by Equations (4.12) and (4.13). Their sum equals

$$u_s^{(1)} = u_0 \left[f^{(1)}(\varphi_1, \varphi_{01}, 2\pi) \frac{e^{i(kr_1 + \pi/4)}}{\sqrt{2\pi kr_1}} + f^{(1)}(\varphi_2, \varphi_{02}, 2\pi) \frac{e^{i(kr_2 + \pi/4)}}{\sqrt{2\pi kr_2}} \right] \qquad (5.23)$$

and

$$u_h^{(1)} = u_0 \left[g^{(1)}(\varphi_1, \varphi_{01}, 2\pi) \frac{e^{i(kr_1 + \pi/4)}}{\sqrt{2\pi kr_1}} + g^{(1)}(\varphi_2, \varphi_{02}, 2\pi) \frac{e^{i(kr_2 + \pi/4)}}{\sqrt{2\pi kr_2}} \right]. \qquad (5.24)$$

As a result of relationships (4.12) and (4.13) and (4.14) and (4.15) these expressions can be written in the form

$$u_s^{(1)} = u_s - u_s^{(0)}, \qquad u_h^{(1)} = u_h - u_h^{(0)}. \qquad (5.25)$$

Therefore, the total first-order scattered field is given by

$$u_s = u_s^{(0)} + u_s^{(1)} = u_0 \left[f(\varphi_1, \varphi_{01}, 2\pi) \frac{e^{i(kr_1 + \pi/4)}}{\sqrt{2\pi kr_1}} + f(\varphi_2, \varphi_{02}, 2\pi) \frac{e^{i(kr_2 + \pi/4)}}{\sqrt{2\pi kr_2}} \right]$$

$$(5.26)$$

and

$$u_h = u_h^{(0)} + u_h^{(1)} = u_0 \left[g(\varphi_1, \varphi_{01}, 2\pi) \frac{e^{i(kr_1 + \pi/4)}}{\sqrt{2\pi kr_1}} + g(\varphi_2, \varphi_{02}, 2\pi) \frac{e^{i(kr_2 + \pi/4)}}{\sqrt{2\pi kr_2}} \right]$$

$$(5.27)$$

with functions f and g defined in Equations (2.62) and (2.64). For the far zone these field expressions are simplified to

$$u_s = u_0 \left[f(1) e^{ika(\sin\phi_0 - \sin\phi)} + f(2) e^{-ika(\sin\phi_0 - \sin\phi)} \right] \frac{e^{i(kr + \pi/4)}}{\sqrt{2\pi kr}} \qquad (5.28)$$

and

$$u_h = u_0 \left[g(1) e^{ika(\sin\phi_0 - \sin\phi)} + g(2) e^{-ika(\sin\phi_0 - \sin\phi)} \right] \frac{e^{i(kr + \pi/4)}}{\sqrt{2\pi kr}}, \qquad (5.29)$$

where

$$f(1) \equiv f(\varphi_1, \varphi_{01}, 2\pi), \qquad f(2) \equiv f(\varphi_2, \varphi_{02}, 2\pi),$$

and (5.30)

$$g(1) \equiv g(\varphi_1, \varphi_{01}, 2\pi), \qquad g(2) \equiv g(\varphi_2, \varphi_{02}, 2\pi).$$

In terms of basic coordinates ϕ and ϕ_0, these functions are determined by

$$\begin{Bmatrix} f(1) \\ g(1) \end{Bmatrix} = \frac{1}{2}\left(-\frac{1}{\sin\dfrac{\phi - \phi_0}{2}} \mp \frac{1}{\cos\dfrac{\phi + \phi_0}{2}} \right), \qquad \text{with } -\pi/2 \le \phi \le 3\pi/2$$

$$(5.31)$$

and

$$\begin{Bmatrix} f(2) \\ g(2) \end{Bmatrix} = \frac{1}{2}\left(\frac{1}{\sin\dfrac{\phi - \phi_0}{2}} \mp \frac{1}{\cos\dfrac{\phi + \phi_0}{2}} \right), \qquad \text{with } -\pi/2 \le \phi \le \pi/2, \quad (5.32)$$

but

$$\begin{Bmatrix} f(2) \\ g(2) \end{Bmatrix} = \frac{1}{2}\left(-\frac{1}{\sin\dfrac{\phi - \phi_0}{2}} \pm \frac{1}{\cos\dfrac{\phi + \phi_0}{2}} \right), \qquad \text{with } \pi/2 \le \phi \le 3\pi/2.$$

$$(5.33)$$

As can be seen, functions $g(2)$ and $\partial f(2)/\partial\phi$ are discontinuous in the direction $\phi = \pi/2$. This is a consequence of the fact that they relate to the field generated by the scattering sources distributed over the whole half-plane $-a < y < \infty$. Functions $g(1)$ and $\partial f(1)/\partial\phi$ are also discontinuous. They have different values in the directions $\phi = -\pi/2$ and $\phi = 3\pi/2$ related to the different sides of the half-plane $-\infty < y < a$ containing the sources of the field.

This discontinuity in the field of the first-order edge waves (5.28) and (5.29) can be eliminated in two ways. First, in the calculation of the field generated by the nonuniform component $j_{s,h}^{(1)}$, one should restrict the integration region by the actual surface of the strip. In other words, one should truncate (outside the strip) the component $j_{s,h}^{(1)}$ related to the semi-infinite half-plane. This approach is presented in Section 5.1.4 (see also Michaeli (1987) and Johansen (1996)). Another remedy is the calculation of multiple edge diffraction (Ufimtsev, 2003).

In view of Equations (5.31) to (5.33), the field expressions (5.28) and (5.29) can be written in the form

$$u_{s,h} = u_0 \Phi_{s,h}(\phi, \phi_0)\frac{e^{i(kr+\pi/4)}}{\sqrt{2\pi kr}}, \qquad (5.34)$$

where

$$\Phi_s(\phi, \phi_0) = -\frac{\cos[ka(\sin\phi_0 - \sin\phi)]}{\cos\dfrac{\phi_0 + \phi}{2}} + i\frac{\sin[ka(\sin\phi_0 - \sin\phi)]}{\sin\dfrac{\phi_0 - \phi}{2}},$$

with $-\pi/2 \le \phi \le \pi/2$, \hfill (5.35)

$$\Phi_s(\phi, \phi_0) = \frac{\cos[ka(\sin\phi_0 - \sin\phi)]}{\sin\dfrac{\phi_0 - \phi}{2}} - i\frac{\sin[ka(\sin\phi_0 - \sin\phi)]}{\cos\dfrac{\phi_0 + \phi}{2}},$$

with $\pi/2 \le \phi \le 3\pi/2$, \hfill (5.36)

and

$$\Phi_h(\phi, \phi_0) = \frac{\cos[ka(\sin\phi_0 - \sin\phi)]}{\cos\dfrac{\phi_0 + \phi}{2}} + i\frac{\sin[ka(\sin\phi_0 - \sin\phi)]}{\sin\dfrac{\phi_0 - \phi}{2}},$$

with $-\pi/2 \le \phi \le \pi/2$, \hfill (5.37)

$$\Phi_h(\phi, \phi_0) = \frac{\cos[ka(\sin\phi_0 - \sin\phi)]}{\sin\dfrac{\phi_0 - \phi}{2}} + i\frac{\sin[ka(\sin\phi_0 - \sin\phi)]}{\cos\dfrac{\phi_0 + \phi}{2}},$$

with $\pi/2 \le \phi \le 3\pi/2$. \hfill (5.38)

As can be seen, these functions (in contrast to functions $\Phi_{s,h}^{(0)}(\phi, \varphi_0)$) do not have exact zeros in the directions determined by Equation (5.22).

In the following we provide the explicit expressions of functions $\Phi_{s,h}$ for certain specific directions of observation. For the forward direction $\phi = \phi_0$,

$$\Phi_s(\phi_0, \phi_0) = i2ka\cos\phi_0 - \frac{1}{\cos\phi_0} \hfill (5.39)$$

and

$$\Phi_h(\phi_0, \phi_0) = i2ka\cos\phi_0 + \frac{1}{\cos\phi_0}, \hfill (5.40)$$

and for the specular direction $\phi = \pi - \phi_0$,

$$\Phi_s(\pi - \phi_0, \phi_0) = i2ka\cos\phi_0 - \frac{1}{\cos\phi_0} \hfill (5.41)$$

and

$$\Phi_h(\pi - \phi_0, \phi_0) = -i2ka\cos\phi_0 - \frac{1}{\cos\phi_0}. \hfill (5.42)$$

In addition, for the backscattering direction $\phi = \pi + \phi_0$,

$$\Phi_s(\pi + \phi_0, \phi_0) = -\cos(2ka\sin\phi_0) + i\frac{\sin(2ka\sin\phi_0)}{\sin\phi_0} \hfill (5.43)$$

and

$$\Phi_h(\pi + \phi_0, \phi_0) = -\cos(2ka \sin \phi_0) - i \frac{\sin(2ka \sin \phi_0)}{\sin \phi_0}. \qquad (5.44)$$

In conformity with Equations (4.20) and (4.21), functions (5.39) to (5.42) are singular under the grazing incidence ($\phi_0 = \pm \pi/2$).

A comparison with Equation (5.14) shows that the first term in Equations (5.39) and (5.40) is caused by the uniform component, and the second term by the nonuniform component of the surface sources. In view of Equation (5.16), this observation gives the impression that the field generated by the nonuniform component does not contribute to the total scattered power. An interesting question arises. How does it happen that the nonuniform component $j^{(1)}$ generates the field (5.23) and (5.24), but does not contribute to the total scattered power? The answer is the following. The field (5.23) and (5.24) does contribute to the total scattered field, but through the high-order edge waves generated due to multiple diffraction of the primary edge waves (5.34).

An additional comment is necessary on the field equations (5.28) and (5.29) and functions f and g involved in these equations. Functions f and g are singular in the forward ($\phi = \phi_0$) and specular ($\phi = \pi - \phi_0$) directions related to the geometrical optics boundaries of the incident and reflected rays. However, these singularities cancel each other in the field Equations (5.28) and (5.29) and provide finite values in these special directions, as is seen in Equations (5.39) to (5.42). This cancellation is due to the fact that the incident wave is a plane wave. Because of this, the geometrical optics boundaries related to the different edges of the strip are parallel to each other (Fig. 5.3) and merge in the far zone from the strip. This case represents an exception as compared to a more general situation when the source of the incident wave can be at a finite distance from the scattering object. For example, in the case of the cylindrical incident wave (Fig. 5.4), the shadow boundaries caused by the strip edges are not parallel and therefore the related singularities of functions f and g are separated in space and cannot cancel each other. In such a case, the traditional PTD procedure consists of the calculation of the fields generated separately by the uniform and nonuniform

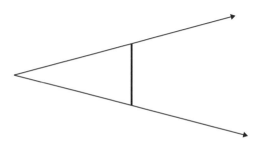

Figure 5.4 Shadow boundaries caused by the strip edges in the field of the incident cylindrical wave. At these boundaries, functions f and g are singular: $f = \infty$ and $g = \infty$.

components of the scattering sources with the subsequent summation of these fields. Each of these fields is finite and hence their sum is finite as well.

5.1.3 Numerical Analysis of the Scattered Field

The relationships $u_s = E_z$ and $u_h = H_z$ exist between the acoustic and electromagnetic fields for this problem.

In this section we present the numerical analysis of the scattered field by utilizing different approaches. Note that the geometry of the problem is shown in Figure 5.1. The soft (1.5) or hard (1.6) boundary conditions are imposed on the strip. The incident wave is the plane wave (5.1). The relationships $u_s = E_z$ and $u_h = H_z$ show the equivalence between the acoustic and electromagnetic fields studied in this section. Equation (5.20) determines the bistatic scattering cross-section. The calculated quantity is the normalized scattering cross-section

$$\frac{\sigma_{s,h}}{kl^2} = \left| \frac{\Phi_{s,h}(\phi, \phi_0)}{kl} \right|^2, \qquad \text{where } l = 2a, \tag{5.45}$$

which is plotted in the decibel scale. For parameters ka and ϕ_0 we take the values $ka = 3\pi$ and $\phi_0 = 45°$, that is, $\varphi_0 = \pi/4$ rad. In this case, the strip width $l = 2a$ equals 3λ and the application of the asymptotic theory is justified. The scattering cross-section is investigated in the directions $\pi/2 \leq \phi \leq 3\pi/2$ around the strip. One should note that the function Φ_s is symmetrical with respect to the strip plane $[\Phi_s(\pi - \phi, \phi_0) = \Phi_s(\phi, \phi_0)]$, and the function Φ_h is anti-symmetrical $[\Phi_h(\pi - \phi, \phi_0) = -\Phi_h(\phi, \phi_0)]$. That is why it is sufficient to calculate these functions only in the interval $\pi/2 \leq \phi \leq 3\pi/2$.

Calculations are performed using the following approaches:

- The PO approximation with the functions $\Phi_{s,h}^{(0)}$ given in Section 5.1.1;
- The first-order PTD approximation with the functions $\Phi_{s,h}$ presented in Section 5.1.2;
- The first-order TED approximation (Ufimtsev, 2003) with the functions

$$\Phi_s(\phi, \phi_0) = \tilde{\Psi}_1(\alpha, \alpha_0) + \tilde{\Psi}_1(-\alpha, -\alpha_0) \tag{5.46}$$

and

$$\Phi_h(\phi, \phi_0) = \tilde{\Phi}_1(\alpha, \alpha_0) + \tilde{\Phi}_1(-\alpha, -\alpha_0), \tag{5.47}$$

where

$$
\tilde{\Psi}_1(\alpha, \alpha_0) = \frac{\sqrt{1-\alpha}\sqrt{1-\alpha_0}}{\alpha + \alpha_0} e^{i\chi(\alpha+\alpha_0)} + \frac{\sqrt{1-\alpha}\sqrt{1+\alpha_0}}{\alpha + \alpha_0}
$$

$$
\times \left[\frac{1-\alpha}{2} \psi(q,\alpha) - \frac{1+\alpha_0}{2} \psi(q,-\alpha_0) \right] e^{iq+i\chi(\alpha-\alpha_0)}
$$

$$
+ \frac{\sqrt{1-\alpha}}{2D_s} e^{i2q} \left[\sqrt{1-\alpha_0} \psi(q,\alpha_0) e^{i\chi\alpha_0} \right.
$$

$$
\left. - \sqrt{1+\alpha_0} \psi(q,-\alpha_0) \psi(q,1) e^{iq-i\chi\alpha_0} \right] \psi(q,\alpha) e^{i\chi\alpha}, \tag{5.48}
$$

$$
\tilde{\Phi}_1(\alpha,\alpha_0) = -\frac{\sqrt{1+\alpha}\sqrt{1+\alpha_0}}{\alpha + \alpha_0} e^{i\chi(\alpha+\alpha_0)} + \frac{1}{\alpha + \alpha_0}
$$

$$
\times \left[(1+\alpha_0)\sqrt{\frac{1+\alpha}{2}} \varphi(q,-\alpha_0) - (1-\alpha)\sqrt{\frac{1-\alpha_0}{2}} \varphi(q,\alpha) \right]
$$

$$
\times e^{iq+i\chi(\alpha-\alpha_0)} - \frac{1}{D_h} e^{i2q} \left[\varphi(q,\alpha_0) e^{i\chi\alpha_0} \right.
$$

$$
\left. - \varphi(q,-\alpha_0)\varphi(q,1) e^{iq-i\chi\alpha_0} \right] \varphi(q,\alpha) e^{i\chi\alpha}. \tag{5.49}
$$

In these expressions,

$$
\alpha = \sin\phi, \qquad \alpha_0 = -\sin\phi_0, \qquad \chi = ka, \qquad q = 2\chi = 2ka, \tag{5.50}
$$

$$
\varphi(q,\alpha) = \frac{2}{\sqrt{\pi}} e^{-i\pi/4} e^{-iq(1+\alpha)} \int_{\sqrt{q(1+\alpha)}}^{\infty} e^{it^2}\, dt, \tag{5.51}
$$

$$
\psi(q,\alpha) = \frac{i}{\sqrt{2(1+\alpha)}} \frac{\partial\varphi(q,\alpha)}{\partial q}, \tag{5.52}
$$

and

$$
D_s = 1 - \psi^2(q,1) e^{i2q}, \qquad D_h = 1 - \varphi^2(q,1) e^{i2q}. \tag{5.53}
$$

The absolute error of the TED approximation (5.46) and (5.47) is equal to

$$
Q_{s,h}(\alpha,\alpha_0) = \tilde{Q}_{s,h}(\alpha,\alpha_0) + \tilde{Q}_{s,h}(-\alpha,-\alpha_0) \tag{5.54}
$$

where

$$
\tilde{Q}_s(\alpha,\alpha_0) = \frac{O(q^{-1/2})}{[1+q(1+\alpha)][1+q(1-\alpha_0)]} \tag{5.55}
$$

and

$$
\tilde{Q}_h(\alpha,\alpha_0) = \frac{\sqrt{1+\alpha}\sqrt{1+\alpha_0}\,O(\sqrt{q})}{[1+q(1+\alpha)][1+q(1+\alpha_0)]} \tag{5.56}
$$

under the conditions $q(1 \pm \alpha) \gg 1$ and $q(1 \pm \alpha_0) \gg 1$. Here, symbol $O(q^m)$ is used to show the asymptotic behavior of quantities $\widetilde{Q}_{s,h}(\alpha, \alpha_0, q)$ when $q \to \infty$. The definition of this symbol has been given above in conjunction with Equation (2.104). Also,

$$\lim \sqrt{1 - \alpha^2}\, \frac{\partial}{\partial \alpha} Q_s(\alpha, \alpha_0) = 0, \qquad \text{with } \alpha \to \pm 1, \tag{5.57}$$

and

$$Q_h(\alpha, \pm 1) = Q_h(\pm 1, \alpha_0) = 0. \tag{5.58}$$

Equations (5.57) and (5.58) are the consequences of the fact that the scattered field (5.46) and (5.47) and its normal derivatives at the plane $x = 0$ are continuous outside the strip. In other words, according to approximation (5.46) and (5.47), no scattering sources exist outside the strip surface.

We have calculated functions (5.46) and (5.47) with the well-established estimations (5.55) and (5.56) to demonstrate the accuracy of the PO and PTD approximations. The results are plotted in Figures 5.5 and 5.6 in the decibel scale as $10 \log(\sigma_{s,h}/kl^2)$. It is seen in these figures that the accuracy of the first-order PTD approximation is higher for the soft boundary condition. Only a small discrepancy with the exact TED curve is observed in the vicinity of the direction $\phi = 90°$, and the PTD and TED curves infact merge in the region $100° < \phi \leq 270°$. The reason for the better approximation provided by the PTD for the soft boundary condition is the faster attenuation of the primary edge waves of the scattering sources $j_s^{(1)}$ compared to the similar waves of $j_h^{(1)}$. In this case, the amplitude of these waves (at the location of the opposite edge of the strip) is of the order $(2ka)^{-3/2}$ for the soft strip, and of the order $(2ka)^{-1/2}$ for the hard strip.

It is also seen in Figures 5.5 and 5.6 that PO cannot provide a reasonable approximation for the scattered field in the vicinity of the minima of the directivity patterns,

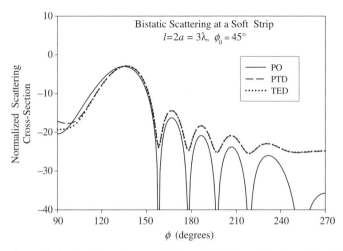

Figure 5.5 Scattering at an acoustically soft strip (E_z-polarization of the electromagnetic field).

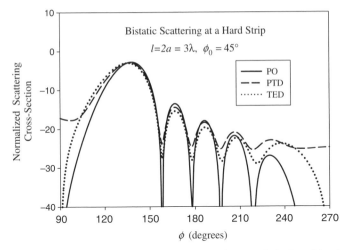

Figure 5.6   Scattering at an acoustically hard strip (H_z-polarization of the electromagnetic field).

where PO predicts wrong pure zeros. The incorrect values of the PTD approximation for the function σ_h in the vicinity of the directions $\phi = 90°$ and $\phi = 270°$ are caused by the fictitious scattering sources $j_h^{(1)}$ distributed outside the strip surface, as has already been discussed at the end of Section 5.1.2. This shortcoming of the first-order PTD is eliminated in the next section.

5.1.4   First-Order PTD with Truncated Scattering Sources $j_h^{(1)}$

> The relationship $u_h = H_z$ exists between the acoustic and electromagnetic fields for this problem.

The geometry of the problem is shown in Figure 5.1 and the incident wave is given by Equation (5.1). To calculate the scattered field u_h, we apply Equation (1.10) and utilize the following observations.

- The symbol r was used in Equation (1.10) for the distance between the integration and observation points. Now we replace it by $\rho = \sqrt{x^2 + (y - \eta)^2 + \zeta^2}$ and retain the symbol r for the polar coordinate $r = \sqrt{x^2 + y^2}$ of the observation point, assuming that $x = r \cos \phi$ and $y = r \sin \phi$.
- The surface of integration in Equation (1.10) covers both faces of the strip: $S = S_- + S_+$ (Fig. 5.7).

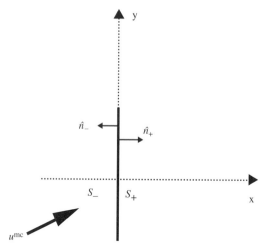

Figure 5.7   The integration surface $S = S_- + S_+$ in Equation (1.10) applied to the strip problem.

- The operator $\dfrac{\partial}{\partial n}$ in Equation (1.10) acts on coordinates of the integration/source point. According to Equations (1.14) and (3.5), it can be replaced by the operators acting on coordinates of the observation point. Namely, $\dfrac{\partial}{\partial n_-} = \dfrac{\partial}{\partial x}$ for the illuminated face S_- of the strip and $\dfrac{\partial}{\partial n_+} = -\dfrac{\partial}{\partial x}$ for the shadowed face S_+, where $\partial/\partial x$ is the differentiation with respect to the observation point (outside the strip).

- The integral in Equation (1.10) over the variable ζ applied to the strip problem can be expressed through the Hankel function $H_0^{(1)}(k\sqrt{x^2 + (y - \eta)^2})$ as shown in Equation (3.7).

- We calculate the scattered field in the far zone where $r \gg ka^2$. For this zone the Hankel function can be approximated by its asymptotic expression

$$H_0^{(1)}\left(k\sqrt{x^2 + (y - \eta)^2}\right) \sim \sqrt{\frac{2}{\pi kr}}\,\mathrm{e}^{ik(r - \eta \sin \phi) - i\pi/4}. \tag{5.59}$$

- We denote the scattering sources as $j_h^{(0)} = u_0 J_h^{(0)}$ and $j_h^{(1)} = u_0 J_h^{(1)}$.

- Then the scattered field in the far zone can be represented in the following form:

$$u_h^{(0)} = u_0 \frac{\mathrm{e}^{i\pi/4}}{2\sqrt{2\pi k}} \frac{\partial}{\partial x} \frac{\mathrm{e}^{ikr}}{\sqrt{r}} \int_{-a}^{a} J_h^{(0)}(\eta, x = -0)\mathrm{e}^{-ik\eta \sin \phi}\,d\eta, \tag{5.60}$$

$$u_h^{(1)} = u_0 \frac{\mathrm{e}^{i\pi/4}}{2\sqrt{2\pi k}} \frac{\partial}{\partial x} \frac{\mathrm{e}^{ikr}}{\sqrt{r}} \left[\int_{-a}^{a} J_h^{(1)}(\eta, x = -0)\mathrm{e}^{-ik\eta \sin \phi}\,d\eta \right.$$

$$\left. - \int_{-a}^{a} J_h^{(1)}(\eta, x = +0)\mathrm{e}^{-ik\eta \sin \phi}\,d\eta \right]. \tag{5.61}$$

Remember that the uniform scattering sources $j_h^{(0)}$ exist only on the illuminated face of the strip where $x = -0$.

- In the far zone approximation one can set

$$\frac{\partial}{\partial x} \frac{e^{ikr}}{\sqrt{r}} \approx ik \cos \phi \frac{e^{ikr}}{\sqrt{r}}. \qquad (5.62)$$

- According to the definition (1.31), one has $J_h^{(0)}(\eta, x = -0) = 2e^{ik\eta \sin \phi_0}$.

- The nonuniform scattering source $j_h^{(1)}$ is found by utilizing the exact solution for the half-plane diffraction problem presented in Section 2.5. According to this solution

  - The function $j_h^{(1)}$ is anti-symmetrical,

  $$j_h^{(1)}(\eta, x = +0) = -j_h^{(1)}(\eta, x = -0) \text{ and consequently}$$

  $$J_h^{(1)}(\eta, x = +0) = -J_h^{(1)}(\eta, x = -0);$$

  - The function $J_h^{(1)}(\eta, x = -0)$ is determined as

  $$J_h^{(1)}(\eta, x = -0) = 2 \left\{ v\left[k(a - \eta), \frac{\pi}{2} - \phi_0 \right] e^{ika \sin \phi_0} \right.$$
  $$\left. + v\left[k(a + \eta), \frac{\pi}{2} + \phi_0 \right] e^{-ika \sin \phi_0} \right\}. \qquad (5.63)$$

- In view of above observations, the functions (5.60) and (5.61) can be approximated by

$$u_h^{(0)} = u_0 i 2 \cos \phi \frac{\sin[ka(\sin \phi_0 - \sin \phi)]}{\sin \phi_0 - \sin \phi} \frac{e^{i(kr + \pi/4)}}{\sqrt{2\pi kr}}, \qquad (5.64)$$

$$u_h^{(1)} = u_0 2ik \cos \phi \frac{e^{i(kr + \pi/4)}}{\sqrt{2\pi kr}} \int_{-a}^{a} \left\{ v\left[k(a - \eta), \frac{\pi}{2} - \phi_0 \right] e^{ika \sin \phi_0} \right.$$
$$\left. + v\left[k(a + \eta), \frac{\pi}{2} + \phi_0 \right] e^{-ika \sin \phi_0} \right\} e^{-ik\eta \sin \phi} \, d\eta. \qquad (5.65)$$

- The asymptotic expression (5.64) was found here by the direct integration of the uniform component of the scattering sources. It is totally identical to the expression (5.11) derived by the summation of two edge waves.

- The field (5.65) is calculated by integration by parts.

- We provide here the final results in terms of the directivity patterns:

$$u_h^{(0,1)} = u_0 \Phi_h^{(0,1)}(\phi, \phi_0) \frac{e^{i(kr + \pi/4)}}{\sqrt{2\pi kr}} \qquad (5.66)$$

where $\Phi_h^{(0)}$ is given by Equation (5.13) and

$$\Phi_h^{(1)}(\phi,\phi_0) = \frac{\cos\phi}{\xi}\left\{-i4\sin(ka\xi) + 2\frac{e^{i3\pi/4}}{\sqrt{\pi}}\left[e^{-ika\xi}\int_0^{\sqrt{2ka(1+\sin\phi_0)}}e^{it^2}\,dt\right.\right.$$

$$\left. - e^{ika\xi}\int_0^{\sqrt{2ka(1-\sin\phi_0)}}e^{it^2}\,dt\right] - 2\frac{e^{i3\pi/4}}{\sqrt{\pi}}\left[\sqrt{\frac{1+\sin\phi_0}{1+\sin\phi}}e^{ika\xi}\right.$$

$$\left.\left. \times\int_0^{\sqrt{2ka(1+\sin\phi)}}e^{it^2}\,dt - \sqrt{\frac{1-\sin\phi_0}{1-\sin\phi}}e^{-ika\xi}\int_0^{\sqrt{2ka(1-\sin\phi)}}e^{it^2}\,dt\right]\right\}$$

$$(5.67)$$

where $\xi = \sin\phi_0 - \sin\phi$.

The directivity pattern of the total scattered field $u_h^{\text{tr}} = u_h^{(0)} + u_h^{(1)}$ (found by truncation of the nonuniform sources) equals $\Phi_h^{\text{tr}} = \Phi_h^{(0)} + \Phi_h^{(1)}$:

$$\Phi_h^{\text{tr}}(\phi,\phi_0) = \frac{\cos\phi}{\xi}\left\{-i2\sin(ka\xi) + 2\frac{e^{i3\pi/4}}{\sqrt{\pi}}\left[e^{-ika\xi}\int_0^{\sqrt{2ka(1+\sin\phi_0)}}e^{it^2}\,dt\right.\right.$$

$$\left. - e^{ika\xi}\int_0^{\sqrt{2ka(1-\sin\phi_0)}}e^{it^2}\,dt\right] - 2\frac{e^{i3\pi/4}}{\sqrt{\pi}}\left[\sqrt{\frac{1+\sin\phi_0}{1+\sin\phi}}e^{ika\xi}\right.$$

$$\left.\left. \times\int_0^{\sqrt{2ka(1+\sin\phi)}}e^{it^2}\,dt - \sqrt{\frac{1-\sin\phi_0}{1-\sin\phi}}e^{-ika\xi}\int_0^{\sqrt{2ka(1-\sin\phi)}}e^{it^2}\,dt\right]\right\}$$

$$(5.68)$$

In the ray region (where $ka(1 \pm \sin\phi_0) \gg 1$ and $ka(1 \pm \sin\phi) \gg 1$), the asymptotic expression of this function is given by

$$\Phi_h^{\text{tr}}(\phi,\phi_0) = g(1)e^{ika(\sin\phi_0-\sin\phi)} + g(2)e^{-ika(\sin\phi_0-\sin\phi)}$$

$$- \cos\phi\left[\frac{e^{ika(\sin\phi_0+\sin\phi)}}{(1+\sin\phi)\sqrt{1+\sin\phi_0}} + \frac{e^{-ika(\sin\phi_0+\sin\phi)}}{(1-\sin\phi)\sqrt{1-\sin\phi_0}}\right]$$

$$\times\frac{e^{i(2ka+\pi/4)}}{\sqrt{2\pi ka}}, \qquad (5.69)$$

where functions $g(1)$ and $g(2)$ are determined in Equations (5.31) and (5.33). These functions relate to the primary edge waves arising at the ends of the strip. The terms in the second line of Equation (5.69) represent the radiation of the primary edge waves when they reach the opposite end of the strip. These terms can be interpreted as the part of the true edge waves arising due to the secondary diffraction. The true secondary edge waves are described by the TED approximation (5.47), which leads to the following asymptotic expression

$$\Phi_h^{TED}(\phi,\phi_0) = g(1)e^{ika(\sin\phi_0-\sin\phi)} + g(2)e^{-ika(\sin\phi_0-\sin\phi)}$$

$$+ \left[\frac{e^{ika(\sin\phi_0+\sin\phi)}}{\sqrt{1+\sin\phi}\sqrt{1+\sin\phi_0}} + \frac{e^{-ika(\sin\phi_0+\sin\phi)}}{\sqrt{1-\sin\phi}\sqrt{1-\sin\phi_0}} \right]$$

$$\times \frac{e^{i(2ka+\pi/4)}}{\sqrt{\pi ka}}. \tag{5.70}$$

The function $\Phi_h^{tr}(\phi,\phi_0)$ contains the terms singular in the specular direction $\phi = \pi - \phi_0$. However, such singular terms cancel each other and generate the finite quantity for this function:

$$\Phi_h^{tr}(\pi - \phi_0, \phi_0)$$

$$= i2ka\cos\phi_0 - \cos\phi_0 \left\{ 4ka \frac{e^{i\pi/4}}{\sqrt{\pi}} \left[\int_0^{\sqrt{2ka(1+\sin\phi_0)}} e^{it^2}\,dt + \int_0^{\sqrt{2ka(1-\sin\phi_0)}} e^{it^2}\,dt \right] \right.$$

$$+ \sqrt{\frac{2ka}{\pi}} e^{i3\pi/4} \left[\frac{e^{i2ka(1+\sin\phi_0)}}{\sqrt{1+\sin\phi_0}} + \frac{e^{i2ka(1-\sin\phi_0)}}{\sqrt{1-\sin\phi_0}} \right] - \frac{e^{i3\pi/4}}{\sqrt{\pi}}$$

$$\times \left. \left[\frac{1}{1+\sin\phi_0} \int_0^{\sqrt{2ka(1+\sin\phi_0)}} e^{it^2}\,dt + \frac{1}{1-\sin\phi_0} \int_0^{\sqrt{2ka(1-\sin\phi_0)}} e^{it^2}\,dt \right] \right\}. \tag{5.71}$$

Expressions (5.68) and (5.71) were used to calculate the normalized scattering cross-section (5.45). The results are plotted in Figure 5.8. It is seen that the truncated PTD is in good agreement with the exact asymptotic theory identified here as the TED (Ufimtsev, 2003). The significant improvement compared to the nontruncated version of PTD (Fig. 5.6) has been achieved in the vicinity of the directions $\phi = 90°$ and $\phi = 270°$.

5.2 DIFFRACTION AT A TRIANGULAR CYLINDER

> The relationships $u_s = E_z$ and $u_h = H_z$ exist between the acoustic and electromagnetic fields for this problem.

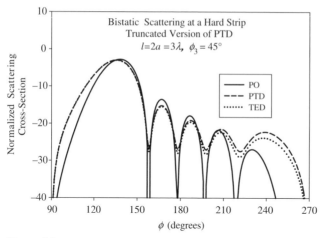

Figure 5.8 Scattering at a hard strip predicted by the truncated version of PTD (H_z-polarization of the electromagnetic field).

For simplicity we consider here the diffraction at a cylinder with the cross-section in the form of an equilateral triangle (Fig. 5.9). Two special cases will be investigated: (a) Symmetric case, when the incident plane wave propagates in the direction parallel to the bisector of the triangle; (b) Backscattering, when the scattered field is evaluated for the direction from which the incident wave comes. First, we study these problems utilizing the PO approximation, and after that it will be corrected by taking into account the first-order edge waves generated by the nonuniform scattering sources $j_{s,h}^{(1)}$.

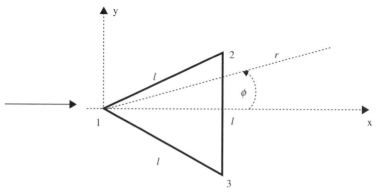

Figure 5.9 Cross-section of the scattering cylinder. Numbers 1, 2, and 3 denote the edges; r and ϕ are polar coordinates of the field point.

5.2.1 Symmetric Scattering: PO Approximation

The incident wave,

$$u^{\text{inc}} = u_0 e^{ikx}, \tag{5.72}$$

generates the identical scattering sources $j_{s,h}^{(0)}$ at faces 1–2 and 1–3 of the cylinder (Fig. 5.9), which are symmetric with respect to the x-axis. The soft (1.5) or hard (1.6) boundary conditions are imposed on the faces. The scattering cylinder is equilateral; the width of each face is equal to l, and each internal angle between faces equals $60°$. The Cartesian coordinates of the edges 1, 2, and 3 are $(0,0)$, (h, a), and $(h, -a)$, respectively, where $h = l\cos(\pi/6)$ and $a = l/2$. Due to the symmetry of the problem, it is sufficient to calculate the scattered field in the directions $0 \leq \phi \leq \pi$.

The traditional integration technique for calculation of the PO scattered field can be avoided here. Indeed, in this approximation, the scattering cylinder can be considered as the combination of the strips 1–2 and 1–3. As was shown in Section 5.1.1, the far field scattered by a strip consists of two edge waves determined by Equations (3.53) and (3.54). We omit all routine calculations related to the transition from the local polar coordinates $(r_{1,2,3}, \varphi_{1,2,3})$ (used for description of individual edge waves) to the basic coordinates (r, ϕ), and provide the final equations for the scattered field in the region $r \gg kl^2, 0 \leq \phi \leq \pi$:

$$u_{s,h}^{(0)} = u_0 \Phi_{s,h}^{(0)}(\phi, 0)\frac{e^{i(kr+\pi/4)}}{\sqrt{2\pi kr}}, \tag{5.73}$$

where

$$\Phi_s^{(0)} = f^{(0)}(1) + f^{(0)}(2)e^{i\psi_2} + f^{(0)}(3)e^{i\psi_3} \tag{5.74}$$

and

$$\Phi_h^{(0)} = g^{(0)}(1) + g^{(0)}(2)e^{i\psi_2} + g^{(0)}(3)e^{i\psi_3}, \tag{5.75}$$

with

$$\psi_2 = kh(1 - \cos\phi) - ka\sin\phi \quad\text{and}\quad \psi_3 = kh(1 - \cos\phi) + ka\sin\phi. \tag{5.76}$$

It is understood here that the uniform sources $j_{s,h}^{(0)}$ induced on each face (1–2 and 1–3), radiate the field in the *entire* surrounding space ($0 \leq \phi \leq 2\pi$).

Functions $f^{(0)}$ and $g^{(0)}$ determine the directivity patterns of the edge waves identified by the corresponding numbers in the argument of these functions. They are given as

$$f^{(0)}(1) = \frac{\sin\dfrac{\pi}{6}}{\cos\left(\dfrac{\pi}{6} - \phi\right) - \cos\dfrac{\pi}{6}} + \frac{\sin\dfrac{\pi}{6}}{\cos\left(\dfrac{\pi}{6} + \phi\right) - \cos\dfrac{\pi}{6}}, \tag{5.77}$$

$$g^{(0)}(1) = \frac{\sin\left(\frac{\pi}{6} - \phi\right)}{\cos\left(\frac{\pi}{6} - \phi\right) - \cos\frac{\pi}{6}} + \frac{\sin\left(\frac{\pi}{6} + \phi\right)}{\cos\left(\frac{\pi}{6} + \phi\right) - \cos\frac{\pi}{6}}, \tag{5.78}$$

$$f^{(0)}(2) = \frac{\sin\frac{\pi}{6}}{\cos\frac{\pi}{6} - \cos\left(\frac{\pi}{6} - \phi\right)}, \qquad g^{(0)}(2) = \frac{\sin\left(\frac{\pi}{6} - \phi\right)}{\cos\frac{\pi}{6} - \cos\left(\frac{\pi}{6} - \phi\right)}, \tag{5.79}$$

$$f^{(0)}(3) = \frac{\sin\frac{\pi}{6}}{\cos\frac{\pi}{6} - \cos\left(\frac{\pi}{6} + \phi\right)}, \qquad g^{(0)}(3) = \frac{\sin\left(\frac{\pi}{6} + \phi\right)}{\cos\frac{\pi}{6} - \cos\left(\frac{\pi}{6} + \phi\right)}. \tag{5.80}$$

All these functions have singularities in the forward direction $\phi = 0$, which cancel each other, being substituted into the functions $\Phi_{s,h}^{(0)}$. This results in the expression

$$\Phi_s^{(0)}(0,0) = \Phi_h^{(0)}(0,0) = ikl. \tag{5.81}$$

A similar situation with compensation of the singularities occurs for the specular direction $\phi = \pi/3$, when

$$\Phi_s^{(0)}\left(\frac{\pi}{3},0\right) = ik\frac{l}{2} + \tan\frac{\pi}{6}e^{i\psi_3}, \qquad \Phi_h^{(0)}\left(\frac{\pi}{3},0\right) = -ik\frac{l}{2} + \frac{1}{\cos\frac{\pi}{6}}e^{i\psi_3}. \tag{5.82}$$

Due to the symmetry of the problem, the same field is scattered in the direction of the specular reflection from the face 1–3:

$$\Phi_s^{(0)}\left(-\frac{\pi}{3},0\right) = \Phi_s^{(0)}\left(\frac{\pi}{3},0\right), \qquad \Phi_h^{(0)}\left(-\frac{\pi}{3},0\right) = \Phi_h^{(0)}\left(\frac{\pi}{3},0\right). \tag{5.83}$$

It is worth noting that the sum of the dominant terms in the reflected fields (5.82) and (5.83) equals $\pm ikl$ and conforms to the field (5.81) scattered in the forward direction. This result is in agreement with the fundamental law that the total power of the reflected field (1.79) is asymptotically (with $kl \gg 1$) equal to the total power of the shadow radiation (1.83).

5.2.2 Backscattering: PO Approximation

Here we assume that the incident wave

$$u^{inc} = u_0 e^{ik(x\cos\phi_0 + y\sin\phi_0)} \tag{5.84}$$

propagates in the direction determined by the angle ϕ_0 given in the interval $-\pi \leq \phi_0 \leq 0$, and the scattered field is evaluated in the opposite direction $\phi = \pi + \phi_0$.

The basic feature of the PO approximation for the backscattering follows from the properties of functions $f^{(0)}$ and $g^{(0)}$ defined in Equations (3.55) to (3.57) through the local polar angles φ and φ_0. These angles determine the directions to the field point and to the source of the incident wave, respectively. For the backscattering direction $\varphi = \varphi_0$, it turns out that $f^{(0)}(\varphi_0, \varphi_0) = -g^{(0)}(\varphi_0, \varphi_0)$. This means that the fields scattered back to the source by the acoustically soft and hard cylinders differ only in sign, and therefore their directivity patterns differ in this way as well:

$$\Phi_s^{(0)} = -\Phi_h^{(0)}. \tag{5.85}$$

It is clear from the previous section that the scattered field consists of the sum of the edge waves. We again omit simple routine calculations of these waves and present the final expressions for the directivity patterns of the total scattered field. We have different expressions for different intervals of observation because of the different number of contributions to the scattered field.

In the interval $0 \leq \phi \leq \pi/6$, only two diffracted waves exist incoming from edges 2 and 3 (Fig. 5.9). In this interval,

$$\Phi_s^{(0)}(\phi) = f^{(0)}(2)e^{i\psi_2} + f^{(0)}(3)e^{i\psi_3} \tag{5.86}$$

where

$$f^{(0)}(2) = -f^{(0)}(3) = -\frac{1}{2}\cot\phi, \tag{5.87}$$

$$\psi_2 = -2k(h\cos\phi + a\sin\phi), \qquad \psi_3 = -2k(h\cos\phi - a\sin\phi) \tag{5.88}$$

with $h = l\cos(\pi/6)$ and $a = l/2$.

Equation (5.86) can be rewritten in the form

$$\Phi_s^{(0)}(\phi) = i\cos\phi \frac{\sin(2ka\sin\phi)}{\sin\phi}e^{-i2kh\cos\phi} \tag{5.89}$$

which predicts the value

$$\Phi_s^{(0)}(0) = i2kae^{-i2kh} = ikle^{-i2kh}. \tag{5.90}$$

In the interval $\pi/6 < \phi < \pi/2$, the scattered field consists of the three edge waves, and respectively

$$\Phi_s^{(0)}(\phi) = f^{(0)}(1) + f^{(0)}(2)e^{i\psi_2} + f^{(0)}(3)e^{i\psi_3}, \tag{5.91}$$

where

$$f^{(0)}(1) = -\frac{1}{2}\tan\left(\frac{\pi}{6} - \phi\right), \tag{5.92}$$

$$f^{(0)}(2) = \frac{1}{2}\left[\tan\left(\frac{\pi}{6} - \phi\right) - \cot\phi\right], \tag{5.93}$$

and

$$f^{(0)}(3) = \frac{1}{2}\cot\phi. \tag{5.94}$$

In the interval $\pi/2 < \phi < 5\pi/6$, only two edge waves contribute to the total scattered field, and

$$\Phi_s^{(0)}(\phi) = f^{(0)}(1) + f^{(0)}(2)e^{i\psi_2}, \tag{5.95}$$

with function $f^{(0)}(1)$ defined in Equation (5.92). The function $f^{(0)}(2)$ in this interval differs from Equation (5.93) and equals

$$f^{(0)}(2) = \frac{1}{2}\tan\left(\frac{\pi}{6} - \phi\right) \tag{5.96}$$

because the face 2–3 is not illuminated by the incident wave and does not generate the scattered field (in the framework of the PO approximation). In the specular direction $\phi = 2\pi/3$, Equation (5.95) determines the value

$$\Phi_s^{(0)}\left(\frac{2\pi}{3}\right) = ikl, \tag{5.97}$$

which agrees with Equation (5.90).

In the interval $5\pi/6 < \phi < \pi$, again all three edges generate the scattered field, and the directivity pattern is determined by Equation (5.91), but with the different functions $f^{(0)}(1)$ and $f^{(0)}(3)$:

$$f^{(0)}(1) = -\frac{1}{2}\left[\tan\left(\frac{\pi}{6} - \phi\right) + \tan\left(\frac{\pi}{6} + \phi\right)\right] \tag{5.98}$$

and

$$f^{(0)}(3) = \frac{1}{2}\tan\left(\frac{\pi}{6} + \phi\right). \tag{5.99}$$

However, the function $f^{(0)}(2)$ is still defined by Equation (5.96). The function (5.98) differs from (5.92) because in this interval of observation, two faces 1–2 and 1–3 are illuminated. The function (5.99) differs from Equation (5.94) because the different faces of edge 3 are illuminated in the intervals $0 \le \phi < \pi/2$ and $5\pi/6 < \phi \le \pi$.

5.2.3 Symmetric Scattering: First-Order PTD Approximation

Here, the acoustic quantity $\Phi_s(\Phi_h)$ is equivalent to the directivity pattern of the $E_z(H_z)$-component of the electromagnetic field

To improve the PO approximation, we include into the scattered field the contributions generated by the nonuniform scattering sources $j_{s,h}^{(1)}$. In the first-order approximation, they have the form of edge waves (4.12) and (4.13). When these waves are added to the PO edge waves (3.53) and (3.54), one can see that (due to relationships (4.14) and (4.15)), the resulting edge waves are defined by the Sommerfeld asymptotics (2.61) and (2.63). Because the acoustically soft and hard cylinders are perfectly reflecting (nontransparent), only those edge waves contribute to the scattered field, which come from the edges visible from the observation point. We also note that in this section we consider the symmetric case when the incident wave is given by Equation (5.72) and propagates along the bisector of the cylinder (Fig. 5.9). The basic polar coordinates r, ϕ are used in the following for the description of the scattered field and the angle ϕ changes in the interval $0 \leq \phi \leq \pi$.

In accordance with these comments, the directivity patterns of the scattered field can be written as follows. In the interval $0 \leq \phi > \pi/6$,

$$\Phi_s(\phi, 0) = f(2)e^{i\psi_2} + f(3)e^{i\psi_3}, \tag{5.100}$$

and

$$\Phi_h(\phi, 0) = g(2)e^{i\psi_2} + g(3)e^{i\psi_3}, \tag{5.101}$$

where

$$\begin{Bmatrix} f(2) \\ g(2) \end{Bmatrix} = \frac{\sin\dfrac{\pi}{n}}{n} \left[\frac{1}{\cos\dfrac{\pi}{n} - \cos\left(\dfrac{\pi}{n} - \dfrac{\phi}{n}\right)} \mp \frac{1}{\cos\dfrac{\pi}{n} + \cos\left(\dfrac{\pi}{5} + \dfrac{\phi}{n}\right)} \right],$$

with $0 \leq \phi \leq \pi$ \hfill (5.102)

and

$$\begin{Bmatrix} f(3) \\ g(3) \end{Bmatrix} = \frac{\sin\dfrac{\pi}{n}}{n} \left[\frac{1}{\cos\dfrac{\pi}{n} - \cos\left(\dfrac{\pi}{n} + \dfrac{\phi}{n}\right)} \mp \frac{1}{\cos\dfrac{\pi}{n} + \cos\left(\dfrac{\pi}{5} - \dfrac{\phi}{n}\right)} \right],$$

with $0 \leq \phi \leq \pi/2$, \hfill (5.103)

where $n = \alpha/\pi = 5/3$ (here $\alpha = 5\pi/3$ is the external angle between the faces of the edge) and $\psi_{2,3}$ is defined in Equation (5.76). The first terms in these functions are singular for the forward direction ($\phi = 0$). However, the singularities of the functions related to the edges 2 and 3 cancel each other, resulting in the expressions

$$\Phi_s(0) = ikl - \frac{1}{n}\cot\frac{\pi}{n} - \frac{\dfrac{2}{n}\sin\dfrac{\pi}{n}}{\cos\dfrac{\pi}{n} + \cos\dfrac{\pi}{5}}, \tag{5.104}$$

and

$$\Phi_h(0) = ikl - \frac{1}{n}\cot\frac{\pi}{n} + \frac{\dfrac{2}{n}\sin\dfrac{\pi}{n}}{\cos\dfrac{\pi}{n} + \cos\dfrac{\pi}{5}}. \tag{5.105}$$

Comparison of these equations with Equation (5.81) shows that the first term here relates to the PO field and the last two terms represent the contributions from the nonuniform component $j_{s,h}^{(1)}$.

In the interval $\pi/6 < \phi < \pi/2$, three edge waves form the total scattered field:

$$\Phi_s(\phi) = f(1) + f(2)e^{i\psi_2} + f(3)e^{i\psi_3}, \tag{5.106}$$

$$\Phi_h(\phi) = g(1) + g(2)e^{i\psi_2} + g(3)e^{i\psi_3}, \tag{5.107}$$

where

$$\begin{Bmatrix} f(1) \\ g(1) \end{Bmatrix} = \frac{\sin\dfrac{\pi}{n}}{n} \left[\frac{1}{\cos\dfrac{\pi}{n} - \cos\left(\dfrac{\pi}{n} - \dfrac{\phi}{n}\right)} \mp \frac{1}{\cos\dfrac{\pi}{n} + \cos\left(\dfrac{\pi}{n} - \dfrac{\phi}{n}\right)} \right],$$

$$\text{with } \pi/6 \le \phi \le \pi. \tag{5.108}$$

There is the specular direction ($\phi = \pi/3$) in this sector of observation. In this direction, the second terms in functions $f(1)$, $g(1)$ and $f(2)$, $g(2)$ are singular. However, these singular terms cancel each other, and the scattered field remains finite:

$$\Phi_s\left(\frac{\pi}{3}\right) = ik\frac{l}{2} + \frac{1}{n}\left(\tan\frac{\pi}{n} + \cot\frac{\pi}{n}\right) + f(3)e^{i\psi_3}, \tag{5.109}$$

$$\Phi_h\left(\frac{\pi}{3}\right) = -ik\frac{l}{2} + \frac{1}{n}\left(\tan\frac{\pi}{n} - \cot\frac{\pi}{n}\right) + g(3)e^{i\psi_3}. \tag{5.110}$$

Due to the symmetry of the problem, these equations also describe the field in the direction of specular reflection from the face 1–3. Here the comments presented at the end of Section 5.2.1 are also pertinent. The power of the total reflected field is asymptotically equal to the power of the shadow radiation determined by the first term in Equations (5.104) and (5.105). It is also clear that the dominant term $\pm ikl/2$ in Equations (5.109) and (5.110) relates to the PO contribution.

In the interval $\pi/2 < \phi < 5\pi/6$, the edge 3 is invisible and the scattered field consists of two edge waves:

$$\Phi_s(\phi) = f(1) + f(2)e^{i\psi_2}, \tag{5.111}$$

$$\Phi_h(\phi) = g(1) + g(2)e^{i\psi_2}. \tag{5.112}$$

In the interval $5\pi/6 < \phi \le \pi$, all three edges are visible and the scattered field is determined again by Equations (5.106) and (5.107) with the functions

$$\begin{Bmatrix} f(3) \\ g(3) \end{Bmatrix} = \frac{\sin\dfrac{\pi}{n}}{n} \left[\frac{1}{\cos\dfrac{\pi}{n} - \cos\left(\dfrac{\pi}{n} - \dfrac{\phi}{n}\right)} \mp \frac{1}{\cos\dfrac{\pi}{n} + \cos\left(\dfrac{\pi}{n} + \dfrac{\phi}{n}\right)} \right],$$

$$\text{with } 5\pi/6 \le \phi \le \pi. \tag{5.113}$$

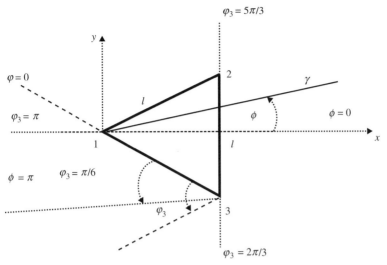

Figure 5.10   Sector $0 \le \varphi_3 \le 5\pi/3$ is the domain of the functions $f(3)$ and $g(3)$. The field point (r, ϕ) is located in the region $0 \le \phi \le \pi$.

This expression is different from Equation (5.103), although both are the exact forms of the same generic expressions (2.62) and (2.64). This difference is due to the fact that the local polar coordinate φ_3 (used in the generic definition of functions $f(3)$ and $g(3)$) cannot be described in the regions $\varphi_3 < \pi/6$ and $\varphi_3 > \pi/6$ (Fig. 5.10) by a single expression in terms of the basic coordinate ϕ under the restriction $0 \le \phi \le \pi$.

Notice also that the asymptotic expressions found above for the quantities u_h and $du_s/d\phi$ are discontinuous in the directions $\phi = \pi/6$, $\phi = \pi/2$, and $\phi = 5\pi/6$, which are the geometrical optics boundaries for the edge waves. These discontinuities can be diminished when the higher-order edge waves (arising due to multiple edge diffraction) are taken into account. However, as shown in Section 5.2.5, such discontinuities are already not significant in the case when $l \ge 3\lambda$.

It is easy to construct similar asymptotics for the bistatic scattering for arbitrary directions of the incident wave. However, their shortcoming is the grazing singularity (4.20) and (4.21), which appears in the case of the grazing directions of the incident wave. This singularity can be removed with application of the uniform theory developed in Section 7.9. An alternative procedure free from the grazing singularity is based on truncation of elementary strips (Michaeli, 1987; Breinbjerg, 1992; Johansen, 1996).

5.2.4 Backscattering: First-Order PTD Approximation

Here, the acoustic quantity $\Phi_s(\Phi_h)$ is equivalent to the directivity pattern of the $E_z(H_z)$-component of the electromagnetic field.

In the previous section it was explained that in the first-order PTD approximation, the field scattered by a triangular cylinder is a linear combination of the edge waves defined in general form by Equations (2.61) and (2.63). Now we apply this PTD theory for the investigation of the backscattering, assuming that the incident wave is given by Equation (5.84) with $-\pi \leq \phi_0 \leq 0$. The scattered field is found in the direction $\phi = \pi + \phi_0$ in the sector $0 \leq \phi \leq \pi$. The geometry of the problem is shown in Figure 5.9. We omit the simple but tedious calculations of the individual edge waves in terms of the basic coordinates r, ϕ and present the final expressions for the directivity patterns of the total scattered field.

In the interval $0 \leq \phi \leq \pi/6$,

$$\Phi_s(\phi) = f(2)e^{ik\psi_2} + f(3)e^{i\psi_3}, \tag{5.114}$$

$$\Phi_h(\phi) = g(2)e^{ik\psi_2} + g(3)e^{i\psi_3}, \tag{5.115}$$

where

$$\begin{Bmatrix} f(2) \\ g(2) \end{Bmatrix} = \frac{\sin\dfrac{\pi}{n}}{n} \left[\frac{1}{\cos\dfrac{\pi}{n} - 1} \mp \frac{1}{\cos\dfrac{\pi}{n} - \cos\left(\dfrac{\pi}{n} + \dfrac{2\phi}{n}\right)} \right],$$

$$\text{with } 0 \leq \phi \leq \pi, \tag{5.116}$$

$$\begin{Bmatrix} f(3) \\ g(3) \end{Bmatrix} = \frac{\sin\dfrac{\pi}{n}}{n} \left[\frac{1}{\cos\dfrac{\pi}{n} - 1} \mp \frac{1}{\cos\dfrac{\pi}{n} - \cos\left(\dfrac{\pi}{n} - \dfrac{2\phi}{n}\right)} \right]$$

$$\text{with } 0 \leq \phi \leq \pi/2. \tag{5.117}$$

Here, the edge parameter equals $n = \alpha/\pi = 5/3$, and the quantities $\psi_{2,3}$ are defined in Equation (5.88). It follows from these equations that for the direction $\phi = 0$,

$$\Phi_s(0) = \left(ikl + \frac{1}{n}\cot\frac{\pi}{n} + \frac{\dfrac{2}{n}\sin\dfrac{\pi}{n}}{\cos\dfrac{\pi}{n} - 1} \right) e^{-i2kh}, \tag{5.118}$$

$$\Phi_h(0) = \left(-ikl - \frac{1}{n}\cot\frac{\pi}{n} + \frac{\dfrac{2}{n}\sin\dfrac{\pi}{n}}{\cos\dfrac{\pi}{n} - 1} \right) e^{-i2kh}, \tag{5.119}$$

where $h = l\cos(\pi/6)$.

In the interval $\pi/6 < \phi \leq \pi/2$, the additional wave appears incoming from edge 1. Hence, in this interval,

$$\Phi_s(\phi) = f(1) + f(2)e^{ik\psi_2} + f(3)e^{i\psi_3} \tag{5.120}$$

and

$$\Phi_h(\phi) = g(1) + g(2)e^{ik\psi_2} + g(3)e^{i\psi_3}, \tag{5.121}$$

where functions $f(2)$, $f(3)$ and $g(2), g(3)$ are defined in Equations (5.116) and (5.117) and

$$\begin{Bmatrix} f(1) \\ g(1) \end{Bmatrix} = \frac{\sin\dfrac{\pi}{n}}{n} \left[\frac{1}{\cos\dfrac{\pi}{n} - 1} \mp \frac{1}{\cos\dfrac{\pi}{n} - \cos\left(\dfrac{\pi}{5} - \dfrac{2\phi}{n}\right)} \right],$$

with $\pi/6 < \phi \le \pi$. \hfill (5.122)

In the interval $\pi/2 < \phi < 5\pi/6$, the diffracted wave from edge 3 disappears, and

$$\Phi_s(\phi) = f(1) + f(2)e^{ik\psi_2} \tag{5.123}$$

and

$$\Phi_h(\phi) = g(1) + g(2)e^{ik\psi_2}. \tag{5.124}$$

There is the specular direction $\phi = 2\pi/3$ in this interval, where

$$\Phi_s\left(\frac{2\pi}{3}\right) = ikl + \frac{1}{n}\cot\frac{\pi}{n} + \frac{\dfrac{2}{n}\sin\dfrac{\pi}{n}}{\cos\dfrac{\pi}{n} - 1} \tag{5.125}$$

and

$$\Phi_h\left(\frac{2\pi}{3}\right) = -ikl - \frac{1}{n}\cot\frac{\pi}{n} + \frac{\dfrac{2}{n}\sin\dfrac{\pi}{n}}{\cos\dfrac{\pi}{n} - 1}. \tag{5.126}$$

Due to the symmetry of the problem, these expressions differ from Equations (5.118) and (5.119) only by the absence of the phase factor e^{-i2kh} (caused by the choice of the incident wave in the form of Equation (5.84) and by the choice of the coordinates origin at the edge 1).

In the interval $5\pi/6 < \phi \le \pi$, all three edges generate the scattered field, which again is determined by Equations (5.120) and (5.121), where the functions $f(3), g(3)$ are defined by the expressions

$$\begin{Bmatrix} f(3) \\ g(3) \end{Bmatrix} = \frac{\sin\dfrac{\pi}{n}}{n} \left(\frac{1}{\cos\dfrac{\pi}{n} - 1} \mp \frac{1}{\cos\dfrac{\pi}{n} + \cos\dfrac{2\phi}{n}} \right), \tag{5.127}$$

different from Equation (5.117). These expressions complete the description of the backscattered field.

5.2.5 Numerical Analysis of the Scattered Field

Here, the scattering cross-section $\sigma_s(\sigma_h)$ for acoustic waves is equal to the scattering cross-section for electromagnetic waves with the component $E_z(H_z)$.

Numerical calculations were performed for the normalized scattering cross-section (5.20)

$$\frac{\sigma_{s,h}}{kl^2} = \left| \frac{\Phi_{s,h}(\phi)}{kl} \right|^2 \tag{5.128}$$

in the decimal scale for the equilateral cylinder with the parameter $kl = 6\pi$ when $l = 3\lambda$. The results are plotted in Figures 5.11 and 5.12 for symmetric scattering (when the incident wave propagates in the direction parallel to the cylinder bisector; Fig. 5.9) and in Figures 5.13 and 5.14 for backscattering.

As is seen in Figures 5.11 and 5.12, PTD significantly improves the PO approximation at the minima of the scattering cross-section. The difference between the PTD and PO data is also appreciable at maxima. This difference is pronounced for the acoustically hard cylinder and it reaches about 6–9 db in the sector $160°–180°$. A similar situation is observed for the backscattering in Figures 5.13 and 5.14. In particular,

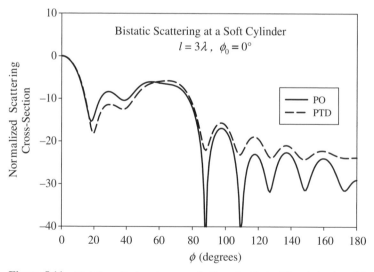

Figure 5.11   Bistatic scattering at an acoustically soft cylinder (E_z-polarization of the electromagnetic wave).

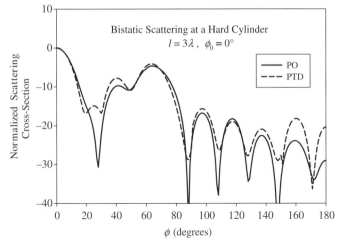

Figure 5.12   Bistatic scattering at an acoustically hard cylinder (H_z-polarization of the electromagnetic wave).

the difference between the PTD and PO curves for the acoustically hard cylinder is about 5–9 db in the directions 50°–70° and 170°–180°.

Notice that more accurate PTD results for triangular cylinders are presented in the paper by Johanson (1996), where the second-order edge waves are partially taken into account.

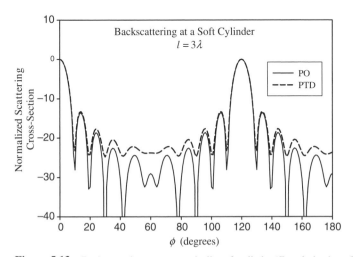

Figure 5.13   Backscattering at an acoustically soft cylinder (E_z-polarization of the electromagnetic wave).

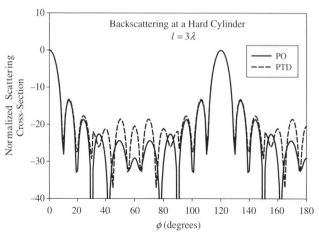

Figure 5.14   Backscattering at an acoustically hard cylinder (H_z-polarization of the electromagnetic wave).

PROBLEMS

5.1   Derive the PO approximation for the field (in the far zone) scattered at a soft strip as shown in Figure 5.1. The incident wave is given by Equation (5.1). Start with the general expressions (1.32), use Equation (3.7) for the Hankel function, apply its asymptotic form (2.29). Represent the scattered field in terms of edge waves and compare the result with Equation (5.4). Verify that they are identical.

5.2   Derive the PO approximation for the field (in the far zone) scattered at a hard strip as shown in Figure 5.1. The incident wave is given by Equation (5.1). Start with the general expressions (1.32), use Equation (3.7) for the Hankel function, apply its asymptotic form (2.29). Represent the scattered field in terms of edge waves and compare the result with Equation (5.5). Verify that they are identical.

5.3   Derive the PO approximation for the field (in the far zone) scattered at a perfectly conducting strip as shown in Figure 5.1. The incident wave is given as

$$E_z^{\text{inc}} = E_{0z}e^{ik(x\cos\varphi_0 + y\sin\varphi_0)}$$

Start with the general expressions (1.87) and (1.97). Use Equation (3.7) for the Hankel function, apply its asymptotic form (2.29). Represent the scattered field in terms of edge waves and compare the result with Equation (5.4). Formulate the relationship between acoustic and electromagnetic diffracted waves.

5.4   Derive the PO approximation for the field (in the far zone) scattered at a perfectly conducting strip as shown in Figure 5.1. The incident wave is given as

$$H_z^{\text{inc}} = H_{0z}e^{ik(x\cos\varphi_0 + y\sin\varphi_0)}.$$

Start with the general expressions (1.88) and (1.97). Use Equation (3.7) for the Hankel function, apply its asymptotic form (2.29). Represent the scattered field in terms of edge waves and compare the result with Equation (5.5). Formulate the relationship between acoustic and electromagnetic diffracted waves.

5.5 Use Equation (5.35). Prove Equation (5.39). Apply Equation (5.16) and calculate the total cross-section of a soft strip.

5.6 Use Equation (5.37). Prove Equation (5.40). Apply Equation (5.16) and calculate the total cross-section of a hard strip.

5.7 Investigate the diffraction of a cylindrical wave at a soft strip (Fig. 5.4). Related coordinates are shown in Figure 5.1. The incident wave is given by

$$u^{\text{inc}} = u_0 \frac{e^{ikr_0}}{\sqrt{kr_0}}, \qquad \text{with } r_0 = \sqrt{(x - x_0)^2 + y^2} \text{ and } k|x_0| \gg 1.$$

Derive the PTD approximation for the scattered field as a sum of two separate components: the PO field $u_s^{(0)}$ and the fringe field $u_s^{(1)}$. Start the calculation of the PO field with the general expressions (1.32), use Equation (3.7) for the Hankel function, apply its asymptotic form (2.29). Represent the far field $u_s^{(0)}$ in integral form. Use Equation (4.12) and write the asymptotic expressions for the edge waves generated by the nonuniform/fringe sources $j_s^{(1)}$. Verify that these waves are finite at the shadow boundaries. Retrace the transition (in the field expressions) from the cylindrical wave excitation to the incident plane wave by setting

$$\lim_{x_0 \to -\infty} u_0 \frac{e^{ikr_0}}{\sqrt{kr_0}} = e^{ikx}.$$

5.8 Investigate the diffraction of a cylindrical wave at a hard strip (Fig. 5.4). Related coordinates are shown in Figure 5.1. The incident wave is given by

$$u^{\text{inc}} = u_0 \frac{e^{ikr_0}}{\sqrt{kr_0}}, \qquad \text{with } r_0 = \sqrt{(x - x_0)^2 + y^2} \text{ and } k|x_0| \gg 1.$$

Derive the PTD approximation for the scattered field as a sum of two separate components: the PO field $u_h^{(0)}$ and the fringe field $u_h^{(1)}$. Start the calculation of the PO field with the general expressions (1.32), use Equation (3.7) for the Hankel function, apply its asymptotic form (2.29). Represent the far field $u_h^{(0)}$ in integral form. Use Equation (4.13) and write the asymptotic expressions for the edge waves generated by the nonuniform/fringe sources $j_h^{(1)}$. Verify that these waves are finite at the shadow boundaries. Retrace the transition (in the field expressions) from the cylindrical wave excitation to the incident plane wave by setting

$$\lim_{x_0 \to -\infty} u_0 \frac{e^{ikr_0}}{\sqrt{kr_0}} = e^{ikx}.$$

5.9 Investigate the diffraction of a cylindrical wave at a perfectly conducting strip (Fig. 5.4). Related coordinates are shown in Figure 5.1. The incident wave is given by

$$E_z^{\text{inc}} = E_{0z} \frac{e^{ikr_0}}{\sqrt{kr_0}}, \qquad \text{with } r_0 = \sqrt{(x - x_0)^2 + y^2} \text{ and } k|x_0| \gg 1.$$

Derive the PTD approximation for the scattered field as a sum of two separate components: the PO field $E_z^{(0)}$ and the fringe field $E_z^{(1)}$. Start the calculation of the PO field
with the general expression (1.87), use Equation (3.7) for the Hankel function, apply its
asymptotic form (2.29). Represent the far field $E_z^{(0)}$ in integral form. Use Equation (4.12)
(adapted for electromagnetic waves) and write the asymptotic expressions for the field
$E_z^{(1)}$. Verify that these waves are finite at the shadow boundaries. Retrace the transition
(in the field expressions) from the cylindrical wave excitation to the incident plane wave
by setting

$$\lim_{x_0 \to -\infty} E_{0z} \frac{e^{ikr_0}}{\sqrt{kr_0}} = e^{ikx}.$$

5.10 Investigate the diffraction of a cylindrical wave at a perfectly conducting strip (Fig. 5.4).
Related coordinates are shown in Figure 5.1. The incident wave is given by

$$H_z^{\text{inc}} = H_{0z} \frac{e^{ikr_0}}{\sqrt{kr_0}}, \qquad \text{with } r_0 = \sqrt{(x - x_0)^2 + y^2} \text{ and } k|x_0| \gg 1.$$

Derive the PTD approximation for the scattered field as a sum of two separate components: the PO field $H_z^{(0)}$ and the fringe field $H_z^{(1)}$. Start the calculation of the PO field
with the general expression (1.88), use Equation (3.7) for the Hankel function, apply
its asymptotic form (2.29). Represent the far field $H_z^{(0)}$ in integral form. Use Equation (4.13) (adapted for electromagnetic waves) and write the asymptotic expressions
for the field $H_z^{(1)}$. Verify that these waves are finite at the shadow boundaries. Retrace the
transition (in the field expressions) from the cylindrical wave excitation to the incident
plane wave by setting

$$\lim_{x_0 \to -\infty} H_{0z} \frac{e^{ikr_0}}{\sqrt{kr_0}} = e^{ikx}.$$

Chapter **6**

Axially Symmetric Scattering of Acoustic Waves at Bodies of Revolution

A similar problem for electromagnetic waves is considered in Chapter 2 of Ufimtsev (2003).

This chapter develops the first-order PTD for acoustic waves scattered at bodies of revolution with sharp edges. Axially symmetric scattering is studied. This situation occurs when an incident plane wave propagates in the direction along the symmetry axis of a body of revolution. An edge of a body of revolution is a circle. When its diameter is much greater than a wavelength, then the nonuniform scattering sources $j_{s,h}^{(1)}$ induced near the edge are asymptotically identical to those near the edge of the tangential conic surface consisting of two parts (Fig. 6.1). Diffraction at this surface is an appropriate canonical problem, which is studied in Section 6.1. Its solution is used in the next sections to determine the field scattered at certain bodies of revolution.

6.1 DIFFRACTION AT A CANONICAL CONIC SURFACE

The geometry of the problem is illustrated in Figures 6.1 and 6.2. The solid lines in Figure 6.1 show a general view of a body of revolution with a circular edge. The dashed tangent lines belong to the tangential conic surface. The cross-section of this surface by the meridian plane and some related denotations are presented in Figure 6.2. Here, ξ is the distance from the edge along the generatrix; r', ϑ', ψ' the spherical coordinates and ρ, ψ' the polar coordinates of the point on the conic surface; R, ϑ, ψ the coordinates of the observation point; φ_0 the angle of incidence measured

Fundamentals of the Physical Theory of Diffraction. By Pyotr Ya. Ufimtsev
Copyright © 2007 John Wiley & Sons, Inc.

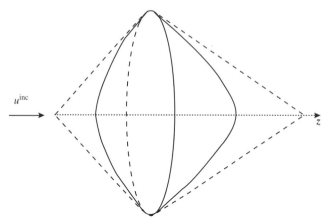

Figure 6.1   A body of revolution (solid lines) with a circular edge and a conic surface (dashed lines) tangential to a body at the edge points.

from the illuminated side of the conic surface; and $\alpha - \varphi_0$ the angle of incidence measured from the shadowed side; the meaning of the angles ω and Ω is clear; the edge points 1 and 2 are symmetrical.

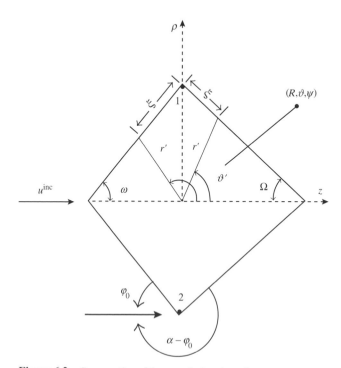

Figure 6.2   Cross-section of the canonical conic surface.

6.1.1 Integrals for the Scattered Field

It is supposed that the incident plane wave

$$u^{\text{inc}} = u_0 e^{ikz} \tag{6.1}$$

propagates along the symmetry axis of the conic surface. The incident wave undergoes diffraction at this surface and creates there the nonuniform scattering sources

$$j_s^{(1)} = u_0 J_s^{(1)}, \qquad j_h^{(1)} = u_0 J_h^{(1)}. \tag{6.2}$$

It is obvious that, due to the symmetry of the problem, these sources do not depend on the polar angle ψ'. The scattered field is described by general Equations (1.16) and (1.17). In the particular case of the conic surface, the quantities involved in these equations are determined by the following expressions. The quantities

$$ds^- = (a - \xi \sin \omega)d\xi \, d\psi'$$

and

$$ds^+ = (a - \xi \sin \Omega)d\xi \, d\psi' \tag{6.3}$$

are the differential elements of the conic surface on its illuminated ($z < 0$) and shadowed ($z > 0$) sides, respectively;

$$\left. \begin{array}{l} r' \sin \vartheta' = a - \xi \sin \omega \\ r' \cos \vartheta' = -\xi \cos \omega \end{array} \right\} \quad \text{for the points with } z < 0, \tag{6.4}$$

$$\left. \begin{array}{l} r' \sin \vartheta' = a - \xi \sin \Omega \\ r' \cos \vartheta' = \xi \cos \Omega \end{array} \right\} \quad \text{for the points with } z > 0; \tag{6.5}$$

$$(\hat{m} \cdot \hat{n})^- = \sin \vartheta \cos \omega \cos(\psi' - \psi) - \sin \omega \cos \vartheta, \quad \text{for the points with } z < 0, \tag{6.6}$$

$$(\hat{m} \cdot \hat{n})^+ = \sin \vartheta \cos \Omega \cos(\psi' - \psi) + \sin \Omega \cos \vartheta, \quad \text{for the points with } z > 0. \tag{6.7}$$

In this section, we use the symbol Ω for the angle shown in Figure 6.2. In Equations (1.16) and (1.17), the same symbol was used for another angle (Fig. 1.2). We note this to avoid possible confusion.

In view of the above comments, the scattered field (1.16–1.17) can be represented in this form:

$$u_s^{(1)} = -u_0 \frac{1}{4\pi} \frac{e^{ikR}}{R} \left[\int_0^{\xi_{eff}} J_s^{(1)}(k\xi, \varphi_0) e^{ik\xi \cos\omega \cos\vartheta} (a - \xi \sin\omega) d\xi \right.$$

$$\times \int_0^{2\pi} e^{-ik\Psi^-(\psi',\psi)} d\psi' + \int_0^{\xi_{eff}} J_s^{(1)}(k\xi, \alpha - \varphi_0) e^{-ik\xi \cos\Omega \cos\vartheta} (a - \xi \sin\Omega) d\xi$$

$$\times \left. \int_0^{2\pi} e^{-ik\Psi^+(\psi',\psi)} d\psi' \right], \tag{6.8}$$

$$u_h^{(1)} = -u_0 \frac{ik}{4\pi} \frac{e^{ikR}}{R}$$

$$\times \left[\int_0^{\xi_{eff}} J_h^{(1)}(k\xi, \varphi_0) e^{ik\xi \cos\omega \cos\vartheta} (a - \xi \sin\omega) d\xi \int_0^{2\pi} e^{-ik\Psi^-(\psi',\psi)} (\hat{m} \cdot \hat{n})^- d\psi' \right.$$

$$+ \int_0^{\xi_{eff}} J_h^{(1)}(k\xi, \alpha - \varphi_0) e^{-ik\xi \cos\Omega \cos\vartheta} (a - \xi \sin\Omega) d\xi$$

$$\times \left. \int_0^{2\pi} e^{-ik\Psi^+(\psi',\psi)} (\hat{m} \cdot \hat{n})^+ d\psi' \right], \tag{6.9}$$

where

$$\Psi^-(\psi',\psi) = (a - \xi \sin\omega) \sin\vartheta \cos(\psi' - \psi)$$

and

$$\Psi^+(\psi',\psi) = (a - \xi \sin\Omega) \sin\vartheta \cos(\psi' - \psi). \tag{6.10}$$

The first integrals in the brackets of Equations (6.8) and (6.9) relate to the illuminated surface ($z < 0$), and the second integrals relate to the shadowed surface ($z > 0$). These integrals are calculated over the interval $0 \leq \xi \leq \xi_{eff}$, as the nonuniform sources $J_{s,h}^{(1)}(k\xi)$ decrease with increasing ξ and can be neglected at a certain distance $\xi > \xi_{eff}$ from the edge. In the next sections we present the asymptotic estimates for the scattered field.

6.1.2 Ray Asymptotics

First we investigate the field in the observation points, which are visible from all edge points ($0 \leq \psi' \leq 2\pi$). This happens for the two intervals of the observation directions: $0 \leq \vartheta \leq \Omega$ and $\pi - \omega \leq \vartheta \leq \pi$. We assume that

$$k(a - \xi_{eff} \sin\omega) \sin\vartheta \gg 1$$

and

$$k(a - \xi_{eff} \sin\Omega) \sin\vartheta \gg 1 \tag{6.11}$$

and apply the stationary phase technique (Copson, 1965; Murray, 1984) to the integrals over the variable ψ'. The stationary points $\psi_{1,2}$ are found from the condition

$$\frac{d\Psi^\pm(\psi',\psi)}{d\psi'} = 0 \tag{6.12}$$

which leads to the equation

$$\frac{d\cos(\psi' - \psi)}{d\psi'} = -\sin(\psi' - \psi) = 0, \qquad \text{at } \psi' = \psi'_{\text{st}}. \tag{6.13}$$

In the interval $0 \le \psi' \le 2\pi$, two stationary points exist:

$$\psi_1 = \psi \qquad \text{and} \qquad \psi_2 = \pi + \psi. \tag{6.14}$$

In accordance with this asymptotic method, the function $\cos(\psi' - \psi)$ contained in $\Psi^\pm$ is approximated by

$$\cos(\psi' - \psi) \approx 1 - \frac{1}{2}(\psi' - \psi_1)^2 \qquad \text{in the vicinity of the point } \psi' = \psi_1 \tag{6.15}$$

and

$$\cos(\psi' - \psi) \approx -1 + \frac{1}{2}(\psi' - \psi_2)^2 \qquad \text{in the vicinity of the point } \psi' = \psi_2. \tag{6.16}$$

The slowly varying factor $(\hat{m} \cdot \hat{n})^\pm$ is approximated by its value at the stationary points. The initial integral over the variable ψ' asymptotically equals the sum of two integrals calculated in the vicinity of each stationary point. The intervals of integration in these integrals are extended from $-\infty$ to $+\infty$. These standard manipulations lead to the asymptotic expression

$$\int_0^{2\pi} e^{-ik\Psi^\pm(\psi',\psi)} (\hat{m} \cdot \hat{n})^\pm d\psi' \sim \sqrt{\frac{2\pi}{k\left[a - \xi \sin\left(\frac{\Omega}{\omega}\right)\right] \sin\vartheta}}$$

$$\times \left\{ (\hat{m} \cdot \hat{n})^\pm \Big|_{\psi'=\psi_1} \cdot e^{-ik\left[a-\xi\sin\left(\frac{\Omega}{\omega}\right)\right]\sin\vartheta} \cdot e^{i\frac{\pi}{4}} \right.$$

$$\left. + (\hat{m} \cdot \hat{n})^\pm \Big|_{\psi'=\psi_2} \cdot e^{ik\left[a-\xi\sin\left(\frac{\Omega}{\omega}\right)\right]\sin\vartheta} \cdot e^{-i\frac{\pi}{4}} \right\} \tag{6.17}$$

where

$$(\hat{m} \cdot \hat{n})^+ \Big|_{\psi'=\psi_1} = \sin(\vartheta + \Omega), \qquad (\hat{m} \cdot \hat{n})^- \Big|_{\psi'=\psi_1} = \sin(\vartheta - \omega),$$

$$(\hat{m} \cdot \hat{n})^+ \Big|_{\psi'=\psi_2} = -\sin(\vartheta - \Omega), \qquad (\hat{m} \cdot \hat{n})^- \Big|_{\psi'=\psi_2} = -\sin(\vartheta + \omega). \tag{6.18}$$

Equation (6.17) allows one to reduce Equations (6.8) and (6.9) to single integrals over the variable ξ. These integrals will contain the factors $\sqrt{a - \xi \sin \omega}$ and $\sqrt{a - \xi \sin \Omega}$, which can be approximated to $\sqrt{a}$ under the condition $a \gg \xi_{\text{eff}}$.

Before we present the resulting expressions for Equations (6.8) and (6.9), we introduce the local polar coordinates φ_1 and φ_2 at the stationary points ψ_1 and ψ_2. In Figure 6.2 these points are denoted as 1 and 2. The local coordinates are shown in Figure 6.3 for point 1 and in Figures 6.4 and 6.5 for point 2.

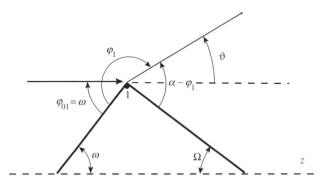

Figure 6.3 Local polar coordinates at the stationary point 1 ($\psi'_{\text{st}} = \psi$).

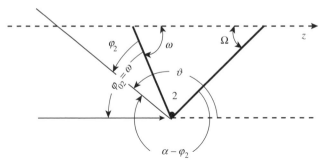

Figure 6.4 Local coordinates at the stationary point 2 ($\psi'_{\text{st}} = \pi + \psi$) for the observation directions in the interval $\pi - \omega \leq \vartheta \leq \pi$.

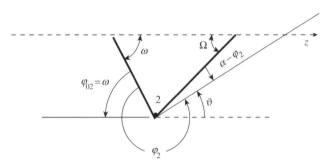

Figure 6.5 Local coordinates at the stationary point 2 ($\psi'_{\text{st}} = \pi + \psi$) for the observation directions in the interval $0 \leq \vartheta \leq \Omega$.

Considering the above relationships between coordinates φ_1, φ_2, and ϑ, and utilizing Equation (6.17), one can obtain the following approximations for Equations (6.8) and (6.9):

$$
u_{\mathrm{s}}^{(1)} = -u_0 \frac{a}{2\sqrt{2\pi k a \sin \vartheta}} \frac{e^{ikR}}{R} \left\{ e^{-ika \sin \vartheta + i\frac{\pi}{4}} \left[\int_0^{\xi_{\mathrm{eff}}} J_{\mathrm{s}}^{(1)}(k\xi, \varphi_0) e^{-ik\xi \cos \varphi_1} \, d\xi \right. \right.
$$

$$
+ \int_0^{\xi_{\mathrm{eff}}} J_{\mathrm{s}}^{(1)}(k\xi, \alpha - \varphi_0) e^{-ik\xi \cos(\alpha - \varphi_1)} \, d\xi \Bigg]
$$

$$
+ e^{ika \sin \vartheta - i\frac{\pi}{4}} \left[\int_0^{\xi_{\mathrm{eff}}} J_{\mathrm{s}}^{(1)}(k\xi, \varphi_0) e^{-ik\xi \cos \varphi_2} \, d\xi \right.
$$

$$
+ \int_0^{\xi_{\mathrm{eff}}} J_{\mathrm{s}}^{(1)}(k\xi, \alpha - \varphi_0) e^{-ik\xi \cos(\alpha - \varphi_2)} \, d\xi \Bigg] \Bigg\} \tag{6.19}
$$

$$
u_{\mathrm{h}}^{(1)} = -u_0 \frac{ika}{2\sqrt{2\pi k a \sin \vartheta}} \frac{e^{ikR}}{R}
$$

$$
\times \left\{ e^{-ika \sin \vartheta + i\frac{\pi}{4}} \left[\sin \varphi_1 \int_0^{\xi_{\mathrm{eff}}} J_{\mathrm{h}}^{(1)}(k\xi, \varphi_0) e^{-ik\xi \cos \varphi_1} \, d\xi \right. \right.
$$

$$
+ \sin(\alpha - \varphi_1) \int_0^{\xi_{\mathrm{eff}}} J_{\mathrm{h}}^{(1)}(k\xi, \alpha - \varphi_0) e^{-ik\xi \cos(\alpha - \varphi_1)} \, d\xi \Bigg]
$$

$$
+ e^{ika \sin \vartheta - i\frac{\pi}{4}} \left[\sin \varphi_2 \int_0^{\xi_{\mathrm{eff}}} J_{\mathrm{h}}^{(1)}(k\xi, \varphi_0) e^{-ik\xi \cos \varphi_2} \, d\xi \right.
$$

$$
+ \sin(\alpha - \varphi_2) \int_0^{\xi_{\mathrm{eff}}} J_{\mathrm{h}}^{(1)}(k\xi, \alpha - \varphi_0) e^{-ik\xi \cos(\alpha - \varphi_2)} \, d\xi \Bigg] \Bigg\} \tag{6.20}
$$

Under the condition $ka \gg 1$, the scattering sources $J_{\mathrm{s,h}}^{(1)}$ near the edge of the conic surface are asymptotically equivalent to those near the edge of the tangential wedge. Hence, the expressions inside the brackets in Equations (6.19) and (6.20) are also asymptotically equivalent to the similar expressions in Equations (4.29) and (4.30), which relate to the wedge diffraction problem. Utilizing this observation, one can rewrite Equations (6.19) and (6.20) as

$$
u_{\mathrm{s}}^{(1)} = \frac{u_0 a}{\sqrt{2\pi k a \sin \vartheta}} \left[f^{(1)}(1) e^{-ika \sin \vartheta + i\frac{\pi}{4}} + f^{(1)}(2) e^{ika \sin \vartheta - i\frac{\pi}{4}} \right] \frac{e^{ikR}}{R}, \tag{6.21}
$$

$$
u_{\mathrm{h}}^{(1)} = \frac{u_0 a}{\sqrt{2\pi k a \sin \vartheta}} \left[g^{(1)}(1) e^{-ika \sin \vartheta + i\frac{\pi}{4}} + g^{(1)}(2) e^{ika \sin \vartheta - i\frac{\pi}{4}} \right] \frac{e^{ikR}}{R}. \tag{6.22}
$$

These expressions can be interpreted as the ray asymptotics for the field $u_{\mathrm{s,h}}^{(1)}$. They show that, under the condition $ka \sin \vartheta \gg 1$, this field consists of two diffracted rays incoming from the stationary points 1 and 2 at the circular edge. As is also seen there, the ray from point 2 undergoes the additional phase shift equal $-\pi/2$ while crossing the focal line along the z-axis.

The above approximations were derived for the regions $0 < \vartheta \leq \Omega$ and $\pi - \omega \leq \vartheta < \pi$ when both stationary points are visible from the observation point. The same technique is used for the calculation of the diffracted field in the region $\Omega \leq \vartheta \leq \pi - \omega$, when the stationary point 2 is not visible and therefore does not contribute to the first-order edge waves. In this case, only the vicinity of the stationary point 1 participates in the calculation that results in

$$u_s^{(1)} = \frac{u_0 a}{\sqrt{2\pi ka \sin \vartheta}} f^{(1)}(1) e^{-ika \sin \vartheta + i\frac{\pi}{4}} \frac{e^{ikR}}{R} \tag{6.23}$$

and

$$u_h^{(1)} = \frac{u_0 a}{\sqrt{2\pi ka \sin \vartheta}} g^{(1)}(1) e^{-ika \sin \vartheta + i\frac{\pi}{4}} \frac{e^{ikR}}{R}. \tag{6.24}$$

Functions $f^{(1)}$ and $g^{(1)}$ are defined by Equations (4.14) and (4.15) with Equations (3.55) to (3.57) and (2.62) and (2.64). According to these equations,

$$f^{(1)}(1) = \frac{\sin \frac{\pi}{n}}{n} \left(\frac{1}{\cos \frac{\pi}{n} - \cos \frac{\pi - \vartheta}{n}} - \frac{1}{\cos \frac{\pi}{n} - \cos \frac{\pi + 2\omega - \vartheta}{n}} \right)$$
$$- \frac{\sin \omega}{\cos \omega - \cos(\omega - \vartheta)} \tag{6.25}$$

and

$$g^{(1)}(1) = \frac{\sin \frac{\pi}{n}}{n} \left(\frac{1}{\cos \frac{\pi}{n} - \cos \frac{\pi - \vartheta}{n}} + \frac{1}{\cos \frac{\pi}{n} - \cos \frac{\pi + 2\omega - \vartheta}{n}} \right)$$
$$- \frac{\sin(\omega - \vartheta)}{\cos \omega - \cos(\omega - \vartheta)}, \tag{6.26}$$

where

$$n = \frac{\alpha}{\pi} = 1 + \frac{\omega + \Omega}{\pi}. \tag{6.27}$$

These expressions for functions $f^{(1)}$ and $g^{(1)}$ are valid in the entire region $0 \leq \vartheta \leq \pi$, although we have two different expressions for functions $f^{(1)}(2)$ and $g^{(1)}(2)$. For the region $0 \leq \vartheta \leq \Omega$,

$$f^{(1)}(2) = \frac{\sin\dfrac{\pi}{n}}{n}\left(\frac{1}{\cos\dfrac{\pi}{n} - \cos\dfrac{\pi + \vartheta}{n}} - \frac{1}{\cos\dfrac{\pi}{n} - \cos\dfrac{\pi + 2\omega + \vartheta}{n}}\right)$$

$$- \frac{\sin\omega}{\cos\omega - \cos(\omega + \vartheta)} \tag{6.28}$$

and

$$g^{(1)}(2) = \frac{\sin\dfrac{\pi}{n}}{n}\left(\frac{1}{\cos\dfrac{\pi}{n} - \cos\dfrac{\pi + \vartheta}{n}} + \frac{1}{\cos\dfrac{\pi}{n} - \cos\dfrac{\pi + 2\omega + \vartheta}{n}}\right)$$

$$- \frac{\sin(\omega + \vartheta)}{\cos\omega - \cos(\omega + \vartheta)}, \tag{6.29}$$

and for the region $\pi - \omega \leq \vartheta \leq \pi$ the corresponding expressions are

$$f^{(1)}(2) = \frac{\sin\dfrac{\pi}{n}}{n}\left(\frac{1}{\cos\dfrac{\pi}{n} - \cos\dfrac{\pi - \vartheta}{n}} - \frac{1}{\cos\dfrac{\pi}{n} - \cos\dfrac{\pi - 2\omega - \vartheta}{n}}\right)$$

$$- \frac{\sin\omega}{\cos\omega - \cos(\omega + \vartheta)} \tag{6.30}$$

and

$$g^{(1)}(2) = \frac{\sin\dfrac{\pi}{n}}{n}\left(\frac{1}{\cos\dfrac{\pi}{n} - \cos\dfrac{\pi - \vartheta}{n}} + \frac{1}{\cos\dfrac{\pi}{n} - \cos\dfrac{\pi - 2\omega - \vartheta}{n}}\right)$$

$$- \frac{\sin(\omega + \vartheta)}{\cos\omega - \cos(\omega + \vartheta)}. \tag{6.31}$$

Some terms in functions $f^{(1)}$ and $g^{(1)}$ are singular in certain directions ϑ; however, such singularities cancel each other and these functions are always finite. For the forward direction $\vartheta = 0$ (that is, the shadow boundary behind the body),

$$f^{(1)}(1) = f^{(1)}(2) = -\frac{\dfrac{1}{n}\sin\dfrac{\pi}{n}}{\cos\dfrac{\pi}{n} - \cos\dfrac{\pi + 2\omega}{n}} - \frac{1}{2n}\cot\dfrac{\pi}{n} + \frac{1}{2}\cot\omega \tag{6.32}$$

and

$$g^{(1)}(1) = g^{(1)}(2) = \frac{\dfrac{1}{n}\sin\dfrac{\pi}{n}}{\cos\dfrac{\pi}{n} - \cos\dfrac{\pi + 2\omega}{n}} - \frac{1}{2n}\cot\dfrac{\pi}{n} - \frac{1}{2}\cot\omega. \tag{6.33}$$

For the specular direction $\vartheta = 2\omega$ (corresponding to the ray reflected from the conic surface),

$$f^{(1)}(1) = \frac{\frac{1}{n}\sin\frac{\pi}{n}}{\cos\frac{\pi}{n} - \cos\frac{\pi - 2\omega}{n}} + \frac{1}{2n}\cot\frac{\pi}{n} + \frac{1}{2}\cot\omega \qquad (6.34)$$

and

$$g^{(1)}(1) = \frac{\frac{1}{n}\sin\frac{\pi}{n}}{\cos\frac{\pi}{n} - \cos\frac{\pi - 2\omega}{n}} - \frac{1}{2n}\cot\frac{\pi}{n} + \frac{1}{2}\cot\omega. \qquad (6.35)$$

Notice also that for the directions $\vartheta = 0$ and $\vartheta = \pi$, the relationships

$$f^{(1)}(1) = f^{(1)}(2) \qquad \text{and} \qquad g^{(1)}(1) = g^{(1)}(2) \qquad (6.36)$$

are valid, due to the symmetry of the problem.

6.1.3 Focal Fields

Focal fields for acoustic and electromagnetic waves generated by the nonuniform sources $j^{(1)}$ are different, due to the vector nature of electromagnetic fields.

Given the incident wave (6.1), every point at the z-axis is a focal point for diffracted edge waves (Fig. 6.2). In the far zone, the position of any focal points with $z > 0$ is determined by the coordinate $\vartheta = 0$, and the points with $z < 0$ are characterized by the coordinate $\vartheta = \pi$. For the focal points, the general field expressions (6.8) and (6.9) are simplified. Under the condition $a \gg \xi_{\text{eff}}$, they can be written as

$$u_s^{(1)} = -u_0 \frac{a}{2} \frac{e^{ikR}}{R} \left[\int_0^{\xi_{\text{eff}}} J_s^{(1)}(k\xi, \varphi_0) e^{\pm ik\xi \cos\omega} \, d\xi \right.$$

$$\left. + \int_0^{\xi_{\text{eff}}} J_s^{(1)}(k\xi, \alpha - \varphi_0) e^{\mp ik\xi \cos\Omega} \, d\xi \right] \qquad (6.37)$$

and

$$u_h^{(1)} = -u_0 \frac{ika}{2} \frac{e^{ikR}}{R} \left[\mp \sin\omega \int_0^{\xi_{\text{eff}}} J_s^{(1)}(k\xi, \varphi_0) e^{\pm ik\xi \cos\omega} \, d\xi \right.$$

$$\left. \pm \sin\Omega \int_0^{\xi_{\text{eff}}} J_s^{(1)}(k\xi, \alpha - \varphi_0) e^{\mp ik\xi \cos\Omega} \, d\xi \right]. \qquad (6.38)$$

The upper (lower) sign here relates to $\vartheta = 0\,(\vartheta = \pi)$. In terms of the local coordinates they look the same for both directions:

$$u_s^{(1)} = -u_0 \frac{a}{2} \frac{e^{ikR}}{R} \left[\int_0^{\xi_{\text{eff}}} J_s^{(1)}(k\xi, \varphi_0) e^{-ik\xi \cos \varphi_1} \, d\xi \right.$$

$$\left. + \int_0^{\xi_{\text{eff}}} J_s^{(1)}(k\xi, \alpha - \varphi_0) e^{-ik\xi \cos(\alpha - \varphi_1)} \, d\xi \right]. \tag{6.39}$$

$$u_h^{(1)} = -u_0 \frac{ika}{2} \frac{e^{ikR}}{R} \left[\sin \varphi_1 \int_0^{\xi_{\text{eff}}} J_s^{(1)}(k\xi, \varphi_0) e^{-ik\xi \cos \varphi_1} \, d\xi \right.$$

$$\left. + \sin(\alpha - \varphi_1) \int_0^{\xi_{\text{eff}}} J_s^{(1)}(k\xi, \alpha - \varphi_0) e^{-ik\xi \cos(\alpha - \varphi_1)} \, d\xi \right], \tag{6.40}$$

with $\varphi_1 = \omega\,(\varphi_1 = \pi + \omega)$ when $\vartheta = \pi\,(\vartheta = 0)$. Then we use Equations (4.29) and (4.30) and find the following expressions for the focal field:

$$u_s^{(1)} = u_0 a f^{(1)}(1) \frac{e^{ikR}}{R} \tag{6.41}$$

and

$$u_h^{(1)} = u_0 a g^{(1)}(1) \frac{e^{ikR}}{R}. \tag{6.42}$$

It is seen that the focal field is $\sqrt{ka}$ times higher in magnitude than the ray fields (6.21) and (6.22). In conjunction with Equations (6.41) and (6.42), we recall relationships (6.36), which are valid for the focal points.

6.1.4 Bessel Interpolations for the Field $u_{s,h}^{(1)}$

In the previous sections, we derived the ray and focal asymptotic expressions for the field $u_{s,h}^{(1)}$. The ray asymptotics (6.21) to (6.24) are valid under the condition $ka \sin \vartheta \gg 1$ (i.e., away from the focal line), and the asymptotic expressions (6.41) and (6.42) determine the field directly on the focal line (where $\sin \vartheta = 0$). Now our goal is to construct such approximations, that would continuously describe the diffracted field both in the ray and focal regions. This can be done utilizing the Bessel functions $J_0(ka \sin \vartheta)$ and $J_1(ka \sin \vartheta)$.

For the large argument ($ka \sin \vartheta \gg 1$), they can be approximated by (Gradshteyn and Ryzhik, 1994)

$$J_0(ka \sin \vartheta) \sim \frac{1}{\sqrt{2\pi ka \sin \vartheta}} \left(e^{ika \sin \vartheta - i\frac{\pi}{4}} + e^{-ika \sin \vartheta + i\frac{\pi}{4}} \right) \tag{6.43}$$

and

$$J_1(ka \sin \vartheta) \sim \frac{1}{\sqrt{2\pi ka \sin \vartheta}} \left(e^{ika \sin \vartheta - i\frac{3\pi}{4}} + e^{-ika \sin \vartheta + i\frac{3\pi}{4}} \right). \tag{6.44}$$

It follows from these expressions that

$$\frac{e^{-ika\sin\vartheta+i\frac{\pi}{4}}}{\sqrt{2\pi ka\sin\vartheta}} \approx \frac{1}{2}[J_0(ka\sin\vartheta)-iJ_1(ka\sin\vartheta)] \tag{6.45}$$

and

$$\frac{e^{ika\sin\vartheta-i\frac{\pi}{4}}}{\sqrt{2\pi ka\sin\vartheta}} \approx \frac{1}{2}[J_0(ka\sin\vartheta)+iJ_1(ka\sin\vartheta)]. \tag{6.46}$$

With these observations, the ray asymptotics (6.21) and (6.22) can be written as

$$u_s^{(1)} = u_0\frac{a}{2}\frac{e^{ikR}}{R}\left\{f^{(1)}(1)[J_0(ka\sin\vartheta)-iJ_1(ka\sin\vartheta)]\right.$$
$$\left.+f^{(1)}(2)[J_0(ka\sin\vartheta)+iJ_1(ka\sin\vartheta)]\right\} \tag{6.47}$$

$$u_h^{(1)} = u_0\frac{a}{2}\frac{e^{ikR}}{R}\left\{g^{(1)}(1)[J_0(ka\sin\vartheta)-iJ_1(ka\sin\vartheta)]\right.$$
$$\left.+g^{(1)}(2)[J_0(ka\sin\vartheta)+iJ_1(ka\sin\vartheta)]\right\}. \tag{6.48}$$

Now we analytically continue these expressions into the focal region. If we take into account Equation (6.36), as well as the relationships $J_0(0)=1$ and $J_1(0)=0$, one can see that expressions (6.47) and (6.48) exactly transform into the focal asymptotics (6.41) and (6.42), when $\vartheta \to 0$ and $\vartheta \to \pi$. This means that the above formulas (6.47) and (6.48) can be considered as the appropriate approximations for the diffracted field in the regions $0 \le \vartheta \le \Omega$ and $\pi - \omega \le \vartheta \le \pi$.

In the region $\Omega \le \vartheta \le \pi - \omega$, where the stationary point 2 is not visible, the modified versions of Equations (6.47) and (6.48),

$$u_s^{(1)} = u_0\frac{a}{2}f^{(1)}(1)[J_0(ka\sin\vartheta)-iJ_1(ka\sin\vartheta)]\frac{e^{ikR}}{R} \tag{6.49}$$

and

$$u_h^{(1)} = u_0\frac{a}{2}g^{(1)}(1)[J_0(ka\sin\vartheta)-iJ_1(ka\sin\vartheta)]\frac{e^{ikR}}{R}, \tag{6.50}$$

represent the analytical continuation of the ray asymptotics (6.23) and (6.24) into the entire region $\Omega \le \vartheta \le \pi - \omega$.

In the next sections, the above results found for the canonical problem are applied for calculation of the field scattered at certain bodies of revolution.

6.2 SCATTERING AT A DISK

The geometry of the problem is shown in Figure 6.6. The incident plane wave is given by Equation (6.1) and propagates in the positive direction of the z-axis. Because of

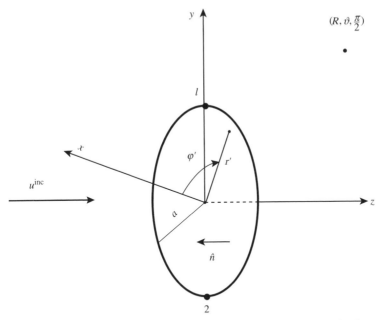

Figure 6.6   An incident wave propagates in the direction along the z-axis and undergoes diffraction at a circular disk of radius a. Edge points 1 and 2 are in the plane $x = 0$.

that, and due to the axial symmetry of the disk, the scattered field is the same in all meridian planes $\varphi = $ const. $(0 \leq \varphi \leq 2\pi)$. In addition, the scattered field is symmetric with respect to the disk plane: $u_s(-z) = u_s(z)$, $u_h(-z) = -u_h(z)$. Therefore, it is sufficient to investigate the field only in the plane $\varphi = \pi/2$ for the directions $0 \leq \vartheta \leq \pi/2$.

6.2.1 Physical Optics Approximation

The relationships $u_s^{(0)} = E_\varphi^{(0)}$, $u_h^{(0)} = H_\varphi^{(0)}$ are valid for the disk diffraction problem under the conditions $u_s^{inc} = E_\varphi^{inc}$, $u_h^{(0)} = H_\varphi^{(0)}$.

Acoustic Waves

In this approximation, the scattered field is considered as the radiation generated by the uniform component (1.31) of the scattering sources, which are induced on the illuminated side $(z = -0)$ of the disk. This field is determined by Equations (1.16) and (1.17), where one should set

$$\vartheta' = \frac{\pi}{2}, \qquad \varphi = \frac{\pi}{2}, \qquad \cos\Omega = \sin\vartheta\sin\varphi', \qquad \hat{m}\cdot\hat{n} = -\frac{z}{r} \approx -\cos\vartheta,$$

$$(6.51)$$

$$j_s = j_s^{(0)} = -2iku_0, \qquad j_h = j_h^{(0)} = 2u_0. \tag{6.52}$$

With these settings, the field is described by

$$u_s^{(0)} = u_0 \frac{ik}{2\pi} \frac{e^{ikR}}{R} \int_0^a r'dr' \int_0^{2\pi} e^{-ikr'\sin\vartheta\sin\varphi'} d\varphi' \tag{6.53}$$

and

$$u_h^{(0)} = u_0 \frac{ik}{2\pi} \cos\vartheta \frac{e^{ikR}}{R} \int_0^a r'dr' \int_0^{2\pi} e^{-ikr'\sin\vartheta\sin\varphi'} d\varphi'. \tag{6.54}$$

According to Gradshteyn and Ryzhik (1994),

$$\frac{1}{2\pi} \int_0^{2\pi} e^{-ix\sin\varphi'} d\varphi' = J_0(x), \qquad \int J_0(x)xdx = xJ_1(x), \tag{6.55}$$

where J_0 and J_1 are the Bessel functions. With these relationships, one can represent Equations (6.53) and (6.54) as

$$u_s^{(0)} = u_0 \frac{ia}{\sin\vartheta} J_1(ka\sin\vartheta) \frac{e^{ikR}}{R} \tag{6.56}$$

and

$$u_h^{(0)} = u_0 \frac{ia}{\sin\vartheta} \cos\vartheta J_1(ka\sin\vartheta) \frac{e^{ikR}}{R}. \tag{6.57}$$

For small arguments ($ka\sin\vartheta \ll 1$), one can replace $J_1(ka\sin\vartheta)$ by $(ka\sin\vartheta)/2$. Then we set $\vartheta = 0, \pi$ and find the field on the focal line:

$$u_s^{(0)} = u_0 \frac{ika^2}{2} \frac{e^{ikR}}{R}, \qquad \text{for } \vartheta = 0 \text{ and } \vartheta = \pi \tag{6.58}$$

and

$$u_h^{(0)} = \pm u_0 \frac{ika^2}{2} \frac{e^{ikR}}{R}, \qquad \text{for } \vartheta = \begin{pmatrix} 0 \\ \pi \end{pmatrix}. \tag{6.59}$$

In the directions away from the focal line where $ka\sin\vartheta \gg 1$, the Bessel function in Equations (6.56) and (6.57) can be replaced by its asymptotic expression (6.44).

Equations we obtain in this way reveal a ray structure of the field, which consists of two diffracted rays coming from the stationary points 1 and 2:

$$u_s^{(0)} = \frac{u_0 a}{\sqrt{2\pi ka \sin \vartheta}} \left[f^{(0)}(1)e^{-ika\sin\vartheta + i\frac{\pi}{4}} + f^{(0)}(2)e^{ika\sin\vartheta - i\frac{\pi}{4}} \right] \frac{e^{ikR}}{R} \qquad (6.60)$$

and

$$u_h^{(0)} = \frac{u_0 a}{\sqrt{2\pi ka \sin \vartheta}} \left[g^{(0)}(1)e^{-ika\sin\vartheta + i\frac{\pi}{4}} + g^{(0)}(2)e^{ika\sin\vartheta - i\frac{\pi}{4}} \right] \frac{e^{ikR}}{R}, \qquad (6.61)$$

where the functions

$$f^{(0)}(1) = -f^{(0)}(2) = -\frac{1}{\sin \vartheta} \qquad (6.62)$$

and

$$g^{(0)}(1) = -g^{(0)}(2) = -\frac{\cos \vartheta}{\sin \vartheta} \qquad (6.63)$$

determine the directivity patterns of diffracted rays in the PO approximation.

Electromagnetic Waves

To show the relationship between the acoustic and electromagnetic diffracted fields, we present here the PO approximation for the field scattered at a perfectly conducting disk (Ufimtsev, 1962):

$$E_\vartheta^{(0)} = Z_0 H_\varphi^{(0)} = -iaZ_0 H_{0x} \sin\varphi \frac{\cos \vartheta}{\sin \vartheta} J_1(ka\sin\vartheta) \frac{e^{ikR}}{R}$$

and

$$E_\varphi^{(0)} = -Z_0 H_\vartheta^{(0)} = -iaZ_0 H_{0x} \cos\varphi \frac{1}{\sin \vartheta} J_1(ka\sin\vartheta) \frac{e^{ikR}}{R}. \qquad (6.64)$$

This field is due to the diffraction of the incident wave

$$E_y^{inc} = -Z_0 H_x^{inc} = -Z_0 H_{0x} e^{ikz} = E_{0y}e^{ikz}, \qquad (6.65)$$

where $Z_0 = \sqrt{\mu_0/\varepsilon_0} = 120\pi$ ohms is the impedance of free space. In view of the equations

$$E_{0\varphi} = E_{0y} \cos\varphi = -Z_0 H_{0x} \cos\varphi, \qquad (6.66)$$

the expressions (6.64) can be rewritten as

$$E_\varphi^{(0)} = E_{0\varphi} \frac{ia}{\sin \vartheta} J_1(ka\sin\vartheta) \frac{e^{ikR}}{R}$$

and

$$H_\varphi^{(0)} = H_{0\varphi} \frac{ia}{\sin\vartheta} \cos\vartheta\, J_1(ka\sin\vartheta)\, \frac{e^{ikR}}{R}. \tag{6.67}$$

Their comparison with Equations (6.56) and (6.57) reveals the following equivalence existing between the acoustic and electromagnetic diffracted fields:

$$
\begin{aligned}
u_{\rm s}^{(0)} &= E_\varphi^{(0)}, &&\text{if } u_0 = E_{0\varphi}, \\
u_{\rm h}^{(0)} &= H_\varphi^{(0)}, &&\text{if } u_0 = H_{0\varphi}.
\end{aligned}
\tag{6.68}
$$

These equations are in an agreement with relationships (1.100) and (1.101).

6.2.2   Field Generated by Nonuniform Scattering Sources

The field generated by the nonuniform scattering sources $j_{\rm s,h}^{(1)}$ was investigated in Section 6.1.4. Here we reproduce the related approximations

$$
\begin{aligned}
u_{\rm s}^{(1)} = u_0 \frac{a}{2} \frac{e^{ikR}}{R} \Big\{ &f^{(1)}(1)[J_0(ka\sin\vartheta) - i\,J_1(ka\sin\vartheta)] \\
&+ f^{(1)}(2)[J_0(ka\sin\vartheta) + i\,J_1(ka\sin\vartheta)] \Big\}
\end{aligned}
\tag{6.69}
$$

and

$$
\begin{aligned}
u_{\rm h}^{(1)} = u_0 \frac{a}{2} \frac{e^{ikR}}{R} \Big\{ &g^{(1)}(1)[J_0(ka\sin\vartheta) - i\,J_1(ka\sin\vartheta)] \\
&+ g^{(1)}(2)[J_0(ka\sin\vartheta) + i\,J_1(ka\sin\vartheta)] \Big\},
\end{aligned}
\tag{6.70}
$$

where $0 \le \vartheta \le \pi$. The ray-type asymptotics of (6.69) and (6.70) are shown in Equations (6.21) and (6.22). The focal asymptotics of (6.69) and (6.7) are determined by Equations (6.41) and (6.42).

General expressions for functions $f^{(1)}(1)$, $f^{(1)}(2)$ and $g^{(1)}(1)$, $g^{(1)}(2)$ are given by Equations (6.25) to (6.31). For the scattering disk, they are written below. Functions $f^{(1)}(1)$ and $g^{(1)}(1)$ are described in the entire region $0 \le \vartheta \le \pi$ by

$$f^{(1)}(1) = -\frac{1}{2}\left(\frac{1}{\sin\dfrac{\vartheta}{2}} + \frac{1}{\cos\dfrac{\vartheta}{2}} \right) + \frac{1}{\sin\vartheta} \tag{6.71}$$

and

$$g^{(1)}(1) = \frac{1}{2}\left(-\frac{1}{\sin\dfrac{\vartheta}{2}} + \frac{1}{\cos\dfrac{\vartheta}{2}}\right) + \frac{\cos\vartheta}{\sin\vartheta}. \tag{6.72}$$

Functions $f^{(1)}(2)$ and $g^{(1)}(2)$ are described by

$$f^{(1)}(2) = \frac{1}{2}\left(\frac{1}{\sin\dfrac{\vartheta}{2}} - \frac{1}{\cos\dfrac{\vartheta}{2}}\right) - \frac{1}{\sin\vartheta} \tag{6.73}$$

and

$$g^{(1)}(2) = \frac{1}{2}\left(\frac{1}{\sin\dfrac{\vartheta}{2}} + \frac{1}{\cos\dfrac{\vartheta}{2}}\right) - \frac{\cos\vartheta}{\sin\vartheta} \tag{6.74}$$

in the region $0 \le \vartheta \le \pi/2$, and by

$$f^{(1)}(2) = \frac{1}{2}\left(-\frac{1}{\sin\dfrac{\vartheta}{2}} + \frac{1}{\cos\dfrac{\vartheta}{2}}\right) - \frac{1}{\sin\vartheta} \tag{6.75}$$

and

$$g^{(1)}(2) = -\frac{1}{2}\left(\frac{1}{\sin\dfrac{\vartheta}{2}} + \frac{1}{\cos\dfrac{\vartheta}{2}}\right) - \frac{\cos\vartheta}{\sin\vartheta} \tag{6.76}$$

in the region $\pi/2 \le \vartheta \le \pi$.

Functions $f^{(1)}(1)$ and $f^{(1)}(2)$ are symmetric with respect to the disk plane,

$$f^{(1)}(1, \pi - \vartheta) = f^{(1)}(1, \vartheta), \qquad f^{(1)}(2, \pi - \vartheta) = f^{(1)}(2, \vartheta), \tag{6.77}$$

but functions $g^{(1)}(1)$ and $g^{(1)}(2)$ are antisymmetric,

$$g^{(1)}(1, \pi - \vartheta) = -g^{(1)}(1, \vartheta), \qquad g^{(1)}(2, \pi - \vartheta) = -g^{(1)}(2, \vartheta). \tag{6.78}$$

Here, the functions with the argument $\pi - \vartheta$ correspond to the left half-space $(z < 0)$.
It follows from Equations (6.71) to (6.76) that

$$f^{(1)}(1) = f^{(1)}(2) = -\frac{1}{2}, \qquad \text{for } \vartheta = 0 \text{ and } \vartheta = \pi, \tag{6.79}$$

and

$$g^{(1)}(1) = g^{(1)}(2) = \pm\frac{1}{2}, \qquad \text{for } \vartheta = \begin{pmatrix} 0 \\ \pi \end{pmatrix}. \tag{6.80}$$

After substitution of these values into Equations (6.41) and (6.42), it follows that

$$u_{\mathrm{s}}^{(1)} = -\frac{u_0 a}{2}\frac{e^{ikR}}{R}, \qquad \text{for } \vartheta = 0 \text{ and } \vartheta = \pi, \tag{6.81}$$

and

$$u_{\mathrm{h}}^{(1)} = \pm\frac{u_0 a}{2}\frac{e^{ikR}}{R}, \qquad \text{for } \vartheta = \begin{pmatrix} 0 \\ \pi \end{pmatrix}. \tag{6.82}$$

It is worth noting that the amplitude of the focal field generated by the nonuniform sources does not depend on frequency.

6.2.3 Total Scattered Field

The sum

$$u_{\mathrm{s,h}} = u_{\mathrm{s,h}}^{(0)} + u_{\mathrm{s,h}}^{(1)} \tag{6.83}$$

provides the first-order PTD approximation for the scattered field:

- Quantities $u_{\mathrm{s,h}}^{(0)}$ and $u_{\mathrm{s,h}}^{(1)}$ are defined by Equations (6.56) and (6.57) and Equations (6.69) and (6.70).
- Their rays-type asymptotics are determined in Equations (6.60), (6.61) and (6.21), (6.22).

  When they are included in Equation (6.83), the functions $f^{(0)}$, $g^{(0)}$ contained both in (6.60) and (6.61) and in (6.21) and (6.22) cancel each other. As a result, the ray asymptotics for the total field contain only the Sommerfeld functions f and g:

$$u_{\mathrm{s}} = \frac{u_0 a}{\sqrt{2\pi ka \sin\vartheta}}\left[f(1)e^{-ika\sin\vartheta + i\frac{\pi}{4}} + f(2)e^{ika\sin\vartheta - i\frac{\pi}{4}}\right]\frac{e^{ikR}}{R} \tag{6.84}$$

and

$$u_{\mathrm{h}} = \frac{u_0 a}{\sqrt{2\pi ka \sin\vartheta}}\left[g(1)e^{-ika\sin\vartheta + i\frac{\pi}{4}} + g(2)e^{ika\sin\vartheta - i\frac{\pi}{4}}\right]\frac{e^{ikR}}{R}. \tag{6.85}$$

Functions $f(1,2)$ and $g(1,2)$ are described by Equations (6.71) to (6.76), where the last terms (being outside the parentheses) should be omitted.

- The focal asymptotics for $u_{s,h}^{(0)}$ and $u_{s,h}^{(1)}$ are presented in Equations (6.58), (6.59) and, (6.81), (6.82). Their summation leads to

$$u_s = u_0 \frac{ika^2}{2}\left(1 + \frac{i}{ka}\right)\frac{e^{ikR}}{R}, \qquad \text{for } \vartheta = 0 \text{ and } \vartheta = \pi, \qquad (6.86)$$

and

$$u_h = \pm u_0 \frac{ika^2}{2}(1 - \frac{i}{ka})\frac{e^{ikR}}{R}, \qquad \text{for } \vartheta = \begin{pmatrix}0\\\pi\end{pmatrix}. \qquad (6.87)$$

Approximation (6.87) for the direction $\vartheta = \pi$ agrees with Equation (14.114) in the work by Bowman *et al.* (1987).

The results of numerical calculations shown in Figures 6.7 and 6.8 supplement the above analysis of the scattered field. The normalized scattering cross-section is the quantity

$$\sigma_{norm} = \frac{\sigma_{s,h}}{\pi a^2 (ka)^2}$$

plotted in the decibel scale. The ordinary scattering cross-section σ is defined by Equation (1.26).

In Figures 6.7 and 6.8, the contribution to the scattered field generated by the nonuniform component of the scattering sources is clearly seen. The field scattered by the hard disk must be equal to zero in the direction $\vartheta = 90°$. The finite value of the PTD field in this direction (Fig. 6.8) is caused by the fictitious nonuniform sources distributed in the plane outside the disk surface. This shortcoming can be removed by the truncation of these fictitious sources, as was shown in Section 5.1.4.

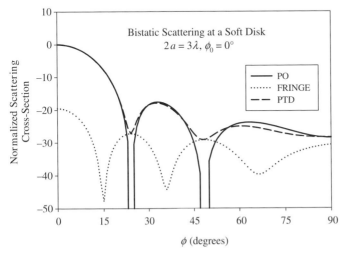

Figure 6.7   Scattering at the acoustically soft disk. The curve "FRINGE" relates to the field generated by the nonuniform ("fringe") scattering sources. According to Equations (1.100) and (1.101), the PO curve here also demonstrates the scattering of electromagnetic waves at a perfectly conducting disk.

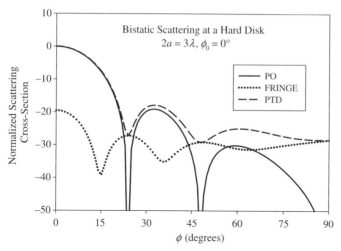

Figure 6.8   Scattering at the acoustically hard disk. The curve "FRINGE" relates to the field generated by the nonuniform ("fringe") scattering sources. According to Equations (1.100) and (1.101), the PO curve here also demonstrates the scattering of electromagnetic waves at a perfectly conducting disk.

6.3 SCATTERING AT CONES: FOCAL FIELD

The PO approximations for acoustic and electromagnetic focal fields in this problem are identical. Asymptotics for the *total* acoustic and electromagnetic fields are different.

6.3.1 Asymptotic Approximations for the Field

The geometry of the problem is shown in Figure 6.9. The incident plane wave is given by Equation (6.1). First, we calculate the focal field radiated by the uniform components $j_{s,h}^{(0)}$ of the induced scattering sources. They are defined by Equation (1.31) and determined on the cone surface by

$$j_s^{(0)} = -2u_0 ik \sin \omega e^{ikz} \qquad \text{and} \qquad j_h^{(0)} = 2u_0 e^{ikz}. \qquad (6.88)$$

The scattered field in the far zone is defined by Equations (1.16) and (1.17), where one should set

$$\hat{m} \cdot \hat{n} = -\sin \omega \cos \vartheta, \qquad \text{with } \vartheta = 0 \text{ or } \vartheta = \pi, \qquad (6.89)$$

and

$$ds = \xi \sin \omega \, d\xi \, d\varphi', \qquad \text{with } \xi = r'. \qquad (6.90)$$

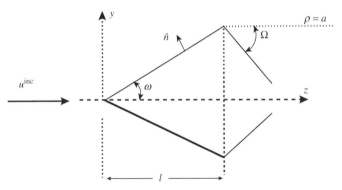

Figure 6.9   Cross-section of a cone by the plane *yoz*. A radius of the edge is *a*.

For the observation points on the focal line ($\vartheta = 0, \pi$), the integral over the variable φ' equals 2π. Hence, the field $u_{\text{s,h}}^{(0)}$ can be represented in the form

$$u_s^{(0)} = u_0 ik \sin^2 \omega \frac{e^{ikR}}{R} \int_0^b e^{ik\xi(1-\cos\vartheta)\cos\omega} \xi \, d\xi \qquad (6.91)$$

and

$$u_h^{(0)} = u_0 ik \sin^2 \omega \cos \vartheta \frac{e^{ikR}}{R} \int_0^b e^{ik\xi(1-\cos\vartheta)\cos\omega} \xi \, d\xi, \qquad (6.92)$$

where $b = a/\sin\omega$ is the length of the cone generatrix and a is the radius of the edge. It follows from these equations that

$$u_s^{(0)} = u_h^{(0)} = u_0 \frac{ika^2}{2} \frac{e^{ikR}}{R} \qquad (6.93)$$

in the forward direction ($\vartheta = 0$), and

$$u_s^{(0)} = -u_h^{(0)} = -u_0 \frac{i}{4k} \tan^2 \omega \frac{e^{ikR}}{R} + u_0 \cdot \left(\frac{i}{4k} \tan^2 \omega + \frac{a}{2} \tan \omega \right) \frac{e^{ikR}}{R} e^{i2kl} \qquad (6.94)$$

in the backscattering direction ($\vartheta = \pi$).

The forward scattering (6.93) is exactly the same as Equations (6.58) and (6.59) for the disk and actually represents the shadow radiation introduced earlier in Section 1.3.4. This result is in complete agreement with the Shadow Contour Theorem presented in Section 1.3.5, since the cone and the disk of radius *a* produce the same shadow and generate the same shadow radiation.

Comparison of the acoustic field, (6.93) and (6.94), with the electromagnetic PO field scattered by a perfectly conducting cone (Equation (2.4.3) in Ufimtsev (2003) reveals the relationships

$$E_x^{(0)} = u_s^{(0)}, \qquad \text{if } E_x^{(0)} = u^{\text{inc}}$$

$$H_y^{(0)} = u_h^{(0)}, \qquad \text{if } H_y^{(0)} = u^{\text{inc}} \tag{6.95}$$

for the directions $\vartheta = 0$ and $\vartheta = \pi$. This result is in complete agreement with the general relationships (1.100) and (1.101).

The field radiated by the nonuniform components of the scattering sources $j_{s,h}^{(1)}$ (induced near the circular edge on the cone) was investigated in Section 6.1. According to Equations (6.41) and (6.42), this field equals

$$u_s^{(1)} = u_0 a f^{(1)} \frac{e^{ikR}}{R} \tag{6.96}$$

and

$$u_h^{(1)} = u_0 a g^{(1)} \frac{e^{ikR}}{R} \tag{6.97}$$

in the forward direction $\vartheta = 0$, and

$$u_s^{(1)} = u_0 a f^{(1)} \frac{e^{ikR}}{R} e^{i2kl} \tag{6.98}$$

and

$$u_h^{(1)} = u_0 a g^{(1)} \frac{e^{ikR}}{R} e^{i2kl} \tag{6.99}$$

in the backscattering direction $\vartheta = \pi$. Here,

$$f^{(1)} = -\frac{\frac{1}{n} \sin \frac{\pi}{n}}{\cos \frac{\pi}{n} - \cos \frac{\pi + 2\omega}{n}} - \frac{1}{2n} \cot \frac{\pi}{n} + \frac{1}{2} \cot \omega \tag{6.100}$$

and

$$g^{(1)} = \frac{\frac{1}{n} \sin \frac{\pi}{n}}{\cos \frac{\pi}{n} - \cos \frac{\pi + 2\omega}{n}} - \frac{1}{2n} \cot \frac{\pi}{n} - \frac{1}{2} \cot \omega \tag{6.101}$$

for the forward direction $\vartheta = 0$, and

$$f^{(1)} = \frac{\sin \frac{\pi}{n}}{n} \left(\frac{1}{\cos \frac{\pi}{n} - 1} - \frac{1}{\cos \frac{\pi}{n} - \cos \frac{2\omega}{n}} \right) - \frac{1}{2} \tan \omega \tag{6.102}$$

and

$$g^{(1)} = \frac{\sin \frac{\pi}{n}}{n} \left(\frac{1}{\cos \frac{\pi}{n} - 1} + \frac{1}{\cos \frac{\pi}{n} - \cos \frac{2\omega}{n}} \right) + \frac{1}{2} \tan \omega \qquad (6.103)$$

for the backscattering direction $\vartheta = \pi$. In these equations, $n = 1 + (\omega + \Omega)/\pi$.

In contrast to the above noted equivalence of the PO acoustic and electromagnetic fields, the acoustic field $u_{s,h}^{(1)}$ generated by the nonuniform scattering sources is different from the electromagnetic field radiated by the nonuniform electric currents. The electromagnetic field is given by Equations (2.3.18) and (2.3.19) in Ufimtsev (2003) and is determined by a linear combination of functions $f^{(1)}$ and $g^{(1)}$, but in the acoustic case the field $u_s^{(1)}$ depends only on $f^{(1)}$ and the field $u_h^{(1)}$ depends only on $g^{(1)}$.

The above equations describe the field $u_{s,h}^{(1)}$ generated *only* by the nonuniform scattering sources induced near the cone edge. There are two other types of nonuniform sources induced on the cone surface, which we have neglected here. The first is caused by the smooth bending of the surface, which is a small quantity inversely proportional to $(k\rho)$ at a distance far from the cone tip. Here ρ is a polar coordinate of the surface ($\rho = a$ at the edge points). The second is the nonuniform component concentrated near the cone tip. It is caused by both the sharp tip and by the large curvature of the cone surface due to its smooth bending near the tip. At a far distance ξ from the tip, it is of the order $1/k\xi$. In the case of electromagnetic diffraction, it was shown (Ufimtsev, 2003) that in the backscattering direction ($\vartheta = \pi$) one can neglect the field radiated by these types of nonuniform scattering sources. The asymptotic analysis of the electromagnetic field scattered by a semi-infinite cone also confirms this observation (Felsen, 1955; Bowman *et al.*, 1987).

Taking these comments into account, the first-order PTD approximation for the backscattered total field (at the focal line $\vartheta = \pi$) can be found by summation of Equations (6.94), (6.98) and (6.99):

$$u_s = u_0 a \left[-\frac{i}{4ka} \tan^2 \omega (1 - e^{i2kl}) \right.$$

$$\left. + \frac{\sin \frac{\pi}{n}}{n} \left(\frac{1}{\cos \frac{\pi}{n} - 1} - \frac{1}{\cos \frac{\pi}{n} - \cos \frac{2\omega}{n}} \right) e^{i2kl} \right] \frac{e^{ikR}}{R} \qquad (6.104)$$

and

$$u_h = u_0 a \left[\frac{i}{4ka} \tan^2 \omega (1 - e^{i2kl}) \right.$$

$$\left. + \frac{\sin \frac{\pi}{n}}{n} \left(\frac{1}{\cos \frac{\pi}{n} - 1} + \frac{1}{\cos \frac{\pi}{n} - \cos \frac{2\omega}{n}} \right) e^{i2kl} \right] \frac{e^{ikR}}{R}. \qquad (6.105)$$

In the limiting case when $\omega \to \pi/2$ and the front part of the object (Fig. 6.9) transforms into the disk, the above equations for the backscattered field are reduced to

$$u_s = u_0 a \left(\frac{ika}{2} + \frac{\frac{1}{n} \sin \frac{\pi}{n}}{\cos \frac{\pi}{n} - 1} + \frac{1}{2n} \cot \frac{\pi}{n} \right) \frac{e^{ikR}}{R} \qquad (6.106)$$

and

$$u_h = u_0 a \left(-\frac{ika}{2} + \frac{\frac{1}{n} \sin \frac{\pi}{n}}{\cos \frac{\pi}{n} - 1} - \frac{1}{2n} \cot \frac{\pi}{n} \right) \frac{e^{ikR}}{R} \qquad (6.107)$$

with $n = (3/2) + (\Omega/\pi)$. Finally, with $\Omega \to \pi/2$ these expressions transform into

$$u_s = u_0 \left(\frac{ika^2}{2} - \frac{a}{2} \right) \frac{e^{ikR}}{R} \qquad (6.108)$$

and

$$u_h = -u_0 \left(\frac{ika^2}{2} + \frac{a}{2} \right) \frac{e^{ikR}}{R}, \qquad (6.109)$$

which exactly coincide with Equations (6.86), (6.87) for the field scattered back from the disk.

6.3.2   Numerical Analysis of Backscattering

For the field represented in the form

$$u_{s,h} = u_0 \Phi_{s,h} \frac{e^{ikR}}{R}, \qquad (6.110)$$

the scattering cross-section is defined according to Equation (1.26) by

$$\sigma_{s,h} = 4\pi \left| \Phi_{s,h} \right|^2. \qquad (6.111)$$

We calculated the normalized scattering cross-section

$$\sigma_{s,h}^{norm} = \frac{\sigma_{s,h}}{\pi a^2} \qquad (6.112)$$

in the decibel scale, that is, the quantity $10 \log(\sigma_{s,h}^{norm})$. Note that this normalized scattering cross-section is different from the one used in Section 6.2.3 for the disk problem.

According to Section 6.3.1, the PO predicts the following approximation

$$\sigma^{(0)} = \sigma_s^{(0)} = \sigma_h^{(0)} = \pi a^2 \left| \frac{i}{2ka} \tan^2 \omega (1 - e^{i2kl}) - \tan \omega \, e^{i2kl} \right|^2. \qquad (6.113)$$

In the limiting case when $\omega \to \pi/2$,

$$\sigma^{(0)} = \pi a^2 (ka)^2. \tag{6.114}$$

Together with the contribution from the nonuniform scattering sources $j_{\mathrm{s,h}}^{(1)}$, the total backscattering cross-section equals

$$\sigma_{\mathrm{s}} = \pi a^2 \left| \frac{i}{2ka} \tan^2 \omega (1 - e^{i2kl}) \right.$$

$$\left. - \frac{2}{n} \left(\sin \frac{\pi}{n} \right) \left(\frac{1}{\cos \dfrac{\pi}{n} - 1} - \frac{1}{\cos \dfrac{\pi}{n} - \cos \dfrac{2\omega}{n}} \right) e^{i2kl} \right|^2 \tag{6.115}$$

and

$$\sigma_{\mathrm{h}} = \pi a^2 \left| \frac{i}{2ka} \tan^2 \omega (1 - e^{i2kl}) \right.$$

$$\left. + \frac{2}{n} \left(\sin \frac{\pi}{n} \right) \left(\frac{1}{\cos \dfrac{\pi}{n} - 1} + \frac{1}{\cos \dfrac{\pi}{n} - \cos \dfrac{2\omega}{n}} \right) e^{i2kl} \right|^2 \tag{6.116}$$

with $n = 1 + (\omega + \Omega)/\pi$.

In the limiting case when $\omega \to \pi/2$,

$$\sigma_{\mathrm{s}} = \pi a^2 \left| ika + \frac{\dfrac{2}{n} \sin \dfrac{\pi}{n}}{\cos \dfrac{\pi}{n} - 1} + \frac{1}{n} \cot \frac{\pi}{n} \right|^2, \tag{6.117}$$

and

$$\sigma_{\mathrm{h}} = \pi a^2 \left| -ika + \frac{\dfrac{2}{n} \sin \dfrac{\pi}{n}}{\cos \dfrac{\pi}{n} - 1} - \frac{1}{n} \cot \frac{\pi}{n} \right|^2 \tag{6.118}$$

with $n = (3/2) + (\Omega/\pi)$.

Finally when both ω and Ω are equal to $\pi/2$, and the cone transforms into the disk,

$$\sigma_{\mathrm{s}} = \pi a^2 |ika - 1|^2 \tag{6.119}$$

and

$$\sigma_{\mathrm{h}} = \pi a^2 |ika + 1|^2. \tag{6.120}$$

The normalized backscattering cross-section (6.112) of the cone was numerically analyzed as the function of three variables: the length l, the angle ω and the angle Ω.

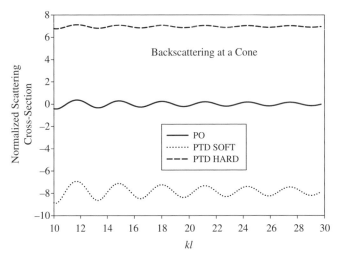

Figure 6.10   Backscattering at a cone: dependence on the cone length l. According to Equation (6.95), the PO curve also represents the scattering of electromagnetic waves from a perfectly conducting cone.

The results are presented in the decibel scale as follows:

- For the calculation of Equation (6.112) as a function of the length l, the variables were set as

$$\omega = 45°, \qquad \Omega = 90°, \qquad a = l \tan \omega = l, \qquad 10 \leq kl < 30.$$

In this case, $1.5\lambda < l < 4.8\lambda$ and $3\lambda < 2a < 9.6\lambda$. The results are demonstrated in Figure 6.10. As is seen here, the data for the hard cone are higher than those for the soft cone. The difference between them is about 15 dB. The PO data are approximately in the middle.

- For the calculation of Equation (6.112) as a function of the angle ω, the variables were set as

$$ka = 3\pi, \qquad 10° \leq \omega \leq 90°, \qquad \Omega = 90°.$$

In this case, $2a = 3\lambda$, $0 \leq l \leq 8.5\lambda$. The results are plotted in Figure 6.11. A big difference can be observed between the soft and hard data here, at about 40 dB for narrow cones.

- For the calculation of Equation (6.112) as a function of the angle Ω (Fig. 6.9), the variables were set as

$$\omega = 10°, \qquad ka = 3\pi, \qquad kl \simeq 17\pi, \qquad 0° \leq \Omega \leq \pi - \omega = 170°.$$

In this case, $2a = 3\lambda$ and $l \simeq 8.5\lambda$. The results are plotted in Figure 6.12. The PO approximation does not depend on the angle Ω, which is why it is represented here by a straight horizontal line. The difference between the soft and hard data is about 42–57 dB. The influence of the cone base shape approaches 11 dB for the soft cone and 16 dB for the hard cone.

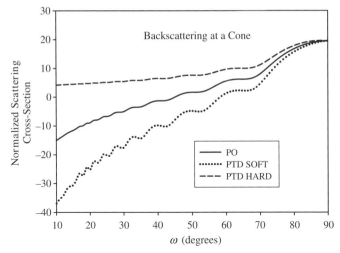

Figure 6.11   Backscattering at a cone: dependence on the vertex angle ω. According to Equation (6.95), the PO curve also represents the scattering of electromagnetic waves from a perfectly conducting cone.

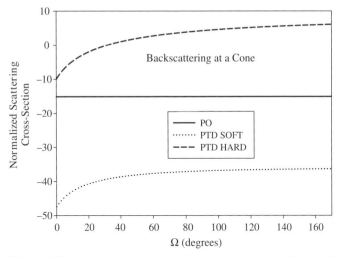

Figure 6.12   Backscattering at a cone: dependence on the angle Ω. According to Equation (6.95), the PO curve also represents the scattering of electromagnetic waves from a perfectly conducting cone.

6.4 BODIES OF REVOLUTION WITH NONZERO GAUSSIAN CURVATURE: BACKSCATTERED FOCAL FIELDS

This section studies symmetrical scattering at bodies of revolution whose illuminated side is an arbitrary smooth convex surface with nonzero Gaussian curvature. A generatrix of such a surface and related geometry are shown in Figure 6.13.

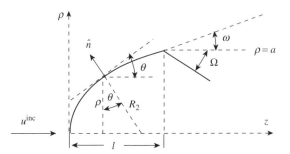

Figure 6.13   Generatrix of the body of revolution.

The incident plane wave (6.1) propagates in the positive direction of the z-axis, which represents the symmetry axis of a scattering object. We use two systems of coordinates: cylindrical coordinates ρ, φ, z and spherical coordinates r, ϑ, φ. The generatrix is given as a function $\rho = \rho(z)$. It is assumed that $\mathrm{d}^2\rho/\mathrm{d}z^2 \neq 0$ for the illuminated side ($0 \leq z \leq l$) of the object. This condition ensures that the Gaussian curvature of this surface is not zero. We also utilize the following denotations related to the edge points ($\rho = a$): $\mathrm{d}\rho/\mathrm{d}z = \tan\omega$ for the illuminated side ($z = l - 0$) and $\mathrm{d}\rho/\mathrm{d}z = -\tan\Omega$ for the shadowed side ($z = l + 0$). The shadowed side is an arbitrary smooth surface with $0 \leq \Omega \leq \pi - \omega$. In the limiting case $\Omega = \pi - \omega$, the scattering object is an infinitely thin (but still perfectly reflecting) screen $\rho = \rho(z)$ with $0 \leq z \leq l$.

The principal radii of curvature of the scattering surface are determined according to the differential geometry (Bronshtein and Semendyaev, 1985)

$$R_1 = \frac{\{1 + [\rho'(z)]^2\}^{3/2}}{|\rho''(z)|} \quad \text{and} \quad R_2 = \rho(z)\sqrt{1 + [\rho'(z)]^2}, \tag{6.121}$$

where $\rho' = \mathrm{d}\rho(z)/\mathrm{d}z$, $\rho'' = \mathrm{d}^2\rho(z)/\mathrm{d}z^2$. The radius R_1 relates to the normal section of the surface by the plane (ρ, z). The radius R_2 relates to the orthogonal normal section. As $\rho' = \tan\theta$ with $\omega \leq \theta \leq \pi/2$, the principal radii can be represented as

$$R_1 = \frac{1}{|\rho''(z)|\cos^3\theta} \quad \text{and} \quad R_2 = \frac{\rho(z)}{\cos\theta}. \tag{6.122}$$

The radius R_2 is shown in Figure 6.13. The Gaussian curvature is determined by

$$k_{\mathrm{G}} = \frac{1}{R_1 R_2} = \frac{|\rho''(z)|}{\rho(z)}\cos^4\theta. \tag{6.123}$$

The above expressions for R_1, R_2, and k_{G} become indefinite at the point $z = 0$. In order to disclose these indefinitenesses, one can use the alternative expressions

$$R_1 = \frac{\{1 + [z'(\rho)]^2\}^{3/2}}{z''(\rho)} \quad \text{and} \quad R_2 = \frac{\rho\sqrt{1 + [z'(\rho)]^2}}{z'(\rho)}, \tag{6.124}$$

where $z = z(\rho)$, $z' = \mathrm{d}z(\rho)/\mathrm{d}\rho$, and $z'' = \mathrm{d}^2 z(\rho)/\mathrm{d}\rho^2$. It follows from these equations that at the point $\rho = z = 0$

$$R_1 = R_2 = \frac{1}{z''(\rho)}. \tag{6.125}$$

The equality of the two principal radii means that the vertex point $\rho = z = 0$ of the scattering surface is an umbilic point.

6.4.1   PO Approximation

This is the field radiated by the uniform components (1.31) of the scattering sources

$$j_{\mathrm{s}}^{(0)} = i u_0 2 k n_z \mathrm{e}^{ikz} \quad \text{and} \quad j_{\mathrm{h}}^{(0)} = u_0 2 \mathrm{e}^{ikz}. \tag{6.126}$$

According to the differential geometry (Bronshtein and Semendyaev, 1985),

$$\hat{n} = \hat{x} \frac{\cos\psi}{\sqrt{1 + [\rho'(z)]^2}} + \hat{y} \frac{\sin\psi}{\sqrt{1 + [\rho'(z)]^2}} - \hat{z} \frac{\rho'(z)}{\sqrt{1 + [\rho'(z)]^2}} \tag{6.127}$$

or

$$\hat{n} = \hat{x} \cos\theta \cos\psi + \hat{y} \cos\theta \sin\psi - \hat{z} \sin\theta. \tag{6.128}$$

Here, we use the letter ψ for the polar coordinate of the scattering surface and retain the letter φ for the polar coordinate of the field point.

The scattered field in the far zone is determined by Equations (1.16) and (1.17), where one should set

$$\mathrm{d}s = \rho(z)\sqrt{1 + (\rho')^2}\,\mathrm{d}z\,\mathrm{d}\psi, \qquad r' \sin\vartheta' = \rho(z), \qquad r' \cos\vartheta' = z \tag{6.129}$$

and

$$\hat{m} = \nabla r = \hat{x} \sin\vartheta \cos\varphi + \hat{y} \sin\vartheta \sin\varphi + \hat{z} \cos\vartheta. \tag{6.130}$$

Due to the axial symmetry of the scattered field, it is sufficient to calculate the field only in the meridian plane $\varphi = \pi/2$. Taking this into account, one has

$$\hat{m} \cdot \hat{n} = \frac{\sin\vartheta \sin\psi - \rho'(z) \cos\vartheta}{\sqrt{1 + [\rho'(z)]^2}}, \tag{6.131}$$

$$u_{\mathrm{s}}^{(0)} = u_0 \frac{ik}{2\pi} \frac{\mathrm{e}^{ikR}}{R} \int_0^l \mathrm{e}^{ikz(1-\cos\vartheta)} \rho(z) \rho'(z) \mathrm{d}z \int_0^{2\pi} \mathrm{e}^{-ik\rho(z) \sin\vartheta \sin\psi} \mathrm{d}\psi \tag{6.132}$$

and

$$u_{\mathrm{h}}^{(0)} = -u_0 \frac{ik}{2\pi} \frac{\mathrm{e}^{ikR}}{R} \int_0^l \mathrm{e}^{ikz(1-\cos\vartheta)} \rho(z) \mathrm{d}z$$

$$\times \int_0^{2\pi} \mathrm{e}^{-ik\rho(z) \sin\vartheta \sin\psi} [\sin\vartheta \sin\psi - \rho'(z) \cos\vartheta] \mathrm{d}\psi \tag{6.133}$$

It follows from these equations that in the forward focal direction ($\vartheta = 0$),

$$u_s^{(0)} = u_h^{(0)} = u_0 \frac{ika^2}{2} \frac{e^{ikR}}{R}. \tag{6.134}$$

For the backscattering direction ($\vartheta = \pi$), the expressions (6.132) and (6.133) are reduced to

$$u_s^{(0)} = -u_h^{(0)} = u_0 ik \frac{e^{ikR}}{R} \int_0^l e^{i2kz} \rho(z)\rho'(z)\mathrm{d}z. \tag{6.135}$$

By integrating by parts, one obtains the following asymptotic estimations:

$$u_s^{(0)} = -u_h^{(0)} = -u_0 \frac{1}{2} \left[\rho(0)\rho'(0) - \rho(l)\rho'(l)e^{i2kl} + O\left(\frac{1}{k}\right) \right] \frac{e^{ikR}}{R}. \tag{6.136}$$

The indefiniteness for the first term in the square brackets is disclosed by the following manipulations:

$$\rho(0)\rho'(0) = \lim_{z \to 0} \rho(z)\frac{\mathrm{d}\rho(z)}{\mathrm{d}z}$$

$$= \lim_{\rho \to 0} \frac{\rho}{\mathrm{d}z(\rho)/\mathrm{d}\rho} = \lim_{\rho \to 0} \frac{1}{\mathrm{d}^2 z(\rho)/\mathrm{d}\rho^2} = \frac{1}{z''(0)}. \tag{6.137}$$

In view of Equation (6.125), this quantity determines the radius of curvature of the scattering surface at the point $z = 0$. Now Equation (6.136) can be written as

$$u_s^{(0)} = -u_h^{(0)} = -u_0 \frac{1}{2} \left[\frac{1}{z''(0)} - a \tan \omega e^{i2kl} \right] \frac{e^{ikR}}{R}. \tag{6.138}$$

Here the first term represents the ordinary ray reflected from the vertex of the body, and the second term is the first-order focal field generated by the uniform components $j_{s,h}^{(0)}$ of the scattering sources induced near the circular edge ($z = l$). Indeed, due to Equations (6.125) and (6.138), the backscattering cross-section (1.26) related to the reflection from the body vertex equals

$$\sigma = \pi \frac{1}{[z''(0)]^2} = \pi R_{1,2}^2 \tag{6.139}$$

and totally agrees with Equation (1.27). We note that, according to Equations (1.100) and (1.101), Equations (6.138) and (6.139) are also valid for electromagnetic waves scattered from perfectly conducting objects.

6.4.2 Total Backscattered Focal Field: First-Order PTD Asymptotics

In this approximation, the total scattered field in the direction $\vartheta = \pi$ is found by summation of its components (6.138) and (6.41), (6.42) generated by the uniform and nonuniform scattering sources $j_{s,h}^{(0)}$ and $j_{s,h}^{(1)}$, respectively. Note that functions $f^{(1)}$ and $g^{(1)}$ in Equations (6.41) and (6.42) are determined for the case $\vartheta = \pi$ by Equations (6.102) and (6.103). The summation results in the following asymptotic expressions:

$$
u_s = -\frac{u_0}{2}\left[\frac{1}{z''(0)} - a\frac{2\sin\dfrac{\pi}{n}}{n}\left(\frac{1}{\cos\dfrac{\pi}{n} - 1} - \frac{1}{\cos\dfrac{\pi}{n} - \cos\dfrac{2\omega}{n}}\right)e^{i2kl}\right]\frac{e^{ikR}}{R}
\tag{6.140}
$$

and

$$
u_h = \frac{u_0}{2}\left[\frac{1}{z''(0)} + a\frac{2\sin\dfrac{\pi}{n}}{n}\left(\frac{1}{\cos\dfrac{\pi}{n} - 1} + \frac{1}{\cos\dfrac{\pi}{n} - \cos\dfrac{2\omega}{n}}\right)e^{i2kl}\right]\frac{e^{ikR}}{R},
\tag{6.141}
$$

with $n = 1 + (\omega + \Omega)/\pi$ where $0 \le \omega \le \pi/2$ and $0 \le \Omega \le \pi - \omega$. In the next two sections we consider the backscattering from two specific bodies of revolution.

6.4.3 Backscattering from Paraboloids

The PO approximations for acoustic and electromagnetic fields in this problem are identical. Focal asymptotics for the *total* acoustic and electromagnetic fields are different.

Asymptotics for the Scattered Field

The illuminated surface of the scattering object is a paraboloid with the generatrix given by the equation

$$
\rho^2(z) = 2pz,
\tag{6.142}
$$

where $0 \le z \le l$. The focus of a paraboloid is located at the point $z = p/2$. According to Equation (6.125), the focal parameter p equals the radius of the paraboloid curvature at the vertex point $z = 0$. The shape of the shadowed side of the scattering object and its other geometric characteristics are shown in Figure 6.13.

Due to Equations (6.138) and (6.140), (6.141), the focal backscattered field is described by the following asymptotic expressions, where the first relates to the PO part of the field:

$$u_{\rm s}^{(0)} = -u_{\rm h}^{(0)} = -\frac{u_0}{2}\left(p - a \tan \omega \, e^{i2kl}\right)\frac{e^{ikR}}{R}. \tag{6.143}$$

The next two expressions represent the first-order PTD approximations:

$$u_{\rm s} = -\frac{u_0}{2}\left[p - a\frac{2\sin\dfrac{\pi}{n}}{n}\left(\frac{1}{\cos\dfrac{\pi}{n} - 1} - \frac{1}{\cos\dfrac{\pi}{n} - \cos\dfrac{2\omega}{n}}\right)e^{i2kl}\right]\frac{e^{ikR}}{R} \tag{6.144}$$

and

$$u_{\rm h} = \frac{u_0}{2}\left[p + a\frac{2\sin\dfrac{\pi}{n}}{n}\left(\frac{1}{\cos\dfrac{\pi}{n} - 1} + \frac{1}{\cos\dfrac{\pi}{n} - \cos\dfrac{2\omega}{n}}\right)e^{i2kl}\right]\frac{e^{ikR}}{R}, \tag{6.145}$$

with $n = 1 + (\omega + \Omega)/\pi$. Comparison with the electromagnetic PO field scattered by a perfectly conducting paraboloid (Equation (2.5.3) in Ufimtsev (2003)) reveals the following relationships:

and
$$E_x^{(0)} = u_{\rm s}^{(0)}, \qquad \text{if } E_x^{(0)} = u^{\rm inc}$$

$$H_y^{(0)} = u_{\rm h}^{(0)}, \qquad \text{if } H_y^{(0)} = u^{\rm inc}. \tag{6.146}$$

This result is in complete agreement with the general relationships (1.100) and (1.101).

Taking into account that $\rho'(l) = p/a = \tan \omega$ and $p = a \tan \omega$, the above expressions can be rewritten as

$$u_{\rm s}^{(0)} = -u_{\rm h}^{(0)} = -u_0\frac{a}{2}\tan\omega(1 - e^{i2kl})\frac{e^{ikR}}{R} \tag{6.147}$$

and

$$u_s = -u_0 \frac{a}{2} \left[\tan \omega - \frac{2 \sin \frac{\pi}{n}}{n} \left(\frac{1}{\cos \frac{\pi}{n} - 1} - \frac{1}{\cos \frac{\pi}{n} - \cos \frac{2\omega}{n}} \right) e^{i2kl} \right] \frac{e^{ikR}}{R},$$

(6.148)

$$u_h = u_0 \frac{a}{2} \left[\tan \omega + \frac{2 \sin \frac{\pi}{n}}{n} \left(\frac{1}{\cos \frac{\pi}{n} - 1} + \frac{1}{\cos \frac{\pi}{n} - \cos \frac{2\omega}{n}} \right) e^{i2kl} \right] \frac{e^{ikR}}{R}.$$

(6.149)

These equations are useful for investigation of the continuous transformation of the paraboloid into the flat disk when $\omega \to \pi/2$. Utilizing the relationship $l = a^2/2p = (a \cot \omega)/2$, one can show that in the limiting case $\omega = \pi/2$

$$u_s^{(0)} = -u_h^{(0)} = u_0 \frac{ika^2}{2} \frac{e^{ikR}}{R}$$

(6.150)

and

$$u_s = u_0 \frac{a}{2} \left(ika + \frac{1}{n} \cot \frac{\pi}{n} + \frac{\frac{2}{n} \sin \frac{\pi}{n}}{\cos \frac{\pi}{n} - 1} \right) \frac{e^{ikR}}{R},$$

(6.151)

$$u_h = -u_0 \frac{a}{2} \left(ika + \frac{1}{n} \cot \frac{\pi}{n} - \frac{\frac{2}{n} \sin \frac{\pi}{n}}{\cos \frac{\pi}{n} - 1} \right) \frac{e^{ikR}}{R}$$

(6.152)

with $n = 3/2 + \Omega/\pi$.

The distinguishing feature of the PO field (6.147) is its oscillations with the pure zeros corresponding to parameters $kl = m\pi$ where $m = 1, 2, 3, \ldots$. Other properties of the scattered field are illustrated in the next section.

Numerical Analysis of Backscattering

Here we calculate the normalized scattering cross-section (6.112), which is determined in terms of the scattered field by Equations (6.110) and (6.111). According to section (Asymptotics for the Scattered field), one can derive the following expressions for the scattering cross-section.

The PO approximation is given by

$$\sigma_s^{(0)} = \sigma_h^{(0)} = \pi a^2 \left| \tan \omega \cdot (1 - e^{i2kl}) \right|^2$$

(6.153)

and the first-order PTD by

$$\sigma_s = \pi a^2 \left| \tan \omega - \frac{2 \sin \frac{\pi}{n}}{n} \left(\frac{1}{\cos \frac{\pi}{n} - 1} - \frac{1}{\cos \frac{\pi}{n} - \cos \frac{2\omega}{n}} \right) e^{i2kl} \right|^2 \qquad (6.154)$$

and

$$\sigma_h = \pi a^2 \left| \tan \omega + \frac{2 \sin \frac{\pi}{n}}{n} \left(\frac{1}{\cos \frac{\pi}{n} - 1} + \frac{1}{\cos \frac{\pi}{n} - \cos \frac{2\omega}{n}} \right) e^{i2kl} \right|^2, \qquad (6.155)$$

with $n = 1 + (\omega + \Omega)/\pi$.

When $l \to 0$ and $\omega \to \pi/2$, the above expressions transform into

$$\sigma^{(0)} = \pi a^2 (ka)^2, \qquad (6.156)$$

$$\sigma_s = \pi a^2 \left| ika + \frac{1}{n} \cot \frac{\pi}{n} + \frac{\frac{2}{n} \sin \frac{\pi}{n}}{\cos \frac{\pi}{n} - 1} \right|^2, \qquad (6.157)$$

and

$$\sigma_h = \pi a^2 \left| ika + \frac{1}{n} \cot \frac{\pi}{n} - \frac{\frac{2}{n} \sin \frac{\pi}{n}}{\cos \frac{\pi}{n} - 1} \right|^2, \qquad (6.158)$$

with $n = 3/2 + \Omega/\pi$.

Here we present three types of calculation similar to those in Section 6.3.2 for the cone. The first type consists of calculations for conformal paraboloids, which differ by their length; the second type relates to the transformation of paraboloids into the disk; and the third type reveals the influence of the shadowed base of paraboloids on backscattering. The calculated quantity is the normalized scattering cross-section (6.112).

- In the study of conformal paraboloids, the focal parameter is set constant ($kp = 3\pi \tan 14°$) and the length of paraboloids changes in the interval $6\pi \leq kl \leq 36$. In this case, the radius of the paraboloid base and its length are in the intervals

$$1.5\lambda \leq a \leq 2\lambda \qquad \text{and} \qquad 3\lambda \leq l \leq 5.8\lambda.$$

For a given frequency ($k = $ const.), the focal parameter p is constant for all paraboloids with different l. This condition actually means that all these paraboloids are just the different sections of the same semi-infinite paraboloid. This is why they are called conformal.

According to the relationship $l = a^2/2p = (a \cot \omega)/2$, the angle ω (Fig. 6.13) is determined by the equation $\cot \omega = \sqrt{2kl/ka}$. For the given values of kl, this angle is in the interval $10.24° \le \omega \le 14°$. For the angle Ω (Fig. 6.13), we take its value as $90°$. The results of calculations are plotted in Figure 6.14. The difference between the soft and hard data approaches 16–19 dB. Figure 6.14 demonstrates the rough PO data, which do not depend at all on the boundary conditions and are totally incorrect in the vicinity of minima.

- The next topic is the transformation of the paraboloid into the disk. In this process, each intermediate shape between the initial parabolid and the final disk is a paraboloid whose focal parameter p depends on its length l. It follows from the equation $l = a^2/2p$ that $p = a^2/2l$. To find the angle ω (Fig. 6.13), we use the additional equation $p = a \tan \omega$ and obtain $\tan \omega = a/2l$, or $\tan \omega = ka/2kl$. For the parameter ka we take the constant value $ka = 3\pi$, which does not depend on the paraboloid length. In this case, the diameters of all the intermediate paraboloids and the final disk equal $2a = 3\lambda$. For the initial paraboloid (which is transformed into the disk), we take $kl = 6\pi$, that is, $l = 3\lambda$. The results are plotted in Figure 6.15.

- Now we consider the influence of the paraboloid shadowed base on backscattering. The illuminated part of the object under investigation is the paraboloid with parameters $ka = 3\pi$ and $kl = 6\pi$, when the base diameter and the length of the paraboloid are equal to each other: $2a = l = 3\lambda$. The critical parameter of the base, which affects the edge wave, is the angle Ω. It changes from zero to $\Omega = \pi - \omega$, where $\omega = \tan^{-1}(ka/2kl) \simeq 14°$. In the limiting

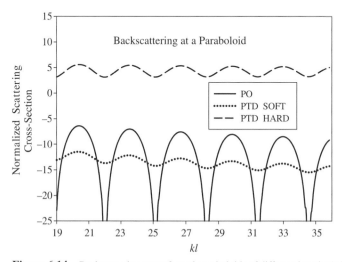

Figure 6.14   Backscattering at conformal paraboloids of different lengths (with constant focal parameter p). According to Equation (6.146), the PO curve also represents scattering of electromagnetic waves from perfectly conducting paraboloids.

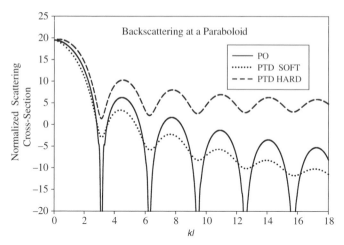

Figure 6.15   Transformation of the paraboloid into a disk with continuous maintaining of the paraboloidial shape. According to Equation (6.146), the PO curve also represents scattering of electromagnetic waves from perfectly conducting paraboloids.

case $\Omega = \pi - \omega$, the scattering object transforms into the perfectly reflecting infinitely thin screen. The results of this investigaton are shown in Figure 6.16.

The PO approximation does not depend on the shape of the shadowed part of the paraboloid. For the chosen parameter $kl = 6\pi$, it predicts the zero value for the scattered field. In the decibel scale it equals minus infinity and it is outside the figure area. As is seen in this figure, the backscattering from a soft paraboloid depends on the angle Ω only a little (the change of scattering

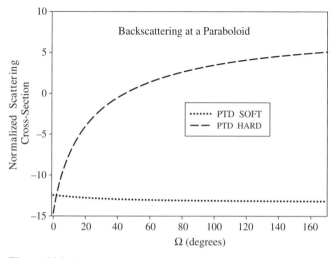

Figure 6.16   Influence of the base shape on backscattering.

cross-section is about 3 dB), although a strong dependence is observed for the hard paraboloid (about 20 dB).

6.4.4 Backscattering from Spherical Segments

The PO approximations for acoustic and electromagnetic fields in this problem are identical. Focal asymptotics for the *total* acoustic and electromagnetic fields are different.

Asymptotics for the Scattered Field

The illuminated surface of the scattering object is a spherical segment whose generatrix is shown in Figure 6.17 and given by the equation

$$z(\rho) = b - \sqrt{b^2 - \rho^2}, \qquad \text{with } 0 \le z \le l, \tag{6.159}$$

where b is the radius of the spherical surface. It follows from Equation (6.159) that

$$z'(\rho) = \frac{dz(\rho)}{d\rho} = \frac{\rho}{\sqrt{b^2 - \rho^2}} = \cot \theta \tag{6.160}$$

and

$$z''(\rho) = \frac{d^2 z(\rho)}{d\rho^2} = \frac{b^2}{(b^2 - \rho^2)^{3/2}}, \qquad z''(0) = \frac{1}{b}. \tag{6.161}$$

The angle $\theta(z)$ in Equation (6.160) is displayed in Figure 6.13. At the point $z = l$, $\rho = a$, this angle equals $\theta(l) = \omega$. For the given quantities b and a, the angle ω and the segment length l are defined by equations

$$\tan \omega = \frac{\sqrt{b^2 - a^2}}{a} \qquad \text{and} \qquad l = b - \sqrt{b^2 - a^2}, \tag{6.162}$$

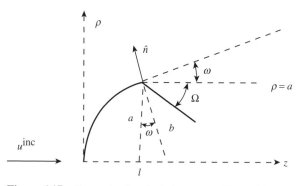

Figure 6.17   Generatrix of a spherical segment with an arbitrary shadowed base.

but for the given ω and a, the segment radius b and the length l are determined as

$$b = \frac{a}{\cos \omega} \qquad l = a \frac{1 - \sin \omega}{\cos \omega}. \qquad (6.163)$$

The last relationships are helpful for the investigation of the continuous transformation of the spherical segment into the flat disk when $\omega \to \pi/2$, $l \to 0$, and $a = $ const.

According to Equations (6.138), (6.161), and (6.163), the field $u_{s,h}^{(0)}$ is determined by

$$u_s^{(0)} = -u_h^{(0)} = -u_0 \frac{1}{2}(b - a \tan \omega \, e^{i2kl}) \frac{e^{ikR}}{R} \qquad (6.164)$$

or by

$$u_s^{(0)} = -u_h^{(0)} = -u_0 \frac{a}{2\cos \omega}(1 - \sin \omega e^{i2kl}) \frac{e^{ikR}}{R}. \qquad (6.165)$$

Comparison with the electromagnetic PO field scattered by a perfectly conducting spherical segment (Equation (2.6.4) in Ufimtsev (2003)) reveals the following relationships:

and

$$E_x^{(0)} = u_s^{(0)}, \qquad \text{if } E_x^{(0)} = u^{\text{inc}}$$

$$H_y^{(0)} = u_h^{(0)}, \qquad \text{if } H_y^{(0)} = u^{\text{inc}}. \qquad (6.166)$$

This result is in a complete agreement with the general relationships (1.100) and (1.101).

In view of Equations (6.140), (6.141), (6.161), and (6.163) the field $u_{s,h} = u_{s,h}^{(0)} + u_{s,h}^{(1)}$ is described by

$$u_s = -u_0 \frac{1}{2}\left[b - a \frac{2 \sin \frac{\pi}{n}}{n}\left(\frac{1}{\cos \frac{\pi}{n} - 1} - \frac{1}{\cos \frac{\pi}{n} - \cos \frac{2\omega}{n}} \right) e^{i2kl} \right] \frac{e^{ikR}}{R}$$

$$(6.167)$$

and

$$u_h = u_0 \frac{1}{2}\left[b + a \frac{2 \sin \frac{\pi}{n}}{n}\left(\frac{1}{\cos \frac{\pi}{n} - 1} + \frac{1}{\cos \frac{\pi}{n} - \cos \frac{2\omega}{n}} \right) e^{i2kl} \right] \frac{e^{ikR}}{R}, \qquad (6.168)$$

or by

$$
u_s = -u_0 \frac{a}{2} \left[\frac{1}{\cos\omega} - \frac{2\sin\frac{\pi}{n}}{n} \left(\frac{1}{\cos\frac{\pi}{n} - 1} - \frac{1}{\cos\frac{\pi}{n} - \cos\frac{2\omega}{n}} \right) e^{i2kl} \right] \frac{e^{ikR}}{R}
$$

(6.169)

and

$$
u_h = u_0 \frac{a}{2} \left[\frac{1}{\cos\omega} + \frac{2\sin\frac{\pi}{n}}{n} \left(\frac{1}{\cos\frac{\pi}{n} - 1} + \frac{1}{\cos\frac{\pi}{n} - \cos\frac{2\omega}{n}} \right) e^{i2kl} \right] \frac{e^{ikR}}{R},
$$

(6.170)

with $n = 1 + (\omega + \Omega)/\pi$, where $0 \leq \omega \leq \pi/2$ and $0 \leq \Omega \leq \pi - \omega$.

In the limiting case when the spherical segment continuously transforms into the flat disk ($\omega \to \pi/2$ and $l \to 0$), Equations (6.164), (6.165), (6.166) and (6.167), (6.168) are exactly reduced to Equations (6.150) and (6.151) and (6.152), respectively.

Numerical Analysis of Backscattering

In this section, we calculate the normalized scattering cross-section (6.112), taking into account Equations (6.110) and (6.111). According to the previous section, the following expressions for the ordinary scattering cross-section are valid. The PO approximation is given by

$$
\sigma_s^{(0)} = \sigma_h^{(0)} = \pi a^2 \left| \frac{1}{\cos\omega} - \tan\omega \cdot e^{i2kl} \right|^2
$$

(6.171)

and the first-order PTD by

$$
\sigma_s = \pi a^2 \left| \frac{1}{\cos\omega} - \frac{2\sin\frac{\pi}{n}}{n} \left(\frac{1}{\cos\frac{\pi}{n} - 1} - \frac{1}{\cos\frac{\pi}{n} - \cos\frac{2\omega}{n}} \right) e^{i2kl} \right|^2
$$

(6.172)

and

$$
\sigma_h = \pi a^2 \left| \frac{1}{\cos\omega} + \frac{2\sin\frac{\pi}{n}}{n} \left(\frac{1}{\cos\frac{\pi}{n} - 1} + \frac{1}{\cos\frac{\pi}{n} - \cos\frac{2\omega}{n}} \right) e^{i2kl} \right|^2 ,
$$

(6.173)

with $n = 1 + (\omega + \Omega)/\pi$.

Two types of calculation are presented here.

- The first is the continuous transformation of the spherical segment into the flat disk in the limiting case $\omega \to \pi/2$. It is assumed that all transition surfaces are

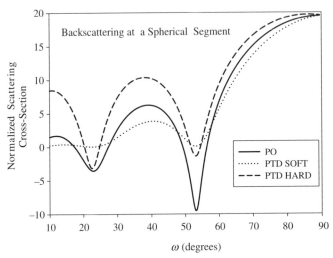

Figure 6.18   Transformation of the spherical segment into a flat disk. According to Equation (6.166), the PO curve also represents the scattering of electromagnetic waves from a perfectly conducting spherical segment.

spherical with the curvature radius $b = a/\cos\omega$. The initial object is given by the parameters $ka = 3\pi$, $kl_0 \simeq 2.5\pi$, $\omega_0 = 10°$, and $\Omega = 90°$. In terms of the wavelength, the base radius and the length of the spherical element are equal to $a = 1.5\lambda$ and $l_0 \simeq 1.26\lambda$. The numerical results for the normalized scattering cross-section are plotted in Figure 6.18. It clearly shows the influence of the

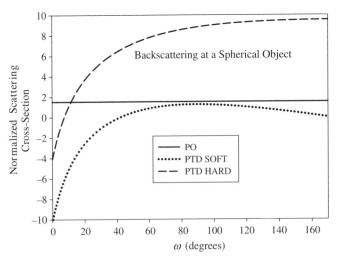

Figure 6.19   Influence of the shadowed part of the spherical segment on backscattering. According to Equation (6.166), the PO curve also represents the scattering of electromagnetic waves from a perfectly conducting spherical segment.

nonuniform component of the scattering sources $(j_{s,h}^{(1)})$ concentrated in the vicinity of the edge.

- In the next calculation we investigate the influence of the shadowed part of the spherical segment on backscattering. The critical parameter of this part is the angle Ω shown in Figure 6.17. It changes from zero to $\pi - \omega$. In the limiting case, when $\Omega = \pi - \omega$, the scattering object transforms into the perfectly reflecting infinitely thin screen. The illuminated spherical part of the segment is determined by the parameters $\omega = 10°$, $ka = 3\pi$, $kl \simeq 2.5\pi$. In terms of the wavelength, $a = 1.5\lambda$ and $l \simeq 1.26\lambda$. The results are plotted in Figure 6.19.

  The PO approximation does not depend on the shape of the shadowed part and it is represented in Figure 6.19 by the horizontal straight line. However, according to PTD, the backscattering increases up to 10 dB for the acoustically soft object and up to 13.5 dB for the acoustically hard object.

6.5 BODIES OF REVOLUTION WITH NONZERO GAUSSIAN CURVATURE: AXIALLY SYMMETRIC BISTATIC SCATTERING

The geometry of the problem is illustrated in Figure 6.20. The incident plane wave (6.1) propagates in the positive direction of the z-axis, which is the symmetry axis of a scattering body of revolution. The generatrix of the illuminated side of this body is given by the equation $\rho = \rho(z)$, with $0 \leq z \leq l$, under the condition $d^2\rho/dz^2 \neq 0$. This condition ensures that the Gaussian curvature of this surface is not zero. The shadowed side is an arbitrary smooth surface with $0 \leq \Omega \leq \pi - \omega$. In the limiting case $\Omega = \pi - \omega$, the scattering object is an infinitely thin perfectly reflecting screen with $\rho = \rho(z)$ and $0 \leq z \leq l$. The tangent to the generatrix forms the angle θ with the z-axis. At the edge points $z = l - 0$, this angle equals $\theta(l) = \omega = \tan^{-1}[d\rho(l)/dz]$. The principal radii of curvature (R_1, R_2) are defined in Equations (6.121) and (6.122), and the Gaussian curvature is given by Equation (6.123). The unit normal $\hat{n}$ to the illuminated surface is defined in Equation (6.128). Due to the axial symmetry of the

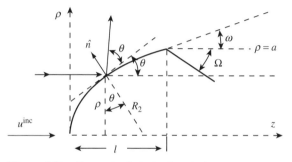

Figure 6.20   Generatrix of a body of revolution.

problem, it is sufficient to calculate the scattered field only in the meridian plane $\varphi = \pi/2$.

6.5.1 Ray Asymptotics for the PO Field

These asymptotics can be derived from the general integral expressions (6.132), (6.133) under the condition $k\rho \sin \vartheta \gg 1$. First we consider the observation points in the region $\pi - \omega < \vartheta \leq \pi$, where the entire illuminated surface of the scattering object is visible (Fig. 6.21).

The integrals in Equations (6.132), (6.133) over the variable ψ are calculated by the stationary phase technique (Copson, 1965; Murray, 1984). The details of this method were briefly explained in Section 6.1.2. The phase function in these integrals has two stationary points: $\psi_1 = \pi/2$ and $\psi_2 = 3\pi/2$. Asymptotic evaluation of these integrals leads to the expressions

$$u_s^{(0)} = u_0 \frac{ik}{\sqrt{2\pi k \sin \vartheta}} \frac{e^{ikR}}{R} \left[e^{i\pi/4} \int_0^l e^{ik\Phi_1(z)} \sqrt{\rho(z)} \rho'(z) dz \right.$$

$$\left. + e^{-i\pi/4} \int_0^l e^{ik\Phi_2(z)} \sqrt{\rho(z)} \rho'(z) dz \right] \tag{6.174}$$

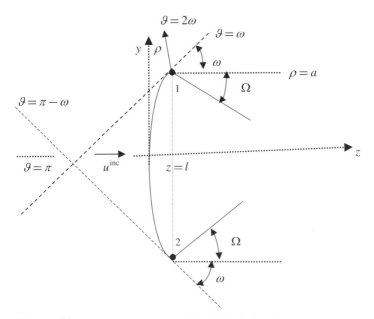

Figure 6.21   Cross-section of a body of revolution by the plane yoz.

and

$$u_h^{(0)} = -u_0 \frac{ik}{\sqrt{2\pi k \sin \vartheta}} \frac{e^{ikR}}{R} \left\{ e^{i\pi/4} \int_0^l e^{ik\Phi_1(z)} \sqrt{\rho(z)} [\sin \vartheta - \rho'(z) \cos \vartheta] dz \right.$$

$$\left. - e^{-i\pi/4} \int_0^l e^{ik\Phi_2(z)} \sqrt{\rho(z)} [\sin \vartheta + \rho'(z) \cos \vartheta] dz \right\} \tag{6.175}$$

where

$$\Phi_1(z) = z(1 - \cos \vartheta) - \rho(z) \sin \vartheta \tag{6.176}$$

and

$$\Phi_2(z) = z(1 - \cos \vartheta) + \rho(z) \sin \vartheta. \tag{6.177}$$

Here the integrals with the factor $\exp[\Phi_1(z)]$ represent the field generated by the vicinity of the stationary line $\psi = \pi/2, 0 < z \le l$, and the integrals with $\exp[\Phi_2(z)]$ describe the field generated by the vicinity of the stationary line $\psi = 3\pi/2, 0 < z \le l$ (Fig. 6.21).

Now we check the functions $\Phi_1(z)$ and $\Phi_2(z)$ for the presence of stationary points z_{st}. It follows from the equation

$$\Phi_1'(z) = 1 - \cos \vartheta - \rho'(z_{st}) \sin \vartheta = 0 \tag{6.178}$$

that

$$\rho'(z_{st}) = d\rho/dz = \tan \theta(z_{st}) = \tan(\vartheta/2) \tag{6.179}$$

and

$$\theta(z_{st}) = \vartheta/2. \tag{6.180}$$

This equation determines the reflection point z_{st} on the scattering surface shown in Figure 6.20. At this point, the tangent to the generatrix $\rho(z)$ forms the angle $\theta = \vartheta/2$ with the z-axis, which agrees with the reflection law.

We then check the function $\Phi_2(z)$ for the stationary point. It follows from the equation

$$\Phi_2'(z) = 1 - \cos \vartheta + \rho'(z_{st}) \sin \vartheta = 0 \tag{6.181}$$

that

$$\rho'(z_{st}) = d\rho/dz = \tan \theta(z_{st}) = -\tan(\vartheta/2) \tag{6.182}$$

and

$$\theta(z_{st}) = -\vartheta/2, \qquad \text{with } -\pi < \vartheta < 0. \tag{6.183}$$

This stationary point relates to the reflected ray in the meridian plane $\varphi = 3\pi/2$. For the same value z_{st} in Equations (6.180) and (6.183), this ray is exactly symmetrical to the reflected ray shown in Figure 6.20. As we consider the scattered field only in the meridian plane $\varphi = \pi/2$, the function $\Phi_2(z)$ does not have any stationary points for the scattering directions in this plane.

By introducing into Equations (6.174) and (6.175) a new integration variable $\xi = z$ for the integrals with function $\Phi_1(z)$, and $\xi = -z$ for the integrals with function $\Phi_2(z)$, one can represent their sum as

$$I(P) = \int_{-l}^{l} F(\xi, P)e^{ik\Phi(\xi,P)}d\xi, \tag{6.184}$$

where symbol P denotes the location of the observation point. For a high frequency of the field (when $k \gg 1$), the factor $\exp[\Phi(\xi, P)]$, being a function of the integration variable ξ, undergoes fast oscillations. Because of this, most differential contributions $F(\xi, P)e^{ik\Phi(\xi,P)}d\xi$ to the integral $I(P)$ asymptotically cancel each other. Only those that are in the vicinity of the stationary point ξ_{st} and in the vicinity of the end points $\xi = -l$ and $\xi = l$ provide substantial contributions to $I(P)$. The contribution of the stationary point is calculated by the stationary phase technique, and the contributions by the end points are found by integrating by parts (Copson, 1965; Murray, 1984). The resulting asymptotic approximation for $I(P)$ is given by

$$\begin{aligned} I(P) \sim & \sqrt{\frac{2\pi}{k\Phi''(\xi_{st}, P)}} F(\xi_{st}, P)e^{ik\Phi(\xi_{st},P)+i\pi/4} \\ & + \frac{1}{ik}\left[\frac{F(l,P)}{\Phi'(l,P)}e^{ik\Phi(l,P)} - \frac{F(-l,P)}{\Phi'(-l,P)}e^{ik\Phi(-l,P)} \right]. \end{aligned} \tag{6.185}$$

The first term in Equation (6.185) represents the contribution from the stationary point, and the rest of the terms provide the contributions from the end points. Only the dominant asymptotic terms for each contribution are retained here.

The outlined procedure was used to represent the scattered field $u_{s,h}^{(0)}$ in the form of three contributions:

$$u_{s,h}^{(0)} = u_{s,h}^{(0)}(z_{st}) + u_{s,h}^{(0)}(1) + u_{s,h}^{(0)}(2), \tag{6.186}$$

where

$$u_s^{(0)}(z_{st}) = -u_h^{(0)}(z_{st}) = -u_0\frac{1}{2}\sqrt{R_1(z_{st})R_2(z_{st})}e^{ik\Phi_1(z_{st})}\frac{e^{ikR}}{R}, \tag{6.187}$$

and

$$u_s^{(0)}(1) + u_s^{(0)}(2) = \frac{u_0 a}{\sqrt{2\pi ka \sin \vartheta}} \left[f^{(0)}(1) e^{-ika \sin \vartheta + i\pi/4} \right.$$

$$\left. + f^{(0)}(2) e^{ika \sin \vartheta - i\pi/4} \right] \frac{e^{ikR}}{R} e^{ikl(1-\cos \vartheta)}, \qquad (6.188)$$

$$u_h^{(0)}(1) + u_h^{(0)}(2) = \frac{u_0 a}{\sqrt{2\pi ka \sin \vartheta}} \left[g^{(0)}(1) e^{-ika \sin \vartheta + i\pi/4} \right.$$

$$\left. + g^{(0)}(2) e^{ika \sin \vartheta - i\pi/4} \right] \frac{e^{ikR}}{R} e^{ikl(1-\cos \vartheta)}. \qquad (6.189)$$

The functions $u_{s,h}^{(0)}(z_{st})$ describe the ordinary ray reflected at the stationary point z_{st} determined by Equations (6.179) and (6.180). The quantities R_1 and R_2 are the principal radii of curvature of the scattering surface. They are defined by Equations (6.121) and (6.122).

Expressions (6.188) and (6.189) determine the sum of two edge-diffracted rays diverging from the edge points 1 and 2 shown in Figure 6.21. The directivity patterns of these rays are defined by the functions

$$f^{(0)}(1) = \frac{\sin \omega}{\cos \omega - \cos(\omega - \vartheta)}, \qquad f^{(0)}(2) = \frac{\sin \omega}{\cos \omega - \cos(\omega + \vartheta)} \qquad (6.190)$$

and

$$g^{(0)}(1) = \frac{\sin(\omega - \vartheta)}{\cos \omega - \cos(\omega - \vartheta)}, \qquad g^{(0)}(2) = \frac{\sin(\omega + \vartheta)}{\cos \omega - \cos(\omega + \vartheta)}. \qquad (6.191)$$

6.5.2   Bessel Interpolations for the PO Field in the Region $\pi - \omega \leq \vartheta \leq \pi$

With the application of relationships (6.45) and (6.46), the above ray asymptotics can be written in the form

$$u_s^{(0)} = u_0 \left[\left[-\frac{1}{2} \sqrt{R_1(z_{st}) R_2(z_{st})} e^{ik\Phi_1(z_{st})} \right. \right.$$

$$+ \frac{a}{2} \{ f^{(0)}(1) [J_0(ka \sin \vartheta) - iJ_1(ka \sin \vartheta)]$$

$$\left. \left. + f^{(0)}(2) [J_0(ka \sin \vartheta) + iJ_1(ka \sin \vartheta)] \} e^{ikl(1-\cos \vartheta)} \right] \right] \frac{e^{ikR}}{R} \qquad (6.192)$$

and

$$u_{\mathrm{h}}^{(0)} = u_0 \left[\!\left[\frac{1}{2}\sqrt{R_1(z_{\mathrm{st}})R_2(z_{\mathrm{st}})}\; \mathrm{e}^{ik\Phi_1(z_{\mathrm{st}})} \right.\right.$$

$$+ \frac{a}{2}\{g^{(0)}(1)[J_0(ka\sin\vartheta) - iJ_1(ka\sin\vartheta)]$$

$$\left.\left. + g^{(0)}(2)[J_0(ka\sin\vartheta) + iJ_1(ka\sin\vartheta)]\} \, \mathrm{e}^{ikl(1-\cos\vartheta)} \right]\!\right] \frac{\mathrm{e}^{ikR}}{R}, \qquad (6.193)$$

where J_0 and J_1 are the Bessel functions. These asymptotics are valid away from the focal line ($\vartheta = \pi$) under the condition $ka\sin\vartheta \gg 1$. The focal field is described by the asymptotics (6.138), which can be rewritten as

$$u_{\mathrm{s}}^{(0)} = -u_{\mathrm{h}}^{(0)} = u_0 \left[-\frac{1}{2}\sqrt{R_1(0)R_2(0)} + \frac{a}{2}\tan\omega\, \mathrm{e}^{i2kl} \right] \frac{\mathrm{e}^{ikR}}{R}, \qquad (6.194)$$

where $R_1(0) = R_2(0) = 1/z''(0)$.

When $\vartheta \to \pi$ the above asymptotics (6.192) and (6.193) exactly transform into the focal asymptotics (6.194). Therefore, the expressions (6.192) and (6.193) can be considered as the appropriate approximations valid in the entire region $\pi - \omega \le \vartheta \le \pi$.

6.5.3   Bessel Interpolations for the PTD Field in the Region $\pi - \omega \le \vartheta \le \pi$

The PTD field consists of the sum of the PO field and the field $u_{\mathrm{s,h}}^{(1)}$ generated by the nonuniform components $j_{\mathrm{s,h}}^{(1)}$ of the scattering sources. The components $j_{\mathrm{s,h}}^{(1)}$ caused by the smooth bending of the scattering surface generate the far field of order k^{-1} (Schensted, 1955), and those caused by the sharp edge create the field of order $k^{-1/2}$ (as shown in Equations (6.21) and (6.22)). Therefore, in the first-order PTD approximation, one can retain only the dominant contributions generated by the edge-type sources $j_{\mathrm{s,h}}^{(1)}$. The uniform asymptotics for these contributions in the region $\pi - \omega \le \vartheta \le \pi$ are given by the expressions (6.47) and (6.48), where one should include the additional factor $\exp[ikl(1 - \cos\vartheta)]$ due to the shift of the coordinates' origin. By the summation of the modified Equations (6.47) and (6.48) with the PO asymptotics (6.192), (6.193), one obtains

$$u_{\mathrm{s}}^{\mathrm{PTD}} = u_0 \left[\!\left[-\frac{1}{2}\sqrt{R_1(z_{\mathrm{st}})R_2(z_{\mathrm{st}})}\; \mathrm{e}^{ik\Phi_1(z_{\mathrm{st}})} \right.\right.$$

$$+ \frac{a}{2}\{f(1)[J_0(ka\sin\vartheta) - iJ_1(ka\sin\vartheta)]$$

$$\left.\left. + f(2)[J_0(ka\sin\vartheta) + iJ_1(ka\sin\vartheta)]\} \, \mathrm{e}^{ikl(1-\cos\vartheta)} \right]\!\right] \frac{\mathrm{e}^{ikR}}{R} \qquad (6.195)$$

and

$$u_h^{PTD} = u_0 \left[\left[\frac{1}{2}\sqrt{R_1(z_{st})R_2(z_{st})}\ e^{ik\Phi_1(z_{st})} \right. \right.$$
$$+ \frac{a}{2}\{g(1)[J_0(ka\sin\vartheta) - iJ_1(ka\sin\vartheta)]$$
$$\left. \left. + g(2)[J_0(ka\sin\vartheta) + iJ_1(ka\sin\vartheta)]\} \ e^{ikl(1-\cos\vartheta)} \right] \frac{e^{ikR}}{R}. \right. \qquad (6.196)$$

The functions $f(1,2)$ and $g(1,2)$ are defined by Equations (6.25), (6.26), and (6.30), and (6.31), where one should omit the last terms, which are exactly cancelled by the terms $f^{(0)}(1,2)$ and $g^{(0)}(1,2)$ during the summation of modified Equations (6.47) and (6.48) with Equations (6.193) and (6.194).

6.5.4 Asymptotics for the PTD Field in the Region $2\omega < \vartheta \le \pi - \omega$ away from the GO Boundary $\vartheta = 2\omega$

In this observation region, the stationary edge point 2 (Fig. 6.21) is not visible (when $\Omega < 2\omega$), and its first-order contribution to the scattered field equals zero. We also assume that the observation directions are far from the last GO ray (reflected at the point 1 and shown in Fig. 6.21), where functions $f(1)$ and $g(1)$ are singular. Under these conditions, one can use the obvious modifications of Equations (6.195) and (6.196) for the scattered field:

$$u_s^{PTD} = u_0 \left[\left[-\frac{1}{2}\sqrt{R_1(z_{st})R_2(z_{st})}\ e^{ik\Phi_1(z_{st})} \right. \right.$$
$$\left. \left. + \frac{a}{2}f(1)[J_0(ka\sin\vartheta) - iJ_1(ka\sin\vartheta)]\ e^{ikl(1-\cos\vartheta)} \right] \frac{e^{ikR}}{R} \right. \qquad (6.197)$$

and

$$u_h^{PTD} = u_0 \left[\left[\frac{1}{2}\sqrt{R_1(z_{st})R_2(z_{st})}\ e^{ik\Phi_1(z_{st})} \right. \right.$$
$$\left. \left. + \frac{a}{2}g(1)[J_0(ka\sin\vartheta) - iJ_1(ka\sin\vartheta)]\ e^{ikl(1-\cos\vartheta)} \right] \frac{e^{ikR}}{R}. \right. \qquad (6.198)$$

6.5.5 Uniform Approximations for the PO Field in the Ray Region $2\omega \le \vartheta \le \pi - \omega$ Including the GO Boundary $\vartheta = 2\omega$

The above ray asymptotics (6.197), (6.198) are not applicable in the vicinity of the geometrical optics boundary $\vartheta = 2\omega$, where the wave field does not have a ray structure. In this so-called transition region, the process of the transverse diffusion of the wave field happens to actually give birth to diffracted rays (Ufimtsev, 2003). The singularities of the functions $f(1)$, $g(1)$ and the singularity of the factor $1/\Phi'(l)$ in Equation (6.185) for the direction $\vartheta \to 2\omega$ (in this case, $\xi_{st} \to l$ and $\Phi'(l, P) \to 0$) are the mathematical evidence of the existence of this process. For the calculation of the field integral (6.184) in this region, one should apply a more accurate method of the stationary phase that allows the approach of the stationary phase to the end point (Felsen and Marcuvitz, 1972). In the following we present the basics of this technique.

First, we modify the canonical integral (6.184). For the sake of simplicity, the symbol P (related to coordinates of the observation point) is omitted. We notice that the edge point 2 (Fig. 6.21) is not visible and therefore its first-order contribution to the scattered field equals zero. In the integral (6.184), this point corresponds to the end point $\xi = -l$. To exclude its contribution, we set $\xi = -\infty$ for the lower limit of integration. Then we introduce a new variable t with the equation

$$\Phi(\xi) = \Phi(\xi_{st}) + t^2. \tag{6.199}$$

Notice that, according to Equation (6.176), the second derivative $\Phi''(\xi_{st}) = -\rho''(\xi_{st}) \sin \vartheta$ is positive and therefore the quantity $\Phi(\xi_{st})$ is the minimum of function $\Phi(\xi)$. Taking this into account, we define the variable t as the continuous and differentiable function of the old variable ξ,

$$t(\xi) = \begin{cases} \sqrt{\Phi(\xi) - \Phi(\xi_{st})}, & \text{for } \xi \geq \xi_{st} \\ -\sqrt{\Phi(\xi) - \Phi(\xi_{st})}, & \text{for } \xi \leq \xi_{st}, \end{cases} \tag{6.200}$$

where the radical is understood in the arithmetic sense. In the vicinity of the stationary point, where $\Phi'(\xi_{st}) = 0$, one can use the Taylor approximations

$$\Phi(\xi) = \Phi(\xi_{st}) + \frac{1}{2!}\Phi''(\xi_{st})(\xi - \xi_{st})^2 + \frac{1}{3!}\Phi'''(\xi_{st})(\xi - \xi_{st})^3 + \cdots \tag{6.201}$$

and

$$t(\xi) = (\xi - \xi_{st})\sqrt{\frac{1}{2}\Phi''(\xi_{st})}\left[1 + \frac{1}{6}\frac{\Phi'''(\xi_{st})}{\Phi''(\xi_{st})}(\xi - \xi_{st})\right]. \tag{6.202}$$

Now the canonical integral (6.184) can be represented in this form

$$I = e^{ik\Phi(\xi_{st})}\int_{-\infty}^{t(l)} e^{ikt^2} G(t)\mathrm{d}t, \tag{6.203}$$

where

$$G(t) = F(\xi)\frac{\mathrm{d}\xi}{\mathrm{d}t} = F(\xi)\frac{2t(\xi)}{\Phi'(\xi)} \tag{6.204}$$

and

$$G(0) = \lim_{\xi \to \xi_{st}}\frac{2t(\xi)}{\Phi'(\xi)}F(\xi) = \sqrt{\frac{2}{\Phi''(\xi_{st})}}F(\xi_{st}). \tag{6.205}$$

The next idea is to extract (in the explicit form!) the Fresnel integral from Equation (6.203). It is accomplished with a simple procedure:

$$I = e^{ik\Phi(\xi_{st})} \left\{ G(0) \int_{-\infty}^{t(l)} e^{ikt^2} dt + \int_{-\infty}^{t(l)} e^{ikt^2} [G(t) - G(0)] dt \right\} \qquad (6.206)$$

or

$$I = \frac{1}{\sqrt{k}} e^{ik\Phi(\xi_{st})} G(0) \int_{-\infty}^{\sqrt{kt(l)}} e^{ix^2} dx + \frac{G[t(l)] - G(0)}{i2kt(l)} e^{ik\Phi(l)} + O\left(\frac{1}{k^2}\right). \qquad (6.207)$$

Under the condition $\sqrt{k}t(l) \gg 1$, when the observation point is far from the geometrical optics boundary $\vartheta = 2\omega$, this expression is reduced asymptotically to the first two terms in Equation (6.185).

When the observation point approaches the boundary ($\vartheta = 2\omega + 0$ and $t = +0$), one should utilize Equations (6.201) and (6.202) and the additional approximations

$$\frac{1}{\Phi'(\xi)} = \frac{1}{(\xi - \xi_{st})\Phi''(\xi_{st})} \left\{ 1 - \frac{1}{2} \frac{\Phi'''(\xi_{st})}{\Phi''(\xi_{st})} (\xi - \xi_{st}) + O\left[(\xi - \xi_{st})^2\right] \right\}, \qquad (6.208)$$

$$\frac{2t(\xi)}{\Phi'(\xi)} = \sqrt{\frac{2}{\Phi''(\xi_{st})}} \left\{ 1 - \frac{1}{3} \frac{\Phi'''(\xi_{st})}{\Phi''(\xi_{st})} (\xi - \xi_{st}) + O[(\xi - \xi_{st})^2] \right\}, \qquad (6.209)$$

$$F(\xi) = F(\xi_{st}) + F'(\xi_{st})(\xi - \xi_{st}) + \cdots, \qquad (6.210)$$

and

$$G(t) - G(0) = (\xi - \xi_{st}) \sqrt{\frac{2}{\Phi''(\xi_{st})}} \left[F'(\xi_{st}) - \frac{1}{3} F(\xi_{st}) \frac{\Phi'''(\xi_{st})}{\Phi''(\xi_{st})} \right]. \qquad (6.211)$$

These relationships lead to the following value of the canonical integral at the boundary $\vartheta = 2\omega + 0$:

$$I = \sqrt{\frac{\pi}{2k\Phi''(l)}} F(l) e^{ik\Phi(l) + i\pi/4}$$

$$+ \frac{1}{ik\Phi''(l)} \left[F'(l) - \frac{1}{3} F(l) \frac{\Phi'''(l)}{\Phi''(l)} \right] e^{ik\Phi(l)} + O\left(\frac{1}{k^2}\right). \qquad (6.212)$$

This technique is applied further to the calculation of the PO field:

$$u_s^{(0)} = u_0 \frac{ik}{\sqrt{2\pi k \sin\vartheta}} e^{i\pi/4} I_s \frac{e^{ikR}}{R} \qquad (6.213)$$

and

$$u_h^{(0)} = -u_0 \frac{ik}{\sqrt{2\pi k \sin\vartheta}} e^{i\pi/4} I_h \frac{e^{ikR}}{R}, \qquad (6.214)$$

where $I_{s,h}$ are defined by Equation (6.203) with

$$F_s(\xi) = \sqrt{\rho(\xi)}\rho'(\xi), \qquad F_h(\xi) = \sqrt{\rho(\xi)}[\sin\vartheta - \rho'(\xi)\cos\vartheta] \qquad (6.215)$$

and

$$\Phi(\xi) = \xi(1 - \cos\vartheta) - \rho(\xi)\sin\vartheta. \qquad (6.216)$$

We omit all intermediate routine manipulations and obtain the final approximations (neglecting the terms of order k^{-2}):

$$u_s^{(0)} = u_0 \frac{e^{ikR}}{R}\left\{ \frac{e^{i3\pi/4}}{2\sqrt{\pi}}\sqrt{R_1(z_{st})R_{st}(z_{st})}e^{ik\Phi(z_{st})}\int_{-\infty}^{\sqrt{k}\,t(l)}e^{ix^2}dx \right.$$
$$\left. + \left[\frac{a}{\sqrt{2\pi ka\sin\vartheta}}f^{(0)}(1) - \frac{\sqrt{R_1(z_{st})R_2(z_{st})}}{4\sqrt{\pi k}\,t(l)}\right]e^{ik\Phi(l)+i\pi/4} \right\} \qquad (6.217)$$

and

$$u_h^{(0)} = u_0 \frac{e^{ikR}}{R}\left\{ -\frac{e^{i3\pi/4}}{2\sqrt{\pi}}\sqrt{R_1(z_{st})R_{st}(z_{st})}e^{ik\Phi(z_{st})}\int_{-\infty}^{\sqrt{k}t(l)}e^{ix^2}dx \right.$$
$$\left. + \left[\frac{a}{\sqrt{2\pi ka\sin\vartheta}}g^{(0)}(1) + \frac{\sqrt{R_1(z_{st})R_2(z_{st})}}{4\sqrt{\pi k}\,t(l)}\right]e^{ik\Phi(l)+i\pi/4} \right\}, \qquad (6.218)$$

where $t(l) = \sqrt{\Phi(l) - \Phi(z_{st})}$; functions $f^{(0)}(1)$ and $g^{(0)}(1)$ are defined in Equations (6.190) and (6.191); and $R_{1,2}$ are the principal radii of curvature of the scattering surface, which are defined in Equation (6.121).

Far from the GO boundary ($\sqrt{k}t(l) \gg 1$), Equations (6.217) and (6.218) transform into

$$u_s^{(0)} = u_0 \frac{e^{ikR}}{R}\left[-\frac{1}{2}\sqrt{R_1(z_{st})R_{st}(z_{st})}e^{ik\Phi(z_{st})} + \frac{a}{\sqrt{2\pi ka\sin\vartheta}}f^{(0)}(1)e^{ik\Phi(l)+i\pi/4} \right]$$

$$(6.219)$$

and

$$u_h^{(0)} = u_0 \frac{e^{ikR}}{R}\left[\frac{1}{2}\sqrt{R_1(z_{st})R_{st}(z_{st})}e^{ik\Phi(z_{st})} + \frac{a}{\sqrt{2\pi ka\sin\vartheta}}g^{(0)}(1)e^{ik\Phi(l)+i\pi/4} \right].$$

$$(6.220)$$

These expressions totally agree with Equations (6.186) to (6.189) (keeping in mind that the edge point 2 is not visible in the region $2\omega \leq \vartheta < \pi - \omega$). The first terms in Equations (6.219) and (6.220) are the ordinary reflected rays and the second terms are the edge-diffracted rays.

Exactly on the geometrical optics boundary ($\vartheta = 2\omega + 0$), Equations (6.217) and (6.218) are reduced to

$$u_s^{(0)} = u_0 \frac{e^{ikR}}{R} \left\{ -\frac{1}{4}\sqrt{R_1(l)R_2(l)} \right.$$

$$\left. + \frac{e^{i\pi/4}}{\sqrt{2\pi k \sin\vartheta}\, \Phi''(l)} \left[F_s'(l) - \frac{1}{3} F_s(l) \frac{\Phi'''(l)}{\Phi''(l)} \right] e^{ik\Phi(l)} \right\} \qquad (6.221)$$

and

$$u_h^{(0)} = u_0 \frac{e^{ikR}}{R} \left\{ \frac{1}{4}\sqrt{R_1(l)R_2(l)} \right.$$

$$\left. - \frac{e^{i\pi/4}}{\sqrt{2\pi k \sin\vartheta}\, \Phi''(l)} \left[F_h'(l) - \frac{1}{3} F_h(l) \frac{\Phi'''(l)}{\Phi''(l)} \right] e^{ik\Phi(l)} \right\}. \qquad (6.222)$$

Note that all the derivatives in these expressions are taken with respect to the variable ξ in the integral (6.184) and the subscripts s and h indicate the type (soft or hard) of the scattering object.

The first-order PTD approximation for the scattered field in this region can be found by the summation of the PO field (6.221), (6.222) with the field $u_{s,h}^{(1)}$ generated by the nonuniform scattering sources $j_{s,h}^{(1)}$. For the field $u_{s,h}^{(1)}$, one can use Equations (6.49) and (6.50), which are not singular at the boundary of reflected rays $\vartheta = 2\omega$. Taking into account the different locations of the coordinates' origin used in Section 6.1 and here, the field $u_{s,h}^{(1)}$ can be written as

$$u_s^{(1)} = u_0 \frac{a}{2} f^{(1)}(1)[J_0(ka\sin\vartheta) - i J_1(ka\sin\vartheta)]e^{ikl(1-\cos\vartheta)} \frac{e^{ikR}}{R} \qquad (6.223)$$

and

$$u_h^{(1)} = u_0 \frac{a}{2} g^{(1)}(1)[J_0(ka\sin\vartheta) - i J_1(ka\sin\vartheta)]e^{ikl(1-\cos\vartheta)} \frac{e^{ikR}}{R}, \qquad (6.224)$$

where functions $f^{(1)}$ and $g^{(1)}$ are defined in Section 6.1.2. Here we note again that the edge point 2 is not visible in the region $2\omega \leq \vartheta < \pi - \omega$ if $\Omega < 2\omega$.

6.5.6 Approximation for the PO Field in the Shadow Region for Reflected Rays

The ordinary rays reflected from the scattering surface do not exist in the shadow region $0 \leq \vartheta < 2\omega$ (Fig. 6.21). This circumstance can be used to obtain a helpful approximation for the PO fields (6.174), (6.175), which otherwise are difficult to calculate. Indeed according to Equation (1.70), the PO field consists of two parts: the reflected field and the so-called shadow radiation. The reflected field (defined by the integral (1.71)) contains all reflected rays and an additional diffracted field that can be neglected in the shadow region, where the basic component of the scattered field is the shadow radiation (1.72). This observation leads directly to the following approximation

$$u_{s,h}^{(0)} \approx u^{sh} = \frac{1}{4\pi} \int_{S_{il}} \left(u^{inc} \frac{\partial}{\partial n} \frac{e^{ikr}}{r} - \frac{\partial u^{inc}}{\partial n} \frac{e^{ikr}}{r} \right) ds, \qquad (6.225)$$

where the integration is performed over the illuminated side of the scattering object. This integral in general is also difficult to calculate. However, the Shadow Contour Theorem developed in Section 1.3.5 greatly simplifies the calculation. According to this theorem, the integral (6.225) is identical to the integral over the black disk located in the plane $z = l$ and having the radius a.

The field scattered by the black disk can be represented in the form (1.73)

$$u^{sh} = \frac{1}{2} \left[u_s^{(0)} + u_h^{(0)} \right]. \qquad (6.226)$$

In the far-field approximation, the quantities $u_s^{(0)}$ and $u_h^{(0)}$ have already been calculated in Section 6.2.1. Utilizing Equations (6.56) and (6.57), and taking into account the shift of the coordinates' origin accepted in the present section, the field (6.226) can be written as

$$u^{sh} = u_0 \frac{ia}{2} \frac{1 + \cos \vartheta}{\sin \vartheta} J_1(ka \sin \vartheta) \frac{e^{ikR}}{R} e^{ikl(1 - \cos \vartheta)}. \qquad (6.227)$$

It is seen that this field really concentrates in the shadow region. It is zero in the backscattering direction ($\vartheta = \pi$), but it is large and exactly equal to the PO field generated by the perfectly reflecting disk in the shadow direction $\vartheta = 0$:

$$u^{sh} = u_0 \frac{ika^2}{2} \frac{e^{ikR}}{R}. \qquad (6.228)$$

The first-order PTD field in the shadow region can be found by the summation of the field (6.227) with the $u_{s,h}^{(1)}$ field generated by the edge sources $j_{s,h}^{(1)}$ and determined in Section 6.1.4. When utilizing the results of that section, one should include the additional factor $\exp[ikl(1 - \cos \vartheta)]$ into Equations (6.47) to (6.50) because of the above-mentioned origin shift.

This section completes the analysis of the field scattered by the bodies of revolution with nonzero Gaussian curvature.

PROBLEMS

6.1   Verify the asymptotic approximation (6.17) for the integral from a function with two stationary points.

6.2   Derive Equations (6.25) and (6.26) for functions $f^{(1)}(1)$, $g^{(1)}(1)$ related to the stationary point 1 (Fig. 6.3 in the canonical problem).

6.3   Derive Equations (6.30) and (6.31) for functions $f^{(1)}(2)$, $g^{(1)}(2)$ related to the stationary point 2 (Fig. 6.4 in the canonical problem).

6.4   Verify Equations (6.34) and (6.35) for functions $f^{(1)}(1)$, $g^{(1)}(1)$ related to the specular direction $\vartheta = 2\omega$ (in the canonical problem).

6.5   Derive the PO approximation (6.64) for the field scattered from a perfectly conducting disk. Start with Equations (1.87) and (1.88), derive Equation (1.96), obtain Equations (1.98) and (1.99), and apply it for the disk problem.

6.6   Derive the focal asymptotics (6.86) and (6.87) for the disk diffraction problem.

6.7   Use Equations (6.104) and (6.105) for the focal field scattered by a cone and prove the limiting form (6.108) and (6.109) when a cone transforms into a disk.

6.8   Equations (6.148) and (6.149) determine the field scattered by a paraboloid. Show that these expressions transform into Equations (6.151) and (6.152) when a paraboloid continuously transforms into a disk.

6.9   Equations (6.169) and (6.170) determine the field scattered by a spherical segment. Show that these expressions transform into Equations (6.151) and (6.152) when a spherical segment continuously transforms into a disk.

Chapter 7

Elementary Acoustic and Electromagnetic Edge Waves

This chapter is based on the papers by Butorin and Ufimtsev (1986), Butorin et al. (1988), and Ufimtsev (1989, 1991, 2006).

The relationships

$$du_s = dE_t \text{ if } u^{inc}(\zeta) = E_t^{inc}(\zeta); \qquad du_h = dH_t \text{ if } u^{inc}(\zeta) = H_t^{inc}(\zeta)$$

exist between acoustic and electromagnetic EEWs propagating in the directions, that belong to the diffraction cone. Here, $\hat{t}$ is the tangent to the scattering edge at the diffraction point ζ.

In the previous sections, it was demonstrated that the edge-diffracted waves provide a significant contribution to the scattered field. These waves represent by themselves the linear superposition of elementary edge waves (EEWs) generated in a certain vicinity of infinitesimal elements of the scattering edge; that is,

$$u = \int_L du(\zeta).$$

Here, L denotes the edge of the scattering object and ζ is the curvilinear coordinate measured along the edge and associated with its length ($d\zeta = dl$). It is supposed that the curvature radius of the edge L is large in terms of the wavelength and it can slowly change along the edge. The angle between the faces of the edge also can slowly change along the edge. The integrand $du(\zeta)$ stands for the field of the EEW.

Our goal is to derive the high-frequency asymptotics for EEWs. Having obtained them, one can calculate the edge waves arising due to diffraction at the large class of objects with arbitrary smoothly curved edges. To achieve this goal, we utilize the asymptotic localization principle. According to this principle, the nonuniform/fringe

---

Fundamentals of the Physical Theory of Diffraction. By Pyotr Ya. Ufimtsev
Copyright © 2007 John Wiley & Sons, Inc.

component $j^{(1)}$ (of the scattering surface sources) induced near the edge is asymptotically (with $k \rightarrow \infty$) equivalent to the component $j_{can}^{(1)}$ induced on the canonical wedge tangent to the real edge. In order to understand what is the appropriate tangency point and what is the appropriate vicinity of this point responsible for the radiation of EEWs, one should appeal to the physical structure of edge waves diffracted at the canonical wedge.

7.1   ELEMENTARY STRIPS ON A CANONICAL WEDGE

In Chapter 4 it was shown that under the oblique incidence on the wedge, the scattered edge waves have the form of conic waves, which can be interpreted as the edge-diffracted rays distributed over the so-called diffraction cone (Fig. 4.4). Hence, the nonuniform component $j_{can}^{(1)}$ induced on the wedge faces is also the ray field running from the edge along the generatrix of this cone. Now it becomes clear how we should choose the appropriate tangency point (of the actual scattering edge L with edge E of the canonical wedge; Fig. 7.1) and how to determine its vicinity responsible for the radiation of the EEWs.

The appropriate tangency point must be the origin of the diffracted ray coming to the observation point on the tangent wedge. We emphasize that in general the tangency point is not the edge point *nearest* to the observation point! What concerns the appropriate vicinity of the tangency point, it must be an infinitely narrow (elementary) strip oriented along the diffracted ray. In Figure 7.1, that is the strip A.

This figure also helps to understand why any other elementary strip, for example strip B, is not acceptable. It is seen that the orientation of such a strip is not consistent with the localization principle. Indeed, the field on this strip does not depend on the local properties of the incident wave and the real scattering edge L at the tangency point. Instead, it consists of the spurious edge waves/rays coming from the fictitious scattering points on the auxiliary edge E, which do not belong to the real scattering edge L.

To complete the definition of the elementary strip, one should determine its length. Although the nonuniform/fringe sources $j^{(1)}$ concentrate near the edge, they

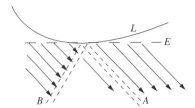

Figure 7.1   Here L is the edge of the actual scattering object and E is the edge of the canonical tangent wedge. The arrows show the edge-diffracted rays diverging from the edge E. The elementary strips A and B belong to the face of the canonical wedge.

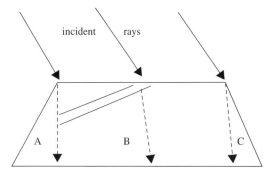

Figure 7.2   Polygonal facet of a scattering object. Only diffracted rays (dotted lines) coming from the upper edge are shown here and discussed in the main text.

are distributed over the whole elementary strip up to its infinite end. By the integration of $j^{(1)}$ over such a semi-infinite strip, one can find the first-order asymptotics for the EEW. However, sometimes it is reasonable to truncate that part of the elementary strip that is outside the real scattering facet (Michaeli, 1987; Breinbjerg, 1992; Johansen, 1996). A similar truncation procedure was considered in Section 5.1.4. From the physical point of view, the truncation results in the additional edge wave (arising at the truncation points), which can be interpreted as a part of the second-order diffraction.

In conclusion we notice a special case when the orientation of elementary strips can be arbitrary. Such a situation is possible for truncated strips on polygonal facets illuminated by a plane wave (Fig. 7.2).

Sections A and C of the polygonal facet are free from diffracted rays, and section B is continuously filled in by such rays. Two parallel thin lines show the elementary strip. All other elementary strips are parallel to this one and have different lengths because they occupy only section B. The field scattered from the facet (i.e., the field radiated by the nonuniform sources distributed in the region B) is determined by the integral over section B. It is clear that the result of integration does not depend on the shape of subsections of the region B, and therefore it does not depend on the orientation of the elementary strips.

7.2 INTEGRALS FOR $j_{s,h}^{(1)}$ ON ELEMENTARY STRIPS

First we choose the elementary strips according to the rule formulated in Section 7.1. They are oriented along the edge-diffracted rays and shown in Figure 7.3. The incident plane wave is given by

$$u^{\text{inc}} = u_0 \, e^{ik\phi^i} \tag{7.1}$$

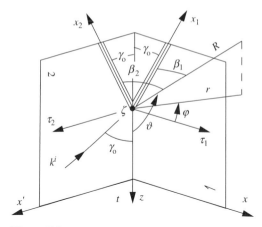

Figure 7.3   Element of the canonical wedge with two elementary strips oriented along the axes x_1 and x_2 with the origin at the point ζ at the edge.

and propagates in the direction $\hat{k}^i = \nabla \phi^i \equiv \mathrm{grad}\ \phi^i$. In the polar cylindrical coordinates $(r,\ \varphi,\ z)$, this wave is determined by the expression

$$u^{\mathrm{inc}} = u_0\, e^{-ikz \cos \gamma_0} \cdot e^{-ik_1 r \cos(\varphi - \varphi_0)}, \tag{7.2}$$

with $k_1 = k \sin \gamma_0, 0 < \gamma_0 < \pi, 0 \leq \varphi_0 \leq \pi$, and $0 \leq \varphi \leq \alpha$. Here, α is the external angle between faces 1 and 2 of the wedge ($\pi \leq \alpha \leq 2\pi$). For the description of the diffracted field at the observation point, we shall use two additional local coordinate systems: the spherical coordinates $(R,\ \vartheta,\ \varphi)$ and the coordinates $(R,\ \beta_1,\ \beta_2)$, where the angles $\beta_{1,2}$ are measured from the elementary strips, that is, from the axes $x_{1,2}$, and do not exceed $180°$ ($0 \leq \beta_{1,2} \leq \pi$).

The local cylindrical coordinates r, φ, z and spherical coordinates R, ϑ, φ are introduced according to the right-hand rule (with respect to their unit vectors, $\hat{r} \times \hat{\varphi} = \hat{z}, \hat{R} \times \hat{\vartheta} = \hat{\varphi}$). One should remember that these coordinates are introduced in such a way that *the angle φ is measured from the illuminated face of the edge and the tangent $\hat{t}$ to the edge is directed along the local polar axis $\hat{z}$ ($\hat{t} = \hat{z}$)*. When both faces are illuminated, one can measure the angle φ from any face, but in this case one should choose the correct direction of the polar axis z and the tangent $\hat{t}$ ($\hat{z} = \hat{t} = \hat{r} \times \hat{\varphi}$). This note is important for correct applications of the theory of EEWs developed here, especially in the case of electromagnetic waves.

In accordance with Sections 2.3 and 4.3, the incident wave (7.2) generates around the wedge the field

$$u_s(r, \varphi) = u_0\, e^{-ikz \cos \gamma_0}[v(k_1 r, \varphi - \varphi_0) - v(k_1 r, \varphi + \varphi_0)] + u_s^{\mathrm{go}}(r, \varphi) \tag{7.3}$$

and

$$u_h(r, \varphi) = u_0 \, e^{-ikz \cos \gamma_0} [v(k_1 r, \varphi - \varphi_0) + v(k_1 r, \varphi + \varphi_0)] + u_h^{go}(r, \varphi), \quad (7.4)$$

where $u_{s,h}^{go}$ is the geometrical optics field (consisting of the incident and reflected plane waves), and

$$v(k_1 r, \psi) = \frac{1}{2\alpha} \int_D \frac{e^{-ik_1 r \cos \eta}}{w(\eta + \psi)} \, d\eta, \quad (7.5)$$

with

$$w(x) = 1 - \exp\left(i\frac{\pi}{\alpha}x\right). \quad (7.6)$$

The integration contour D is shown in Figure 2.5.

We are interested in the nonuniform scattering sources $j_{s,h}^{(1)}$ induced on the elementary strips. In view of Equations (1.11) and (1.31), they are determined as

$$j_s^{(1)} = \frac{\partial u_s(r, \varphi)}{\partial n} - \frac{\partial u_s^{go}(r, \varphi)}{\partial n}, \qquad \text{at } \varphi = 0, \alpha \quad (7.7)$$

and

$$j_h^{(1)} = u_h(r, \varphi) - u_h^{go}(r, \varphi), \qquad \text{at } \varphi = 0, \alpha. \quad (7.8)$$

The normal derivatives in Equation (7.7) are defined as

$$\frac{\partial}{\partial n} = \frac{1}{r}\frac{\partial}{\partial \varphi}, \qquad \text{at } \varphi = 0 \qquad \text{and} \qquad \frac{\partial}{\partial n} = -\frac{1}{r}\frac{\partial}{\partial \varphi}, \qquad \text{at } \varphi = \alpha. \quad (7.9)$$

The differentiation of the function $v(k_1 r, \varphi \mp \varphi_0)$ is carried out by parts:

$$\frac{\partial}{\partial \varphi}v(k_1 r, \varphi \mp \varphi_0) = \frac{1}{2\alpha}\int_D e^{-ik_1 r \cos \eta}\frac{\partial}{\partial \varphi}w^{-1}(\eta + \varphi \mp \varphi_0)d\eta$$

$$= \frac{1}{2\alpha}\int_D e^{-ik_1 r \cos \eta}\frac{\partial}{\partial \eta}w^{-1}(\eta + \varphi \mp \varphi_0)d\eta$$

$$= \frac{1}{2\alpha}\left.\frac{e^{-ik_1 r \cos \eta}}{w(\eta + \varphi \mp \varphi_0)}\right|_{-\frac{3\pi}{2}+i\infty}^{-\frac{\pi}{2}-i\infty} + \frac{1}{2\alpha}\left.\frac{e^{-ik_1 r \cos \eta}}{w(\eta + \varphi \mp \varphi_0)}\right|_{\frac{3\pi}{2}-i\infty}^{\frac{\pi}{2}+i\infty}$$

$$- \frac{ik_1 r}{2\alpha}\int_D \frac{e^{-ik_1 r \cos \eta}\sin \eta}{w(\eta + \varphi \mp \varphi_0)}d\eta$$

$$= -\frac{ik_1 r}{2\alpha}\int_D \frac{e^{-ik_1 r \cos \eta}\sin \eta}{w(\eta + \varphi \mp \varphi_0)}d\eta. \quad (7.10)$$

Here, the terms related to the ends of the contour D are equal to zero. Notice also that the relationships

$$z = \zeta - \xi_{1,2}\cos \gamma_0 \qquad \text{and} \qquad r = \xi_{1,2}\sin \gamma_0 \quad (7.11)$$

are valid for the observation points $x_{1,2} = \xi_{1,2}$ on the elementary strips along the axes $x_{1,2}$.

The above observations result in the following integral expressions for $j_{s,h}^{(1)}$:

$$
j_s^{(1)} = u_0 \frac{ik_1}{2\alpha} e^{-ik(\zeta - \xi_1 \cos\gamma_0)\cos\gamma_0}
$$

$$
\times \int_D e^{-ik_1\xi_1 \sin\gamma_0 \cos\eta} [w^{-1}(\eta + \varphi_0) - w^{-1}(\eta - \varphi_0)] \sin\eta \, d\eta \qquad (7.12)
$$

and

$$
j_h^{(1)} = u_0 \frac{1}{2\alpha} e^{-ik(\zeta - \xi_1 \cos\gamma_0)\cos\gamma_0} \int_D e^{-ik_1\xi_1 \sin\gamma_0 \cos\eta} [w^{-1}(\eta + \varphi_0)
$$

$$
+ w^{-1}(\eta - \varphi_0)] d\eta \qquad (7.13)
$$

on strip 1 ($\varphi = 0$), and

$$
j_s^{(1)} = u_0 \frac{ik_1}{2\alpha} e^{-ik(\zeta - \xi_2 \cos\gamma_0)\cos\gamma_0}
$$

$$
\times \int_D e^{-ik_1\xi_2 \sin\gamma_0 \cos\eta} [w^{-1}(\eta + \alpha - \varphi_0) - w^{-1}(\eta + \alpha + \varphi_0)] \sin\eta \, d\eta
$$

$$
(7.14)
$$

and

$$
j_h^{(1)} = u_0 \frac{1}{2\alpha} e^{-ik(\zeta - \xi_2 \cos\gamma_0)\cos\gamma_0}
$$

$$
\times \int_D e^{-ik_1\xi_2 \sin\gamma_0 \cos\eta} [w^{-1}(\eta + \alpha - \varphi_0) + w^{-1}(\eta + \alpha + \varphi_0)] d\eta \qquad (7.15)
$$

on strip 2 ($\varphi = \alpha$). The identity

$$
w(\eta + \alpha + \varphi_0) = 1 - e^{i\frac{\pi}{\alpha}(\eta + \alpha + \varphi_0)}
$$

$$
= 1 - e^{i\frac{\pi}{\alpha}(2\alpha + \eta - \alpha + \varphi_0)}
$$

$$
= 1 - e^{i\frac{\pi}{\alpha}(\eta - \alpha + \varphi_0)}
$$

$$
= w(\eta - \alpha + \varphi_0)
$$

allows one to rewrite Equations (7.14) and (7.15) in the more convenient form

$$
j_s^{(1)} = u_0 \frac{ik_1}{2\alpha} e^{-ik(\zeta - \xi_2 \cos\gamma_0)\cos\gamma_0}
$$

$$
\times \int_D e^{-ik_1\xi_2 \sin\gamma_0 \cos\eta} [w^{-1}(\eta + \alpha - \varphi_0) - w^{-1}(\eta - \alpha + \varphi_0)] \sin\eta \, d\eta
$$

$$
(7.16)
$$

and

$$j_h^{(1)} = u_0 \frac{1}{2\alpha} e^{-ik(\zeta - \xi_2 \cos \gamma_0) \cos \gamma_0}$$

$$\times \int_D e^{-ik_1 \xi_2 \sin \gamma_0 \cos \eta} [w^{-1}(\eta + \alpha - \varphi_0) + w^{-1}(\eta - \alpha + \varphi_0)] \, d\eta. \quad (7.17)$$

Now it can be seen, that Equations (7.16) and (7.17) related to $j_{s,h}^{(1)}$ on the strip 2 follow from Equations (7.12) and (7.13) for $j_{s,h}^{(1)}$ on strip 1 after the formal replacement of φ_0 by $\alpha - \varphi_0$ and ξ_1 by ξ_2. In the next section, these expressions for $j_{s,h}^{(1)}$ are utilized to calculate the field of the EEWs.

7.3 TRIPLE INTEGRALS FOR ELEMENTARY EDGE WAVES

The field $du_{s,h}^{(1)}$ of the EEW generated by the source $j_{s,h}^{(1)}$ (induced on elementary strip 1, on the face $\varphi = 0$, and on elementary strip 2, on the face $\varphi = \alpha$) is determined in accordance with Equation (1.10), where the differential element of the surface is defined as $ds_{1,2} = \sin \gamma_0 \, d\zeta \, d\xi_{1,2}$.

In applications of the theory of EEWs, one should remember:

- The differential element of the edge is always positive ($d\zeta > 0$);
- The coordinate ζ is associated with the arc length of the edge; and
- The tangent $\hat{t}$ to the edge (which plays the role of a local polar axis $\hat{z}$) is defined as $\hat{t} = d\vec{r}(\zeta)/d\zeta$ where $\vec{r}(\zeta)$ is the position vector of the edge point ζ.

Thus,

$$du_s^{(1)} = du_1^{(1)} + du_2^{(1)} \quad \text{and} \quad du_h^{(1)} = dv_1^{(1)} + dv_2^{(1)}, \quad (7.18)$$

where

$$du_{1,2}^{(1)} = -d\zeta \frac{\sin \gamma_0}{4\pi} \int_0^\infty j_{s1,s2}^{(1)} \frac{e^{ikr_{1,2}}}{r_{1,2}} \, d\xi_{1,2} \quad (7.19)$$

and

$$dv_{1,2}^{(1)} = d\zeta \frac{\sin \gamma_0}{4\pi} \int_0^\infty j_{h1,h2}^{(1)} \frac{\partial}{\partial n} \frac{e^{ikr_{1,2}}}{r_{1,2}} \, d\xi_{1,2}. \quad (7.20)$$

The quantities $du_1^{(1)}$, $dv_1^{(1)}$ and $du_2^{(1)}$, $dv_2^{(1)}$ describe the field generated by the elementary strips 1 and 2, respectively. The quantities $r_1 = \sqrt{(x_1 - \xi_1)^2 + h_1^2}$ and

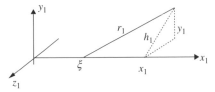

Figure 7.4   Local Cartesian coordinates (x_1, y_1, z_1) with axes x_1 and z_1 placed in wedge face 1 $(\varphi = 0)$, where the axis y_1 is normal to this face $(\hat{y}_1 = \hat{n})$. Similar Cartesian coordinates (x_2, y_2, z_2) are also introduced with axes x_2 and z_2 in face 2 $(\varphi = \alpha)$, and with $\hat{y}_2 = \hat{n}$.

$r_2 = \sqrt{(x_2 - \xi_2)^2 + h_2^2}$ determine the distances between the observation and integration points (Fig. 7.4).

For the Green function of the free space, we use Equation (6.616-3) in Gradshteyn and Ryzhik (1994):

$$\frac{e^{ikr_{1,2}}}{r_{1,2}} = \frac{i}{2} \int_{-\infty}^{\infty} e^{ip(x_{1,2} - \xi_{1,2})} H_0^{(1)}(qh_{1,2}) dp, \tag{7.21}$$

where $q = \sqrt{k^2 - p^2}$, with $\text{Im}(q) \geq 0$, and $h_{1,2} = \sqrt{y_{1,2}^2 + z_{1,2}^2} \geq 0$. The square root q is a two-valued function. In order to make it a single-valued function we introduce two branch cuts $(-\infty \leq \text{Re}(p) \leq -k$ and $k \leq \text{Re}(p) \leq \infty)$ in the complex plane (p). The integration contour in Equation (7.21) is located on the upper side of the left cut and skirts over the branch point $p = -k$, then it follows along the real axis to the right cut and skirts under the branch point $p = k$. After that, the integration contour follows along the lower side of the right cut.

The normal derivatives of the Green function are defined by

$$\frac{\partial}{\partial n} \frac{e^{ikr_{1,2}}}{r_{1,2}} = \nabla' \frac{e^{ikr_{1,2}}}{r_{1,2}} \cdot \hat{y}_{1,2} = -\nabla \frac{e^{ikr_{1,2}}}{r_{1,2}} \cdot \hat{y}_{1,2} = -\frac{\partial}{\partial y_{1,2}} \frac{e^{ikr_{1,2}}}{r_{1,2}}$$

$$= \frac{iy_{1,2}}{2h_{1,2}} \int_{-\infty}^{\infty} e^{ip(x_{1,2} - \xi_{1,2})} q H_1^{(1)}(qh_{1,2}) dp. \tag{7.22}$$

Note that the differential operator ∇' (∇) performs the differentiation with respect to the coordinates of the integration (observation) point. The quantities $y_{1,2}$ are expressed in terms of the spherical coordinates (R, ϑ, φ) as

$$y_1 = R \sin \vartheta \sin \varphi \qquad \text{and} \qquad y_2 = R \sin \vartheta \sin(\alpha - \varphi), \tag{7.23}$$

and the quantities $h_{1,2}$ are determined in terms of the coordinates (R, β_1, β_2) as

$$h_1 = R \sin \beta_1 \qquad \text{and} \qquad h_2 = R \sin \beta_2. \tag{7.24}$$

After substituting the sources $j_{s,h}^{(1)}$ and the above expressions into the field formulas (7.19) and (7.20), one obtains

$$du_1^{(1)} = u_0 e^{-ik\zeta \cos \gamma_0} \, d\zeta \, \frac{k \sin^2 \gamma_0}{16\pi\alpha} I_u(x_1, h_1, \varphi_0) \qquad (7.25)$$

and

$$dv_1^{(1)} = u_0 e^{-ik\zeta \cos \gamma_0} \, d\zeta \, \frac{i}{16\pi\alpha} \frac{\sin \gamma_0 \sin \vartheta \sin \varphi}{\sin \beta_1} I_v(x_1, h_1, \varphi_0) \qquad (7.26)$$

with

$$I_u(x_1, h_1, \varphi_0) = \int_0^\infty e^{ik\xi_1 \cos^2 \gamma_0} \, d\xi_1 \int_{-\infty}^\infty e^{ip(x_1 - \xi_1)} H_0^{(1)}(qh_1) dp$$

$$\times \int_D e^{-ik\xi_1 \sin^2 \gamma_0 \cos \eta} [w^{-1}(\eta + \varphi_0) - w^{-1}(\eta - \varphi_0)] \sin \eta \, d\eta, \qquad (7.27)$$

$$I_v(x_1, h_1, \varphi_0) = \int_0^\infty e^{ik\xi_1 \cos^2 \gamma_0} \, d\xi_1 \int_{-\infty}^\infty e^{ip(x_1 - \xi_1)} q H_1^{(1)}(qh_1) dp$$

$$\times \int_D e^{-ik\xi_1 \sin^2 \gamma_0 \cos \eta} [w^{-1}(\eta + \varphi_0) + w^{-1}(\eta - \varphi_0)] d\eta \qquad (7.28)$$

and

$$du_2^{(1)} = u_0 e^{-ik\zeta \cos \gamma_0} \, d\zeta \, \frac{k \sin^2 \gamma_0}{16\pi\alpha} I_u(x_2, h_2, \alpha - \varphi_0), \qquad (7.29)$$

$$dv_2^{(1)} = u_0 e^{-ik\zeta \cos \gamma_0} \, d\zeta \, \frac{i}{16\pi\alpha} \frac{\sin \gamma_0 \sin \vartheta \sin(\alpha - \varphi)}{\sin \beta_2} I_v(x_2, h_2, \alpha - \varphi_0). \qquad (7.30)$$

It may be seen that Equations (7.29) and (7.30) for the fields generated by strip 2 can be obtained from Equations (7.25) and (7.26) for the fields from strip 1 by the formal replacements

$$x_1 \longrightarrow x_2, \qquad h_1 \longrightarrow h_2, \qquad \beta_1 \longrightarrow \beta_2 \cdot \varphi_0 \longrightarrow \alpha - \varphi_0, \qquad \varphi \longrightarrow \alpha - \varphi. \qquad (7.31)$$

For this reason, further calculations are carried out only for the fields $du_1^{(1)}$, $dv_1^{(1)}$, and the final expressions for the fields $du_2^{(1)}$, $dv_2^{(1)}$ will be obtained using the relationships (7.31).

The next section deals with the analytical work on the integrals I_u and I_v.

7.4 TRANSFORMATION OF TRIPLE INTEGRALS INTO ONE-DIMENSIONAL INTEGRALS

First, we change in Equations (7.27) and (7.28) the order of integration:

$$
I_u = \int_{-\infty}^{\infty} e^{ipx_1} H_0^{(1)}(qh_1)\, dp \int_D [w^{-1}(\eta + \varphi_0) - w^{-1}(\eta - \varphi_0)] \sin \eta \, d\eta
$$

$$
\times \int_0^{\infty} e^{-i\xi_1(p_2 + k_2 \cos \eta)} \, d\xi_1, \tag{7.32}
$$

$$
I_v = \int_{-\infty}^{\infty} e^{ipx_1} q H_1^{(1)}(qh_1)\, dp \int_D [w^{-1}(\eta + \varphi_0) + w^{-1}(\eta - \varphi_0)]\, d\eta
$$

$$
\times \int_0^{\infty} e^{-i\xi_1(p_2 + k_2 \cos \eta)} \, d\xi_1, \tag{7.33}
$$

where

$$
p_2 = p - k \cos^2 \gamma_0 \quad \text{and} \quad k_2 = k \sin^2 \gamma_0. \tag{7.34}
$$

The integral over the variable ξ is calculated under the condition $\text{Im}(p_2 + k_2 \cos \eta) < 0$ to ensure its convergence. Then,

$$
I_u = \frac{1}{i} \int_{-\infty}^{\infty} e^{ipx_1} H_0^{(1)}(qh_1)\, dp \int_D \frac{[w^{-1}(\eta + \varphi_0) - w^{-1}(\eta - \varphi_0)] \sin \eta}{p_2 + k_2 \cos \eta} \, d\eta, \tag{7.35}
$$

and

$$
I_v = \frac{1}{i} \int_{-\infty}^{\infty} e^{ipx_1} q H_1^{(1)}(qh_1)\, dp \int_D \frac{[w^{-1}(\eta + \varphi_0) + w^{-1}(\eta - \varphi_0)]}{p_2 + k_2 \cos \eta} \, d\eta. \tag{7.36}
$$

The above condition for integral convergence is fulfilled for all points on the contour D (with the exception of points $\eta = \pm \pi$), because of the inequality $\text{Im}(k_2 \cos \eta) < 0$ that is valid there. For the points $\eta = \pm \pi$ and $p \neq k$, the convergence can be achieved by the *temporary* assumption that the wave number has a small imaginary part, $k = k' + ik''$ with $k' > 0$ and $0 < k'' \ll 1$. This assumption means a small attenuation of acoustic waves due to a small *temporary* admitted viscosity of the medium. After the transition to Equations (7.35) and (7.36), this assumption can be omitted. A special situation happens when $\eta = \pm \pi$ and the point p approaches the point $p = k$. As was explained above, the integration contour in Equation (7.21) skirts under the branch point $p = k$. This means that under this branch point, the integration point is complex, and $p = p' + ip''$ with $p' = k$ and $p'' = \text{Im}(p) < 0$. Thus, in this special case, the above integral over the variable ξ_1 is convergent due to the inequality $\text{Im}(p) < 0$.

The next work consists of a thorough analysis of the integral over the contour D. Its integrand possesses poles of two types. There are two poles $\eta_1 = \varphi_0$, $\eta_2 = -\varphi_0$

(related to the functions $w^{-1}(\eta \mp \varphi_0)$) and two other poles $\eta_3 = \sigma$, $\eta_4 = -\sigma$ that are the zeros of the denominator $p_2 + k_2 \cos \eta$, when

$$\cos \sigma = -\frac{p_2}{k_2} \quad \text{and} \quad \sigma = \arccos\left(-\frac{p_2}{k_2}\right). \tag{7.37}$$

It is clear that

$$\sigma = \pi - \arccos(p_2/k_2), \quad \text{if } |p_2| \le k_2, \tag{7.38}$$

where for the inverse cosine we take its principal values, $0 \le \arccos(x) \le \pi$. The condition $|p_2| \le k_2$ is satisfied for the points p on the real axis in the interval $k \cos(2\gamma_0) \le p \le k$. In this case, $0 \le \sigma \le \pi$.

In order to define σ for the values $|p_2| > k_2$, one should use the Euler formula for the cosine function. Together with Equation (7.37) they lead to the equation

$$\frac{1}{2}(e^{i\sigma} + e^{-i\sigma}) = -\frac{p_2}{k_2}. \tag{7.39}$$

By using the replacement $e^{-i\sigma} = t$, it is reduced to the quadratic equation with the solution

$$\sigma = i \ln \frac{-p_2 \pm \sqrt{p_2^2 - k_2^2}}{k_2}. \tag{7.40}$$

Now it is necessary to define the appropriate branches for the square root $\sqrt{p_2^2 - k_2^2}$ and for the logarithm function. The square root has two branch points, $p_2 = \pm k_2$. To make this function single-valued, we introduce two branch cuts $(-\infty \le p_2 \le -k_2$ and $k_2 \le p_2 \le \infty)$ and choose the branch where $\sqrt{p_2^2 - k_2^2} \ge 0$ for the points on the upper (lower) side of the left (right) branch cut.

To define the appropriate sign in front of the square root in Equation (7.40) and to define the appropriate branch of the logarithm, we use the conditions $\sigma = \pi$ for $p_2 = k_2$ and $\sigma = 0$ for $p_2 = -k_2$. These conditions lead to the function

$$\sigma = i \ln \frac{-p_2 + \sqrt{p_2^2 - k_2^2}}{k_2} \tag{7.41}$$

where $\ln(-1) = -i\pi$ and $\ln(1) = 0$.

It follows from this equation that

$$\sigma = \pi + i \ln \frac{p_2 - \sqrt{p_2^2 - k_2^2}}{k_2}, \quad \text{with } p_2 \ge k_2, \tag{7.42}$$

and

$$\sigma = i \ln \frac{-p_2 + \sqrt{p_2^2 - k_2^2}}{k_2}, \quad \text{with } p_2 \le -k_2. \tag{7.43}$$

Because the quantity p_2 is real and the argument of the logarithm in Equations (7.42) and (7.43) is always positive, we take the regular arithmetic branch for this logarithm, where $\ln(0) = -\infty$, $\ln(1) = 0$, and $\ln(\infty) = +\infty$.

Now one can trace the position of the pole σ in the complex plane (η) as the function of the variable p in Equations (7.35) and (7.36). When this variable changes from $p = -\infty$ to $p = k\cos(2\gamma_0)$, the pole $\eta_3 = \sigma$ runs in the complex plane (η) along the imaginary axis $\mathrm{Im}(\eta)$ from $\eta = +i\infty$ to $\eta = 0$. When the point p moves from $p = k\cos(2\gamma_0)$ to $p = k$, the pole $\eta_3 = \sigma$ runs along the real axis $\mathrm{Re}(\eta)$ from $\eta = 0$ to $\eta = \pi$. When the point p moves from $p = k$ to $p = +\infty$, the pole $\eta_3 = \sigma$ runs down along the vertical line from $\eta = \pi$ to $\eta = \pi - i\infty$. The location of the pole $\eta_3 = \sigma$ in the complex plane (η) is shown in Figure 7.5 by the thick solid line. The dashed line shows the location of the pole $\eta_4 = -\sigma$.

To calculate the integrals over the contour D, we connect its branches with the additional contours F_+ and F_- (Fig. 7.5), where $\mathrm{Im}(\eta) = A$ and $\mathrm{Im}(\eta) = -A$, respectively. First, we place these contours at a large *finite* distance from the real axis ($A \gg 1$) and apply the Cauchy residue theorem to the integrals over the closed contour $C = D + F_+ + F_-$. Then the results of this theorem are extended to the case when the contours F_+ and F_- are shifted to the *infinite* distance ($A \to \infty$). This procedure is realized in the following for the integrals

$$J_u(C) = \int_C \frac{w^{-1}(\eta + \varphi_0) - w^{-1}(\eta - \varphi_0)}{p_2 + k_2 \cos \eta} \sin \eta \, d\eta \tag{7.44}$$

and

$$J_v(C) = \int_C \frac{w^{-1}(\eta + \varphi_0) + w^{-1}(\eta - \varphi_0)}{p_2 + k_2 \cos \eta} \, d\eta. \tag{7.45}$$

The basic details of this procedure are the same for both integrals. That is why we demonstrate these details only for the integral J_u and then bring the final result for J_v.

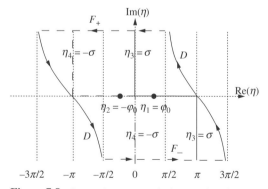

Figure 7.5   Integration contours in the complex plane (η).

When all poles of the integrand of J_u are inside the region closed by the contour $C = D + F_+ + F_-$, the Cauchy theorem states that

$$J_u(C) = 2\pi i \sum_{m=1}^{4} \text{Res}_m. \tag{7.46}$$

Here, the quantities Res_1 and Res_2 are the residues at the poles $\eta_1 = \varphi_0$ and $\eta_2 = -\varphi_0$:

$$\text{Res}_1 = \text{Res}_2 = \frac{\alpha}{i\pi} \cdot \frac{\varepsilon(\varphi_0)\sin\varphi_0}{p_2 + k_2 \cos\varphi_0}, \qquad \text{with} \tag{7.47}$$

$$\varepsilon(x) = \begin{cases} 1, & \text{if} \quad 0 \le x \le \pi \\ 0, & \text{if} \quad \pi < x, \end{cases} \tag{7.48}$$

and

$$\text{Res}_3 = \frac{1}{k_2}[w^{-1}(\sigma - \varphi_0) - w^{-1}(\sigma + \varphi_0)], \tag{7.49}$$

$$\text{Res}_4 = \frac{1}{k_2}[w^{-1}(-\sigma - \varphi_0) - w^{-1}(-\sigma + \varphi_0)] \tag{7.50}$$

are the residues at the poles η_3 and η_4. In addition,

$$\text{Res}_3 + \text{Res}_4 = \frac{i}{k_2}\left[\cot\frac{\pi(\sigma - \varphi_0)}{2\alpha} - \cot\frac{\pi(\sigma + \varphi_0)}{2\alpha}\right]. \tag{7.51}$$

One can verify that the residues Res_3 and Res_4 tend to zero when $\text{Im}(\sigma) \to \pm\infty$.

Notice also that for certain large values of the parameter $|p|$, the poles η_3 and η_4 can reach the contours F_+ and F_-. When they are exactly on these lines, the integral $J_u(C)$ is determined as

$$J_u(C) = 2\pi i\left[\text{Res}_1 + \text{Res}_2 + \frac{1}{2}(\text{Res}_3 + \text{Res}_4)\right]. \tag{7.52}$$

Our goal is to calculate the integral $J_u(D)$ over the contour $D = C - (F_+ + F_-)$ when the contours F_+ and F_- are shifted to infinity ($\text{Im}(\eta) = \pm\infty$). According to Equations (7.46) and (7.52), the integral $J_u(D)$ is determined by

$$J_u(D) = 2\pi i \sum_{m=1}^{4} \text{Res}_m - J_u(F_+ + F_-), \qquad \text{if} \ |\text{Im}(\sigma)| < A, \tag{7.53}$$

and

$$J_u(D) = 2\pi i\left[\text{Res}_1 + \text{Res}_2 + \frac{1}{2}(\text{Res}_3 + \text{Res}_4)\right] - J_u(F_+ + F_-),$$

$$\text{if} \ |\text{Im}(\sigma)| = A, \tag{7.54}$$

where

$$J_u(F_+ + F_-) = J_u(F_+) + J_u(F_-) \tag{7.55}$$

and

$$J_u(F_+) = \int_{\pi/2+iA}^{-3\pi/2+iA} \frac{w^{-1}(\eta + \varphi_0) - w^{-1}(\eta - \varphi_0)}{p_2 + k_2 \cos \eta} \sin \eta \, d\eta, \tag{7.56}$$

$$J_u(F_-) = \int_{-\pi/2-iA}^{3\pi/2-iA} \frac{w^{-1}(\eta + \varphi_0) - w^{-1}(\eta - \varphi_0)}{p_2 + k_2 \cos \eta} \sin \eta \, d\eta. \tag{7.57}$$

When the poles η_3 and η_4 are inside the region closed by the contour $D + F_+ + F_-$ and $A \to \infty$, the integrals $J_u(F_\pm)$ equal zero due to the relationships

$$w^{-1}(\eta \pm \varphi_0) \longrightarrow \begin{cases} 1, & \text{with Im}(\eta) \longrightarrow +\infty \\ 0, & \text{with Im}(\eta) \longrightarrow -\infty. \end{cases} \tag{7.58}$$

$$w^{-1}(\eta + \varphi_0) - w^{-1}(\eta - \varphi_0) \longrightarrow 0, \qquad \text{with Im}(\eta) \longrightarrow \pm\infty, \tag{7.59}$$

and

$$\tan \eta \longrightarrow \begin{cases} i, & \text{with Im}(\eta) \longrightarrow +\infty \\ -i, & \text{with Im}(\eta) \longrightarrow -\infty. \end{cases} \tag{7.60}$$

In the case when the poles η_3 and η_4 are exactly on the contours F_+ and F_-,

$$J_u(F_+) = \lim_{\sigma \to +i\infty} 2\pi i \cdot \left[\frac{1}{2}(\text{Res}_3 + \text{Res}_4) \right]$$

$$+ \lim_{A \to \infty} \lim_{\delta \to 0} \left[\int_{\pi/2+iA}^{\delta+iA} + \int_{-\delta+iA}^{-\pi+\delta+iA} + \int_{-\pi-\delta+iA}^{-3\pi/2+iA} \right]$$

$$\times \frac{w^{-1}(\eta + \varphi_0) - w^{-1}(\eta - \varphi_0)}{p_2 + k_2 \cos \eta} \sin \eta \, d\eta' \tag{7.61}$$

and

$$J_u(F_-) = \lim_{\sigma \to -i\infty} 2\pi i \cdot \left[\frac{1}{2}(\text{Res}_3 + \text{Res}_4) \right]$$

$$+ \lim_{A \to \infty} \lim_{\delta \to 0} \left[\int_{-\pi/2-iA}^{-\delta-iA} + \int_{+\delta-iA}^{\pi-\delta-iA} + \int_{\pi+\delta-iA}^{3\pi/2-iA} \right]$$

$$\times \frac{w^{-1}(\eta + \varphi_0) - w^{-1}(\eta - \varphi_0)}{p_2 + k_2 \cos \eta} \sin \eta \, d\eta'. \tag{7.62}$$

Here, the terms with the double limits represent the principal values of the integrals $J_u(F_\pm)$. In view of Equations (7.59) and (7.60) and the relationship $\text{Res}_{3,4} \to 0$ (when $\text{Im}(\eta) \to \pm i\infty$), the integrals $J_u(F_\pm)$ again equal zero.

Thus, for any positions of η_3 and η_4, it turns out that $J_u(F_+ + F_-) = 0$ with $A \to \infty$. Because of that, and due to Equations (7.53) and (7.54), we obtain the following rigorous result

$$J_u(D) = 2\pi i \sum_{m=1}^{4} \text{Res}_m, \qquad (7.63)$$

where the residues are defined above by Equations (7.47) and (7.49), (7.50).

By application of the same procedure to the integral $J_v(D)$, one can show that it also can be expressed in the form of Equation (7.63), but with other expressions for the residues. We omit all routine intermediate manipulations and bring the final expressions for the field integrals (7.32) and (7.33):

$$I_u = \frac{1}{i} \int_{-\infty}^{\infty} e^{ipx_1} H_0^{(1)}(qh_1) \left\{ \frac{4\alpha\varepsilon(\varphi_0)\sin\varphi_0}{p_2 + k_2\cos\varphi_0} \right. $$
$$\left. + \frac{2\pi}{k_2}\left[\cot\frac{\pi(\sigma + \varphi_0)}{2\alpha} - \cot\frac{\pi(\sigma - \varphi_0)}{2\alpha}\right]\right\}dp \qquad (7.64)$$

and

$$I_v = \frac{1}{i} \int_{-\infty}^{\infty} e^{ipx_1} qH_1^{(1)}(qh_1) \left\{ -\frac{4\alpha\varepsilon(\varphi_0)}{p_2 + k_2\cos\varphi_0} \right.$$
$$\left. + \frac{2\pi}{k_2\sin\sigma}\left[\cot\frac{\pi(\sigma + \varphi_0)}{2\alpha} + \cot\frac{\pi(\sigma - \varphi_0)}{2\alpha}\right]\right\}dp. \qquad (7.65)$$

Integrals $I_{u,v}$ determine the fields (7.25), (7.26) generated by elementary strip 1. In the next section we develop the asymptotic estimations for these fields.

7.5 GENERAL ASYMPTOTICS FOR ELEMENTARY EDGE WAVES

A main contribution to the integrals $I_{u,v}$ is given by the vicinity of the stationary point p_{st}. To determine this point we use the asymptotic expressions for Hankel functions

$$H_0^{(1)}(qh_1) \sim \sqrt{\frac{2}{\pi qh_1}}e^{iqh_1 - i\pi/4}, \qquad H_1^{(1)}(qh_1) \sim \sqrt{\frac{2}{\pi qh_1}}e^{iqh_1 - i3\pi/4} \qquad (7.66)$$

with $qh_1 \gg 1$.

The integrands of $I_{u,v}$ have the fast oscillating factor $\exp[i\Phi(p)] = \exp[i(px_1 + qh_1)]$. The stationary point p_{st} is found from the condition $d\Phi(p)/dp = 0$, which leads to the equation

$$x_1\sqrt{k^2 - p_{\text{st}}^2} - h_1 p_{\text{st}} = 0. \tag{7.67}$$

By substituting here $x_1 = R\cos\beta_1$ and $h_1 = R\sin\beta_1$, one obtains $p_{\text{st}} = k\cos\beta_1$. Now one could apply the standard procedure of the stationary phase technique to derive the asymptotic expressions for $I_{u,v}$, but we use the following idea, which leads to the same result, but faster.

Let us consider the integral

$$I_0 = \int_{-\infty}^{\infty} e^{ipx_1} H_0^{(1)}(qh_1)F(p)dp, \tag{7.68}$$

where $F(p)$ is a slowly varying function. It is clear that

$$I_0 \approx F(p_{\text{st}}) \int_{-\infty}^{\infty} e^{ipx_1} H_0^{(1)}(qh_1)dp, \tag{7.69}$$

and, according to Equation (7.21),

$$I_0 \approx -2iF(p_{\text{st}})\frac{e^{ikR}}{R}, \tag{7.70}$$

where $R = \sqrt{x_1^2 + h_1^2}$.

In a similar way one can evaluate the integral

$$I_1 = \int_{-\infty}^{\infty} e^{ipx_1} q H_1^{(1)}(qh_1)F(p)dp$$

$$= -\frac{d}{dh_1}\int_{-\infty}^{\infty} e^{ipx_1} H_0^{(1)}(qh_1)F(p)dp$$

$$\sim 2iF(p_{\text{st}})\frac{d}{dh_1}\frac{e^{ikR}}{R} \approx -2F(p_{\text{st}})k\frac{h_1}{R}\frac{e^{ikR}}{R}$$

$$= -2k\sin\beta_1 F(p_{\text{st}})\frac{e^{ikR}}{R}, \tag{7.71}$$

with $kR \gg 1$.

There is another advantage of this approach. It is valid under the condition $kR \gg 1$ and is free from the restriction $q_{\text{st}}h_1 = kR\sin\beta_1 \gg 1$ in the standard stationary technique.

The above estimations allow one to obtain the asymptotic expressions for the integrals $I_{u,v}$ defined by Equations (7.64) and (7.65):

$$I_u = -2\frac{e^{ikR}}{R}\left\{\frac{4\alpha\varepsilon(\varphi_0)\sin\varphi_0}{p_{2,st}+k_2\cos\varphi_0}\right.$$
$$\left. +\frac{2\pi}{k_2}\left[\cot\frac{\pi(\sigma_1+\varphi_0)}{2\alpha}-\cot\frac{\pi(\sigma_1-\varphi_0)}{2\alpha}\right]\right\}, \tag{7.72}$$

$$I_v = 2ik\sin\beta_1\frac{e^{ikR}}{R}\left\{-\frac{4\alpha\varepsilon(\varphi_0)}{p_{2,st}+k_2\cos\varphi_0}\right.$$
$$\left. +\frac{2\pi}{k_2\sin\sigma_1}\left[\cot\frac{\pi(\sigma_1+\varphi_0)}{2\alpha}+\cot\frac{\pi(\sigma_1-\varphi_0)}{2\alpha}\right]\right\}, \tag{7.73}$$

where

$$p_{2,st} = p_{st} - k\cos^2\gamma_0 = k(\cos\beta_1 - \cos^2\gamma_0) \tag{7.74}$$

and

$$\cos\beta_1 = \sin\gamma_0\sin\vartheta\cos\varphi - \cos\gamma_0\cos\vartheta. \tag{7.75}$$

Parameter σ_1 is determined according to Equations (7.38) and (7.42), (7.43) as

$$\sigma_1 = \pi - \arccos\left[\frac{\cos\beta_1 - \cos^2\gamma_0}{\sin^2\gamma_0}\right], \quad \text{with } 0 \leq \beta_1 \leq \beta_k, \tag{7.76}$$

and

$$\sigma_1 = i\ln\left\{\cos^2\gamma_0 - \cos\beta_1 + \sqrt{(\cos^2\gamma_0 - \cos\beta_1)^2 - \sin^4\gamma_0}\right\}$$
$$- 2i\ln(\sin\gamma_0), \quad \text{with } \beta_k \leq \beta_1 \leq \pi, \tag{7.77}$$

where

$$\beta_k = \begin{cases} 2\gamma_0, & \text{with } 0 \leq \gamma_0 \leq \pi/2 \\ 2(\pi - \gamma_0), & \text{with } \pi/2 \leq \gamma_0 \leq \pi. \end{cases} \tag{7.78}$$

By substituting Equations (7.72) and (7.73) into Equations (7.25) and (7.26) we find the asymptotic estimations for the field generated by elementary strip 1:

$$du_1^{(1)} = u_0 e^{-ik\zeta\cos\gamma_0}\frac{d\zeta}{2\pi}F_{s,1}^{(1)}(\vartheta,\varphi)\frac{e^{ikR}}{R}, \tag{7.79}$$

$$dv_1^{(1)} = u_0 e^{-ik\zeta\cos\gamma_0}\frac{d\zeta}{2\pi}F_{h,1}^{(1)}(\vartheta,\varphi)\frac{e^{ikR}}{R} \tag{7.80}$$

where $kR \gg 1$ and

$$F_{s,1}^{(1)} = -U(\sigma_1, \varphi_0) \sin^2 \gamma_0, \tag{7.81}$$

$$F_{h,1}^{(1)} = -V(\sigma_1, \varphi_0) \sin \gamma_0 \sin \vartheta \sin \varphi, \tag{7.82}$$

with

$$U(\sigma_1, \varphi_0) = U_t(\sigma_1, \varphi_0) - U_0(\beta_1, \varphi_0), \tag{7.83}$$

$$V(\sigma_1, \varphi_0) = V_t(\sigma_1, \varphi_0) - V_0(\beta_1, \varphi_0), \tag{7.84}$$

$$U_t(\sigma_1, \varphi_0) = \frac{\pi}{2\alpha \sin^2 \gamma_0} \left[\cot \frac{\pi(\sigma_1 + \varphi_0)}{2\alpha} - \cot \frac{\pi(\sigma_1 - \varphi_0)}{2\alpha} \right], \tag{7.85}$$

$$U_0(\beta_1, \varphi_0) = -\frac{\varepsilon(\varphi_0) \sin \varphi_0}{\cos \beta_1 - \cos^2 \gamma_0 + \sin^2 \gamma_0 \cos \varphi_0}, \tag{7.86}$$

$$V_t(\sigma_1, \varphi_0) = \frac{\pi}{2\alpha \sin^2 \gamma_0 \sin \sigma_1} \left[\cot \frac{\pi(\sigma_1 + \varphi_0)}{2\alpha} + \cot \frac{\pi(\sigma_1 - \varphi_0)}{2\alpha} \right], \tag{7.87}$$

and

$$V_0(\beta_1, \varphi_0) = \frac{\varepsilon(\varphi_0)}{\cos \beta_1 - \cos^2 \gamma_0 + \sin^2 \gamma_0 \cos \varphi_0}. \tag{7.88}$$

Here, the quantities U_t, V_t relate to the field generated by the total scattering sources $j_{s,h}^{tot} = j_{s,h}^{(1)} + j_{s,h}^{(0)}$, and U_0, V_0 represent the field radiated by their uniform components $j_{s,h}^{(0)}$. The quantity $\varepsilon(\varphi_0)$ is determined using Equation (7.48).

Asymptotic expressions (7.79) and (7.80) describe the field generated by the scattering sources $j_{s,h}^{(1)}$ induced on strip 1 located at the wedge face $\varphi = 0$. Utilizing Equations (7.79) and (7.80) and the relationships (7.31), one can easily obtain asymptotics for the field $du_2^{(1)}$, $dv_2^{(1)}$ generated by elementary strip 2 belonging to the wedge face $\varphi = \alpha$. The total field of the EEW created by both strips equals

$$du_s^{(1)} = u^{inc}(\zeta) \frac{d\zeta}{2\pi} F_s^{(1)}(\vartheta, \varphi) \frac{e^{ikR}}{R}, \tag{7.89}$$

$$du_h^{(1)} = u^{inc}(\zeta) \frac{d\zeta}{2\pi} F_h^{(1)}(\vartheta, \varphi) \frac{e^{ikR}}{R}, \tag{7.90}$$

where $kR \gg 1$ and

$$F_s^{(1)}(\vartheta, \varphi) = -[U(\sigma_1, \varphi_0) + U(\sigma_2, \alpha - \varphi_0)] \sin^2 \gamma_0, \tag{7.91}$$

$$F_h^{(1)}(\vartheta, \varphi) = -[V(\sigma_1, \varphi_0) \sin \varphi + V(\sigma_2, \alpha - \varphi_0) \sin(\alpha - \varphi)] \sin \gamma_0 \sin \vartheta \tag{7.92}$$

with

$$U(\sigma_2, \alpha - \varphi_0) = U_t(\sigma_2, \alpha - \varphi_0) - U_0(\beta_2, \alpha - \varphi_0) \tag{7.93}$$

and

$$V(\sigma_2, \alpha - \varphi_0) = V_t(\sigma_2, \alpha - \varphi_0) - V_0(\beta_2, \alpha - \varphi_0). \qquad (7.94)$$

Here, the quantity $u^{\text{inc}}(\zeta)$ stands for the incident field at the point ζ on the scattering edge. The parameter σ_2 is defined by Equations (7.76) and (7.77) after the substitution $\beta_1 \rightarrow \beta_2$. The angle β_2 is determined by the expression

$$\cos \beta_2 = \sin \gamma_0 \sin \vartheta \cos(\alpha - \varphi) - \cos \gamma_0 \cos \vartheta, \qquad (7.95)$$

which follows from Equation (7.75) after the substitution of φ by $\alpha - \varphi$.

In view of the observation following Equations (7.87) and (7.88), the EEWs generated by the total scattering sources $j_{\text{s,h}}^{(t)} = j_{\text{s,h}}^{(1)} + j_{\text{s,h}}^{(0)}$ can be obtained by a simple modification of Equations (7.89) and (7.90):

$$du_s^{(t)} = u^{\text{inc}}(\zeta) \frac{d\zeta}{2\pi} F_s^{(t)}(\vartheta, \varphi) \frac{e^{ikR}}{R}, \qquad (7.96)$$

$$du_h^{(t)} = u^{\text{inc}}(\zeta) \frac{d\zeta}{2\pi} F_h^{(t)}(\vartheta, \varphi) \frac{e^{ikR}}{R}, \qquad (7.97)$$

where

$$F_s^{(t)}(\vartheta, \varphi) = -[U_t(\sigma_1, \varphi_0) + U_t(\sigma_2, \alpha - \varphi_0)] \sin^2 \gamma_0 \qquad (7.98)$$

and

$$F_h^{(t)}(\vartheta, \varphi) = -[V_t(\sigma_1, \varphi_0) \sin \varphi + V_t(\sigma_2, \alpha - \varphi_0) \sin(\alpha - \varphi)] \sin \gamma_0 \sin \vartheta. \quad (7.99)$$

As was expected, the elementary edge wave (EEW) (at large distances ($kR \gg 1$) from its origin on the edge) is a spherical wave with the directivity pattern described by Equations (7.91), (7.92) and (7.98), (7.99). This wave can also be interpreted as a set of *elementary edge-diffracted rays*. In the next section we study the analytical properties of EEWs.

7.6 ANALYTIC PROPERTIES OF ELEMENTARY EDGE WAVES

- It is seen that, for real parameters $\sigma_{1,2}$, the directivity patterns $F_{\text{s,h}}^{(1)}$ of EEWs are real functions. One can verify that they also remain real for imaginary parameters $\sigma_{1,2}$. Thus, the functions $F_{\text{s,h}}^{(1)}$ are always real, although their arguments can be complex quantities.

- The field $du_s^{(1)}$ of EEWs for acoustically soft scatters equals zero for the grazing incidence ($\varphi_0 = 0$ or $\varphi_0 = \alpha$). Consider for example the case $\varphi_0 = 0$. Indeed, according to Equations (7.85) and (7.86), $U_t(\sigma_1, 0) = 0$ and $U_0(\beta_1, 0) = 0$. The function $U_t(\sigma_2, \alpha)$ is also equal to zero because

$$\cot \frac{\pi(\sigma_2 + \alpha)}{2\alpha} = \cot \frac{\pi(2\alpha + \sigma_2 - \alpha)}{2\alpha}$$

$$= \cot \frac{\pi(\sigma_2 - \alpha)}{2\alpha}. \tag{7.100}$$

This property is the result of the fact that the incident plane wave cannot propagate along the wedge face. Due to the boundary condition, it is completely cancelled by the reflected plane wave. Therefore, the field $u^{\text{inc}}(\zeta)$ incident on the edge equals zero – no incident field, no diffracted field.

- It is obvious that functions $U_t(\sigma_1, \varphi_0)$, $V_t(\sigma_1, \varphi_0)$ and $U_t(\sigma_2, \alpha - \varphi_0)$, $V_t(\sigma_2, \alpha - \varphi_0)$ are singular at the points $\sigma_1 = \varphi_0$ and $\sigma_2 = \alpha - \varphi_0$, respectively. At these points, functions $U_0(\beta_1, \varphi_0)$, $V_0(\beta_1, \varphi_0)$ and $U_0(\beta_2, \alpha - \varphi_0)$, $V_0(\beta_2, \alpha - \varphi_0)$ are also singular, because in this case $\beta_1 = \beta_{10}$ for $\sigma_1 = \varphi_0$ and $\beta_2 = \beta_{20}$ for $\sigma_2 = \alpha - \varphi_0$, where

$$\cos \beta_{10} = \cos^2 \gamma_0 - \sin^2 \gamma_0 \cos \varphi_0 \tag{7.101}$$

and

$$\cos \beta_{20} = \cos^2 \gamma_0 - \sin^2 \gamma_0 \cos(\alpha - \varphi_0). \tag{7.102}$$

One can show that the singular terms of functions U_t, V_t are cancelled by the singular terms of functions U_0, V_0, and as a result the functions $U = U_t - U_0$, $V = V_t - V_0$ remain finite. We demonstrate this property for the functions $U(\sigma_1, \varphi_0) = U_t(\sigma_1, \varphi_0) - U_0(\beta_1, \varphi_0)$ and $V(\sigma_1, \varphi_0) = V_t(\sigma_1, \varphi_0) - V_0(\beta_1, \varphi_0)$.

According to Equations (7.37) and (7.74),

$$\cos \sigma_1 = \frac{\cos^2 \gamma_0 - \cos \beta_1}{\sin^2 \gamma_0} \quad \text{and} \quad \cos \beta_1 = \cos^2 \gamma_0 - \sin^2 \gamma_0 \cos \sigma_1.$$

$$\tag{7.103}$$

Utilizing the last equality, one can represent the functions $U_0(\beta_1, \varphi_0)$ and $V_0(\beta_1, \varphi_0)$ in the form

$$U_0(\beta_1, \varphi_0) = \frac{1}{2 \sin^2 \gamma_0} \left[\cot \frac{\sigma_1 + \varphi_0}{2} - \cot \frac{\sigma_1 - \varphi_0}{2} \right] \tag{7.104}$$

and

$$V_0(\beta_1, \varphi_0) = \frac{1}{2 \sin^2 \gamma_0 \sin \sigma_1} \left[\cot \frac{\sigma_1 + \varphi_0}{2} + \cot \frac{\sigma_1 - \varphi_0}{2} \right]. \tag{7.105}$$

Now it is easy to prove that

$$U(\varphi_0, \varphi_0) = \lim_{\sigma_1 \to \varphi_0} [U_t(\sigma_1, \varphi_0) - U_0(\beta_1, \varphi_0)]$$

$$= \frac{\pi}{2\alpha \sin^2 \gamma_0} \cot \frac{\pi \varphi_0}{\alpha} - \frac{1}{2 \sin^2 \gamma_0} \cot \varphi_0 \tag{7.106}$$

and

$$V(\varphi_0, \varphi_0) = \lim_{\sigma_1 \to \varphi_0} [V_t(\sigma_1, \varphi_0) - V_0(\beta_1, \varphi_0)]$$

$$= U(\varphi_0, \varphi_0) / \sin \varphi_0. \tag{7.107}$$

These equations also determine the limiting values

$$U(\alpha - \varphi_0, \alpha - \varphi_0) = \frac{\pi}{2\alpha \sin^2 \gamma_0} \cot \frac{\pi(\alpha - \varphi_0)}{\alpha} - \frac{1}{2 \sin^2 \gamma_0} \cot(\alpha - \varphi_0) \tag{7.108}$$

and

$$V(\alpha - \varphi_0, \alpha - \varphi_0) = U(\alpha - \varphi_0, \alpha - \varphi_0) / \sin(\alpha - \varphi_0). \tag{7.109}$$

It follows from these equations that under the grazing incidence ($\varphi_0 = 0$ or $\varphi_0 = \alpha$),

$$\lim_{\varphi_0 \to 0} U(\varphi_0, \varphi_0) = \lim_{\varphi_0 \to \alpha} U(\alpha - \varphi_0, \alpha - \varphi_0) = 0 \tag{7.110}$$

and

$$\lim_{\varphi_0 \to 0} V(\varphi_0, \varphi_0) = \lim_{\varphi_0 \to \alpha} V(\alpha - \varphi_0, \alpha - \varphi_0) = \frac{1}{6 \sin^2 \gamma_0} \left[1 - \left(\frac{\pi}{\alpha} \right)^2 \right]. \tag{7.111}$$

However, expressions (7.106), (7.107) and (7.108), (7.109) remain singular in the grazing directions $\varphi_0 = \pi$ and $\varphi_0 = \alpha - \pi$. The reason for this singularity was explained earlier in conjunction with Equations (4.20) and (4.21) and it is discussed later in Section 7.9, where a new version of PTD is developed that is free from the grazing singularity. This singularity can be also eliminated by the truncation of the elementary strips introduced in Section 7.1 (Johansen, 1996). An example of a similar truncation of scattering sources was considered in Section 5.1.4.

- Notice that the elementary edge-diffracted rays in the directions $\beta_1 = \beta_{10}$ and $\beta_2 = \beta_{20}$ defined by Equations (7.101) and (7.102) form two conic surfaces with axes along the elementary strips 1 and 2 (axes x_1 and x_2 in Fig. 7.3). According to Equation (7.75), the conic surface $\beta_1 = \beta_{10}$ contains the diffracted rays in the directions $\vartheta = \pi - \gamma_0, \varphi = \pi + \varphi_0$ and $\vartheta = \pi - \gamma_0, \varphi = \pi - \varphi_0$, which are on the shadow boundary of the incident and reflected geometrical optics rays, respectively (Fig. 2.7). In view of Equations (7.95) and (7.101), (7.102), the conic surface $\beta_2 = \beta_{20}$ contains the diffracted ray on the boundary of the geometrical optics rays ($\vartheta = \pi - \gamma_0, \varphi = 2\alpha - \pi - \varphi_0$) reflected from the wedge face $\varphi = \alpha$ (Fig. 2.8).

- **EEWs in the directions of the diffraction cone** (Fig. 4.4).   These directions
are determined by $\vartheta = \pi - \gamma_0$. According to Equations (7.37), (7.74), (7.75),
and (7.95),

$$\cos \sigma_1 = -\cos \varphi \quad \text{and} \quad \cos \sigma_2 = -\cos(\alpha - \varphi). \qquad (7.112)$$

The quantities $\sigma_{1,2}$ belong to the interval $0 \le \sigma_{1,2} \le \pi$. Therefore,

$$\sigma_1 = \begin{cases} \pi - \varphi, & \text{for } 0 \le \varphi \le \pi \\ \varphi - \pi, & \text{for } \pi \le \varphi \le 2\pi \end{cases} \qquad (7.113)$$

and

$$\sigma_2 = \begin{cases} \pi - (\varphi - \alpha), & \text{for } -\pi \le \alpha - \varphi \le 0 \\ \pi - (\alpha - \varphi), & \text{for } 0 \le \alpha - \varphi \le \pi \\ \alpha - \varphi - \pi, & \text{for } \pi \le \alpha - \varphi \le 2\pi. \end{cases} \qquad (7.114)$$

The first row in Equation (7.114) relates to the region inside the wedge ($\alpha \le \varphi \le 2\pi$). Utilizing these relationships one can show that the directivity patterns $F_{s,h}^{(1)}$ transform into

$$F_s^{(1)}(\pi - \gamma_0, \varphi) = f^{(1)}(\varphi, \varphi_0, \alpha) \quad \text{and} \quad F_h^{(1)}(\pi - \gamma_0, \varphi) = g^{(1)}(\varphi, \varphi_0, \alpha) \qquad (7.115)$$

outside the wedge ($0 \le \varphi \le \alpha$), and into

$$F_s^{(1)}(\pi - \gamma_0, \varphi) = -f^{(0)}(\varphi, \varphi_0, \alpha) \quad \text{and}$$

$$F_h^{(1)}(\pi - \gamma_0, \varphi) = -g^{(0)}(\varphi, \varphi_0, \alpha) \qquad (7.116)$$

inside the wedge ($\alpha < \varphi < 2\pi$).

The last equation indicates that the total field of EEWs inside the wedge equals zero:

$$F_{s,h}^{(1)}(\pi - \gamma_0, \varphi) + F_{s,h}^{(0)}(\pi - \gamma_0, \varphi) = 0. \qquad (7.117)$$

Here, functions

$$F_s^{(0)}(\pi - \gamma_0, \varphi) = f^{(0)}(\varphi, \varphi_0, \alpha) \quad \text{and} \quad F_h^{(0)}(\pi - \gamma_0, \varphi) = g^{(0)}(\varphi, \varphi_0, \alpha)$$

relate to EEWs radiated by the uniform scattering sources $j_{s,h}^{(0)}$. This is the result of the screening of the region $\alpha < \varphi < 2\pi$ by the perfectly reflecting facets of the wedge.

The functions $f^{(1)}$, $g^{(1)}$ are defined in Equations (4.14) and (4.15). According to Equations (3.56) and (3.57) the functions $f^{(0)}$, $g^{(0)}$ are determined by

$$f^{(0)}(\varphi, \varphi_0, \alpha) = \frac{\varepsilon(\varphi_0) \sin \varphi_0}{\cos \varphi + \cos \varphi_0} + \frac{\varepsilon(\alpha - \varphi_0) \sin(\alpha - \varphi_0)}{\cos(\alpha - \varphi) + \cos(\alpha - \varphi_0)} \qquad (7.118)$$

and

$$g^{(0)}(\varphi, \varphi_0, \alpha) = -\frac{\varepsilon(\varphi_0)\sin\varphi}{\cos\varphi + \cos\varphi_0} - \frac{\varepsilon(\alpha - \varphi_0)\sin(\alpha - \varphi)}{\cos(\alpha - \varphi) + \cos(\alpha - \varphi_0)}, \quad (7.119)$$

with $\varepsilon(x)$ defined in Equation (7.48).

The functions f, g and $f^{(0)}, g^{(0)}$ are singular at the boundaries of the incident and reflected geometric optics rays, that is, in the directions $\varphi = \pi \pm \varphi_0$, $\varphi = 2\alpha - \pi - \varphi_0$. As shown in Section 4.1, these singularities completely cancel each other and functions $f^{(1)} = f - f^{(0)}$, $g^{(1)} = g - g^{(0)}$ are always finite.

Notice also that the EEWs in the directions $\vartheta = \pi - \gamma_0$ generated by the total scattering sources $j_{s,h}^{(t)} = j_{s,h}^{(0)} + j_{s,h}^{(1)}$ satisfy the reciprocity principle. Their directivity patterns $F_{s,h}^{(t)}$ do not change after the permutations $\vartheta \leftrightarrow \gamma_0, \varphi \leftrightarrow \varphi_0$. The situation with this principle in the general case is discussed in Chapter 8.

In conclusion, we present the acoustic EEWs for the directions belonging to the diffraction cone ($\vartheta = \pi - \gamma_0$):

$$\left. \begin{matrix} du_s^{(0)} \\ du_s^{(1)} \\ du_s^{(t)} \end{matrix} \right\} = u^{\text{inc}}(\zeta)\frac{d\zeta}{2\pi} \left\{ \begin{matrix} f^{(0)}(\varphi, \varphi_0, \alpha) \\ f^{(1)}(\varphi, \varphi_0, \alpha) \\ f(\varphi, \varphi_0, \alpha) \end{matrix} \right\} \frac{e^{ikR}}{R}, \quad (7.120)$$

and

$$\left. \begin{matrix} du_h^{(0)} \\ du_h^{(1)} \\ du_h^{(t)} \end{matrix} \right\} = u^{\text{inc}}(\zeta)\frac{d\zeta}{2\pi} \left\{ \begin{matrix} g^{(0)}(\varphi, \varphi_0, \alpha) \\ g^{(1)}(\varphi, \varphi_0, \alpha) \\ g(\varphi, \varphi_0, \alpha) \end{matrix} \right\} \frac{e^{ikR}}{R}. \quad (7.121)$$

7.7 NUMERICAL CALCULATIONS OF ELEMENTARY EDGE WAVES

In this section we present the results of numerical calculations of the elementary edge-diffracted waves radiated by the nonuniform scattering sources $j_{s,h}^{(1)}$. The analytical expressions for the directivity patterns of these waves are given in the previous sections. The quantities $10\log\left|F_{s,h}^{(1)}\right|$ are calculated for the parameters $\alpha = 315°$, $\gamma_0 = 45°$, and $\varphi_0 = 45°$. Figure 7.6 shows the directivity patterns in the plane perpendicular to the edge ($\vartheta = 90°, 0° \le \varphi \le 360°$).

Figure 7.7 demonstrates the directivity patterns in the bisecting plane containing the edge. In this figure, the polar angle θ is defined through the spherical coordinate ϑ as

$$\theta = \begin{cases} \vartheta, & \text{for } \varphi = \alpha/2 \\ 2\pi - \vartheta, & \text{for } \varphi = \alpha/2 + 180°. \end{cases}$$

This means that $0° \leq \theta \leq 180°$ when $\varphi = \alpha/2$ and $180° \leq \theta \leq 360°$ when $\varphi = \alpha/2 + 180°$.

Figures 7.6 and 7.7 show the elementary edge waves both outside and inside the wedge. The presence of elementary waves inside the wedge does not contradict the fact that the wedge is nontransparent. The PTD is based on the Helmholtz equivalency principle (Equations (1.10), (1.54), and (1.59)). According to this principle, a real perfectly reflecting object is replaced by the equivalent scattering sources distributed (*in free space* (!)) over the geometrical surface conformal to the actual scattering surface. These sources generate the field everywhere, including the region inside the object, where this field completely cancels the incident field and ensures the zero total field there. The nonzero elementary edge waves inside the wedge also cancel each other.

Figure 7.8 demonstrates the elementary edge waves propagating in the directions along the diffraction cone.

The following comments are pertinent regarding these calculations:

- As $\varphi_0 = 45°$, only the wedge face $\varphi = 0$ is illuminated. Because of this, only the functions $U_t(\sigma_1, \varphi_0)$, $V_t(\sigma_1, \varphi_0)$, $U_0(\beta_1, \varphi_0)$, and $V_0(\beta_1, \varphi_0)$ are singular,

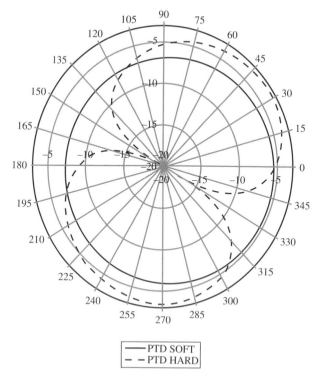

PTD SOFT
- - PTD HARD

Figure 7.6   Directivity patterns of elementary edge waves (radiated by the **nonuniform/fringe** sources $j_{s,h}^{(1)}$) in the plane $\vartheta = 90°$. The interval $315° \leq \varphi \leq 360°$ relates to the region inside the wedge.

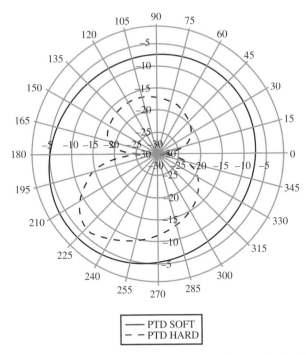

- —— PTD SOFT
- - - PTD HARD

Figure 7.7 Directivity patterns of elementary edge waves (radiated by the **nonuniform/fringe** sources) in the bisecting plane. The interval $180° \leq \theta \leq 360°$ relates to the region inside the wedge.

and their linear combinations in the functions $U(\sigma_1, \varphi_0)$ and $V(\sigma_1, \varphi_0)$ are finite. To treat these singularities, one should apply expressions (7.104) and (7.105), keeping in mind that now $\gamma_0 = 45°$ and $2\sin^2 \gamma_0 = 1$. Together with Equations (7.85) and (7.87), they lead to the following Taylor approximations:

$$U(\sigma_1, \varphi_0) \approx \frac{\pi}{\alpha} \cot \frac{\pi(\sigma_1 + \varphi_0)}{2\alpha} - \cot \frac{\sigma_1 + \varphi_0}{2} + A - B \qquad (7.122)$$

and

$$V(\sigma_1, \varphi_0) \approx \frac{\pi}{\alpha \sin \sigma_1} \cot \frac{\pi(\sigma_1 + \varphi_0)}{2\alpha} - \frac{1}{\sin \sigma_1} \cot \frac{\sigma_1 + \varphi_0}{2} - \frac{1}{\sin \sigma_1}(A - B), \qquad (7.123)$$

where

$$A = \frac{\pi}{\alpha} \left[\frac{\pi(\sigma_1 - \varphi_0)}{6\alpha} + \frac{1}{45} \frac{\pi^3(\sigma_1 - \varphi_0)^3}{8\alpha^3} \right] \qquad (7.124)$$

and

$$B = \frac{\sigma_1 - \varphi_0}{6} + \frac{1}{45} \frac{(\sigma_1 - \varphi_0)^3}{8}. \qquad (7.125)$$

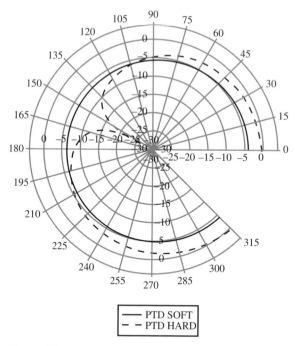

Figure 7.8   Directivity patterns of elementary edge waves radiated by the **nonuniform/fringe** sources in the directions of the diffraction cone ($\vartheta = \pi - \gamma_0, 0 \leq \varphi \leq \alpha$).

- For calculation of the field at the diffraction cone ($\vartheta = \pi - \gamma_0$), the following relationships are useful:

$$\cos \sigma_1 = -\cos \varphi, \qquad \cos \sigma_2 = -\cos(\alpha - \varphi),$$

and

$$\sin \sigma_1 = |\sin \varphi|, \qquad \sin \sigma_2 = |\sin(\alpha - \varphi)|. \qquad (7.126)$$

They allow one to simplify functions $V_t(\sigma_1, \varphi_0) \sin \varphi$ and $V_t(\sigma_2, \alpha - \varphi_0) \sin(\alpha - \varphi)$, taking into account that

$$\sin \varphi / \sin \sigma_1 = \mathrm{sgn}(\sin \varphi) \text{ and } \sin(\alpha - \varphi) / \sin \sigma_2 = \mathrm{sgn}(\sin(\alpha - \varphi)). \qquad (7.127)$$

- The numerical calculations confirm that in the region inside the wedge ($315° \leq \varphi \leq 360°$) the total field $F_{\mathrm{s,h}}^{(t)} = F_{\mathrm{s,h}}^{(1)} + F_{\mathrm{s,h}}^{(0)}$ equals zero in the directions of the diffraction cone ($\vartheta = \pi - \gamma_0$).

7.8   ELECTROMAGNETIC ELEMENTARY EDGE WAVES

The theory of acoustic EEWs presented in Sections 7.1 to 7.7 was extended in the work of Butorin et al. (1987) and Ufimtsev (1991) for electromagnetic waves diffracted at perfectly conducting objects. The basic elements of this extended theory are similar to those in the case of acoustic waves.

The uniform current induced by the incident wave on strips 1 and 2 (Fig. 7.3) is determined according to PO as

$$\vec{j}^{(0)} = 2\{\varepsilon(\varphi_0)[\hat{n}_1 \times \vec{H}^{\text{inc}}] + \varepsilon(\alpha - \varphi_0)[\hat{n}_2 \times \vec{H}^{\text{inc}}]\}, \qquad (7.128)$$

where $\hat{n}_1$ and $\hat{n}_2$ are unit vectors normal to the edge faces 1 and 2, respectively. The nonuniform current $\vec{j}^{(1)}$ is defined as the difference $\vec{j}^{(1)} = \vec{j}^{(t)} - j^{(0)}$. Here, $\vec{j}^{(t)}$ is the total surface current induced on the tangential perfectly conducting wedge. This current is determined by the exact solution of the wedge diffraction problem presented in Sections 2.1 to 2.4 and adapted for electromagnetic waves. The explicit expressions for the current $\vec{j}^{(1)}$ induced on the elementary strips 1 and 2 (Fig. 7.3) are given below in the Problems 7.14 and 7.15. The field radiated by the nonuniform current is found by integration over the elementary strips 1 and 2. The basic integrals are the same as those for the case of acoustic waves. We omit all intermediate calculations and give here the final results.

It is supposed that the incident electromagnetic wave,

$$\vec{E}^{\text{inc}} = \vec{E}_0 e^{ik\phi^i}, \qquad \vec{H}^{\text{inc}} = \vec{H}_0 e^{ik\phi^i}, \qquad (7.129)$$

propagates in the direction $\nabla\phi^i$ and undergoes diffraction at a perfectly conducting object with a curved edge (Fig. 8.1). The nonuniform current $\vec{j}^{(1)}$ induced near the edge radiates the EEWs. Away from the diffraction point ζ at the edge ($kR \gg 1$), their high-frequency asymptotics are expressed as

$$d\vec{E}^{(1)} = \frac{d\zeta}{2\pi}\vec{\mathcal{E}}^{(1)}(\vartheta,\varphi)\frac{e^{ikR}}{R} \quad \text{and} \quad d\vec{H}^{(1)} = [\nabla R \times d\vec{E}^{(1)}]/Z_0. \qquad (7.130)$$

Here $Z_0 = \sqrt{\mu_0/\varepsilon_0} = 120\pi$ ohms is the vacuum impedance; the differential element of the edge is positive ($d\zeta > 0$) and is measured *in the positive direction of the local polar axis* $\hat{z} = \hat{t}$. The quantity

$$\vec{\mathcal{E}}^{(1)}(\vartheta,\varphi) = [E_{0t}(\zeta)\vec{F}^{(1)}(\vartheta,\varphi) + Z_0 H_{0t}(\zeta)\vec{G}^{(1)}(\vartheta,\varphi)]e^{ik\phi^i(\zeta)} \qquad (7.131)$$

is the directivity pattern of the EEWs.

Here, it is also necessary to repeat the note from Section 7.2:

The local cylindrical coordinates r, φ, z and spherical coordinates R, ϑ, φ are introduced according to the right-hand rule (with respect to their unit vectors, $\hat{r} \times \hat{\varphi} = \hat{z}, \hat{R} \times \hat{\vartheta} = \hat{\varphi}$). One should remember that we introduce these coordinates in such a way that *the angle φ is measured from the illuminated face of the edge and the tangent $\hat{t}$ to the edge is directed along the local polar axis $\hat{z}$ ($\hat{t} = \hat{z}$).* When both faces are illuminated, one can measure the angle φ from any face, but in this case one should choose the correct direction of the polar axis z and the tangent $\hat{t}$ ($\hat{z} = \hat{t} = \hat{r} \times \hat{\varphi}$). This note is important for correct applications of the theory of electromagnetic EEWs.

The quantities $E_{0t}(\zeta) \exp[ik\phi^i(\zeta)]$ and $H_{0t}(\zeta) \exp[ik\phi^i(\zeta)]$ are the components of the incident field that are tangential to the edge. Vectors $\vec{F}^{(1)}$ and $\vec{G}^{(1)}$ are specified by their spherical components:

$$F_{\vartheta}^{(1)}(\vartheta, \varphi) = [U(\sigma_1, \varphi_0) + U(\sigma_2, \alpha - \varphi_0)] \sin \vartheta, \qquad F_{\varphi}^{(1)}(\vartheta, \varphi) = 0 \quad (7.132)$$

and

$$
\begin{aligned}
G_{\vartheta}^{(1)}(\vartheta, \varphi) = {} & \frac{\sin \vartheta \cos \gamma_0}{\sin^2 \gamma_0}[\varepsilon(\varphi_0) - \varepsilon(\alpha - \varphi_0)] \\
& + (\sin \gamma_0 \cos \vartheta \cos \varphi - \cos \gamma_0 \sin \vartheta \cos \sigma_1) \, V(\sigma_1, \varphi_0) \\
& - [\sin \gamma_0 \cos \vartheta \cos(\alpha - \varphi) - \cos \gamma_0 \sin \vartheta \cos \sigma_2] \, V(\sigma_2, \alpha - \varphi_0),
\end{aligned}
$$
$$(7.133)$$

$$G_{\varphi}^{(1)}(\vartheta, \varphi) = -[V(\sigma_1, \varphi_0) \sin \varphi + V(\sigma_2, \alpha - \varphi_0) \sin(\alpha - \varphi)] \sin \gamma_0. \quad (7.134)$$

All functions and parameters in Equations (7.132) to (7.134) are the same as those introduced in the previous sections for acoustic waves.

The EEWs radiated by the total current $\vec{j}^{(t)} = \vec{j}^{(1)} + \vec{j}^{(0)}$ are determined by

$$d\vec{E}^{(t)} = \frac{d\zeta}{2\pi} \vec{\mathcal{E}}^{(t)}(\vartheta, \varphi) \frac{e^{ikR}}{R} \qquad \text{and} \qquad d\vec{H}^{(t)} = [\nabla R \times d\vec{E}^{(t)}]/Z_0, \quad (7.135)$$

with

$$\vec{\mathcal{E}}^{(t)}(\vartheta, \varphi) = [E_{0t}(\zeta)\vec{F}^{(t)}(\vartheta, \varphi) + Z_0 H_{0t}(\zeta)\vec{G}^{(t)}(\vartheta, \varphi)]e^{ik\phi^i(\zeta)}, \quad (7.136)$$

where

$$F_{\vartheta}^{(t)}(\vartheta, \varphi) = [U_t(\sigma_1, \varphi_0) + U_t(\sigma_2, \alpha - \varphi_0)] \sin \vartheta, \qquad F_{\varphi}^{(t)}(\vartheta, \varphi) = 0, \quad (7.137)$$

$$
\begin{aligned}
G_{\vartheta}^{(t)}(\vartheta, \varphi) = {} & (\sin \gamma_0 \cos \vartheta \cos \varphi - \cos \gamma_0 \sin \vartheta \cos \sigma_1) V_t(\sigma_1, \varphi_0) \\
& - [\sin \gamma_0 \cos \vartheta \cos(\alpha - \varphi) - \cos \gamma_0 \sin \vartheta \cos \sigma_2] V_t(\sigma_2, \alpha - \varphi_0),
\end{aligned}
$$
$$(7.138)$$

and

$$G_{\varphi}^{(t)}(\vartheta, \varphi) = -[V_t(\sigma_1, \varphi_0) \sin \varphi + V_t(\sigma_2, \alpha - \varphi_0) \sin(\alpha - \varphi)] \sin \gamma_0. \quad (7.139)$$

The difference

$$d\vec{E}^{(0)} = d\vec{E}^{(t)} - d\vec{E}^{(1)}, \qquad d\vec{H}^{(0)} = d\vec{H}^{(t)} - d\vec{H}^{(1)} \quad (7.140)$$

is the electromagnetic field of EEWs generated by the uniform current $\vec{j}^{(0)}$.

For the scattering directions $\vartheta = \pi - \gamma_0$, which relate to the diffraction cone, the functions $\vec{F}^{(1,t)}$ and $\vec{G}^{(1,t)}$ reduce to

$$F_\vartheta^{(1)}(\pi - \gamma_0, \varphi) = -\frac{1}{\sin\gamma_0} f^{(1)}(\varphi, \varphi_0, \alpha), \tag{7.141}$$

$$F_\vartheta^{(t)}(\pi - \gamma_0, \varphi) = -\frac{1}{\sin\gamma_0} f(\varphi, \varphi_0, \alpha), \tag{7.142}$$

$$G_\varphi^{(1)}(\pi - \gamma_0, \varphi) = \frac{1}{\sin\gamma_0} g^{(1)}(\varphi, \varphi_0, \alpha), \tag{7.143}$$

$$G_\varphi^{(t)}(\pi - \gamma_0, \varphi) = \frac{1}{\sin\gamma_0} g(\varphi, \varphi_0, \alpha), \tag{7.144}$$

$$G_\vartheta^{(1)}(\pi - \gamma_0, \varphi) = [\varepsilon(\varphi_0) - \varepsilon(\alpha - \varphi_0)]\cot\gamma_0, \tag{7.145}$$

and

$$G_\vartheta^{(t)}(\pi - \gamma_0, \varphi) = 0. \tag{7.146}$$

The properties of functions $U(\sigma, \psi)$ and $V(\sigma, \psi)$ are described in Sections 7.5 and 7.6.

According to these equations, one can represent the electromagnetic EEWs for the directions $\vartheta = \pi - \gamma_0$ as

$$\left.\begin{array}{l} dE_\vartheta^{(0)} = Z_0 dH_\varphi^{(0)} \\ dE_\vartheta^{(1)} = Z_0 dH_\varphi^{(1)} \\ dE_\vartheta^{(t)} = Z_0 dH_\varphi^{(t)} \end{array}\right\} = -E_t^{\text{inc}}(\zeta)\frac{d\zeta}{2\pi}\frac{1}{\sin\gamma_0} \left\{\begin{array}{l} f^{(0)}(\varphi, \varphi_0, \alpha) \\ f^{(1)}(\varphi, \varphi_0, \alpha) \\ f(\varphi, \varphi_0, \alpha) \end{array}\right\} \frac{e^{ikR}}{R}, \tag{7.147}$$

$$\left.\begin{array}{l} dH_\vartheta^{(0)} = -Y_0 dE_\varphi^{(0)} \\ dH_\vartheta^{(1)} = -Y_0 dE_\varphi^{(1)} \\ dH_\vartheta^{(t)} = -Y_0 dE_\varphi^{(t)} \end{array}\right\} = -H_t^{\text{inc}}(\zeta)\frac{d\zeta}{2\pi}\frac{1}{\sin\gamma_0} \left\{\begin{array}{l} g^{(0)}(\varphi, \varphi_0, \alpha) \\ g^{(1)}(\varphi, \varphi_0, \alpha) \\ g(\varphi, \varphi_0, \alpha) \end{array}\right\} \frac{e^{ikR}}{R}, \tag{7.148}$$

where $Y_0 = 1/Z_0$ and $\hat{t}$ is the unit vector tangent to the scattering edge at the diffraction point ζ. Notice also that the radial components of the far field are of the order $1/R^2$ and they are neglected here. Because of that, in general,

$$E_\vartheta = -E_t/\sin\vartheta, \qquad H_\vartheta = -H_t/\sin\vartheta \tag{7.149}$$

and

$$dE_\vartheta = -dE_t/\sin\vartheta, \qquad dH_\vartheta = -dH_t/\sin\vartheta, \tag{7.150}$$

Taking into account these equations and the condition $\vartheta = \pi - \gamma_0$, one can rewrite the above asymptotics (7.147), (7.148) in the form

$$
\left.\begin{array}{l} dE_t^{(0)} \\ dE_t^{(1)} \\ dE_t^{(t)} \end{array}\right\} = E_t^{\text{inc}}(\zeta) \frac{d\zeta}{2\pi} \left\{\begin{array}{l} f^{(0)}(\varphi, \varphi_0, \alpha) \\ f^{(1)}(\varphi, \varphi_0, \alpha) \\ f(\varphi, \varphi_0, \alpha) \end{array}\right\} \frac{e^{ikR}}{R},
\tag{7.151}
$$

$$
\left.\begin{array}{l} dH_t^{(0)} \\ dH_t^{(1)} \\ dH_t^{(t)} \end{array}\right\} = H_t^{\text{inc}}(\zeta) \frac{d\zeta}{2\pi} \left\{\begin{array}{l} g^{(0)}(\varphi, \varphi_0, \alpha) \\ g^{(1)}(\varphi, \varphi_0, \alpha) \\ g(\varphi, \varphi_0, \alpha) \end{array}\right\} \frac{e^{ikR}}{R}.
\tag{7.152}
$$

Comparison of these equations with Equations (7.120), (7.121) allows one to establish the following relationships between the acoustic and electromagnetic EEWs for the directions belonging to the diffraction cone:

$$
dE_t = du_s, \qquad \text{if } E_t^{\text{inc}}(\zeta) = u^{\text{inc}}(\zeta),
\tag{7.153}
$$

$$
dH_t = du_h, \qquad \text{if } H_t^{\text{inc}}(\zeta) = u^{\text{inc}}(\zeta).
\tag{7.154}
$$

Notice also that the paper by Ufimtsev (1991) investigates in detail the ray, caustic, and focal asymptotics of electromagnetic EEWs, as well as their multiple and slope diffraction. The results of this investigation are presented below in Chapters 8 to 10.

7.9 IMPROVED THEORY OF ELEMENTARY EDGE WAVES: REMOVAL OF THE GRAZING SINGULARITY

The central idea of PTD is the separation of the surface scattering sources into the uniform and nonuniform components in such a way that they would be the most appropriate for calculation of the scattered field. In the case of scattering objects with edges, the nonuniform component is defined as that part of the field that concentrates near edges. Equations (7.7) and (7.8) determine it as the difference between the total field on the tangential wedge and its geometrical optics part. The latter is considered as the uniform component. As is shown in the present book and in other publications, the utilization of these components is really helpful for investigation of many scattering problems.

However, this is not the case for forward scattering in the directions grazing the edge faces (Fig. 7.9), where the above theory of elementary edge waves predicts infinite values for the functions $f^{(1)} = f - f^{(0)}$ and $g^{(1)} = g - g^{(0)}$. It turns out that for these directions, either function f or $f^{(0)}$ (either g or $g^{(0)}$) becomes singular. Which of them becomes singular depends on how the grazing direction is approached: $\varphi = 0$ and then $\varphi_0 \to \pi$, or $\varphi_0 = \pi$ and then $\varphi \to 0$. These singularities indicate that the

Figure 7.9   Grazing incidence on the wedge with $\varphi_0 = \pi$ and grazing scattering in the direction $\varphi = 0$. No singularity exists in the reverse situation, when $\varphi_0 = 0$ and $\varphi = \pi$.

definitions (1.31) and (7.7), (7.8) for uniform and nonuniform components are not adequate for the actual surface field in this case. In particular, component $j_h^{(1)}$ does not vanish away from the edge, but instead it contains a plane wave there. Besides, component $j_h^{(0)}$ includes the *absent* reflected wave.

To remove the grazing singularity we introduce, in the following, new definitions for components $j^{(0)}$ and $j^{(1)}$, which are more appropriate, when the incident wave illuminates both faces of the edge and propagates in the direction close to the face orientation.

7.9.1 Acoustic EEWs

This theory is based on the paper by Ufimtsev (2006a).

The appropriate candidate for a new definition of $j_{h,s}^{(0)}$ is the field $j_{h,s}^{hp}$ induced by the incident wave on the illuminated side of the tangential half-plane. On strip 1 (Fig. 7.3) of the half-plane $\varphi = 0$, these components are determined as

$$j_h^{(0)} \equiv j_h^{hp} = 2u_0\, e^{-ik\zeta \cos \gamma_0}\, e^{ik\xi_1 \cos^2 \gamma_0}$$

$$\times \left[-e^{-ik\xi_1 \sin^2 \gamma_0 \cos \varphi_0} \frac{e^{-\pi/4}}{\sqrt{\pi}} \cdot \int_{\sqrt{2k\xi_1}\, \sin \gamma_0 \cos \frac{\varphi_0}{2}}^{\infty} e^{it^2}\, dt + e^{-ik\xi_1 \sin^2 \gamma_0 \cos \varphi_0} \right],$$

$$(7.155)$$

and

$$j_s^{(0)} \equiv j_s^{hp} = 2u_0\, e^{-ik\zeta \cos \gamma_0}\, e^{ik\xi_1 \cos^2 \gamma_0}$$

$$\times \left[ik \sin \gamma_0 \sin \varphi_0\, e^{-ik\xi_1 \sin^2 \gamma_0 \cos \varphi_0} \frac{e^{-i\pi/4}}{\sqrt{\pi}} \cdot \int_{\sqrt{2k\xi_1}\, \sin \gamma_0 \cos \frac{\varphi_0}{2}}^{\infty} e^{it^2}\, dt \right.$$

$$\left. + \sqrt{\frac{k}{2}} \sin \frac{\varphi_0}{2} \frac{e^{-i\pi/4}}{\sqrt{\pi}} \frac{e^{ik\xi_1 \sin^2 \gamma_0}}{\sqrt{\xi_1}} - ik \sin \gamma_0 \sin \varphi_0\, e^{-ik\xi_1 \sin^2 \gamma_0 \cos \varphi_0} \right].$$

$$(7.156)$$

The last terms in these equations (together with the factor 2 in front of the brackets) relate to the sum of the incident and reflected plane waves. When $\varphi_0 \to \pi$, these equations transform into

$$j_{\mathrm{h}}^{(0)} = u_0 \, e^{-ik\zeta \cos \gamma_0} e^{ik\xi_1} \tag{7.157}$$

and

$$j_{\mathrm{s}}^{(0)} = u_0 \, e^{-ik\zeta \cos \gamma_0} \sqrt{\frac{2k}{\pi \xi_1}} e^{i(k\xi_1 - \pi/4)}, \tag{7.158}$$

where $j_{\mathrm{h}}^{(0)}$ is actually the grazing incident wave, and $j_{\mathrm{s}}^{(0)}$ represents the edge wave. The quantities $j_{\mathrm{h,s}}^{(0)}$ on strip 2 (of the half-plane $\varphi = \alpha$) can be found from Equations (7.157) and (7.158) with the replacement of φ_0 by $\alpha - \varphi_0$ and ξ_1 by ξ_2.

The new nonuniform components $j_{\mathrm{h,s}}^{(1)}$ are defined as the difference between the total scattering source $j_{\mathrm{h,s}}$ (induced on the wedge faces) and the uniform component $j_{\mathrm{h,s}}^{(0)} \equiv j_{\mathrm{h,s}}^{\mathrm{hp}}$ (induced on the tangential half-planes). These nonuniform components are determined by Equations (7.12) and (7.13), where one should replace

$$\alpha w(\eta + \psi) \qquad \text{by} \qquad \alpha w(\eta + \psi) - 2\pi w_{hp}(\eta + \psi) \tag{7.159}$$

with

$$w_{hp}(x) = 1 - e^{ix/2}. \tag{7.160}$$

The field generated by the new component $j_{\mathrm{h,s}}^{(1)}$ is found with a procedure completely similar to that described in Sections 7.3 to 7.5. The associated field integrals are defined by Equations (7.44) and (7.45), where one should apply the replacement (7.159). It is worth noting that now the contribution to these integrals caused by the residues at the poles $\eta = \pm \varphi_0$ is equal to zero. The field of EEWs found in this way can be represented again in the form of Equations (7.89) and (7.90), with the new functions $F_{\mathrm{h,s}}^{(1)}$:

$$du_{\mathrm{h}}^{(1)} = u^{\mathrm{inc}}(\zeta) \frac{d\zeta}{2\pi} F_{\mathrm{h}}^{(1)}(\vartheta, \varphi) \frac{e^{ikR}}{R}, \tag{7.161}$$

$$du_{\mathrm{s}}^{(1)} = u^{\mathrm{inc}}(\zeta) \frac{d\zeta}{2\pi} F_{\mathrm{s}}^{(1)}(\vartheta, \varphi) \frac{e^{ikR}}{R}, \tag{7.162}$$

and

$$F_{\mathrm{h}}^{(1)}(\vartheta, \varphi) = \{[-V_t(\sigma_1, \varphi_0) + V_t^{\mathrm{hp}}(\sigma_1, \varphi_0)] \sin \varphi + [-V_t(\sigma_2, \alpha - \varphi_0)$$
$$+ \varepsilon(\alpha - \varphi_0) V_t^{\mathrm{hp}}(\sigma_2, \alpha - \varphi_0)] \sin(\alpha - \varphi)\} \sin \gamma_0 \sin \vartheta, \tag{7.163}$$

$$F_{\mathrm{s}}^{(1)} = \{[-U_t(\sigma_1, \varphi_0) + U_t^{\mathrm{hp}}(\sigma_1, \varphi_0)]$$
$$+ [-U_t(\sigma_2, \alpha - \varphi_0) + \varepsilon(\alpha - \varphi_0) U_t^{\mathrm{hp}}(\sigma_2, \alpha - \varphi_0)]\} \sin^2 \gamma_0. \tag{7.164}$$

Here, we assume that $0 \leq \varphi_0 \leq \pi$ and utilize the designation (7.48). Functions V_t, U_t are defined in Equations (7.87) and (7.85), but V_t^{hp} and U_t^{hp} are the similar functions associated with the tangential half-plane:

$$V_t^{hp}(\sigma, \psi) = \frac{1}{4 \sin^2 \gamma_0 \sin \sigma} \left[\cot \frac{\sigma + \psi}{4} + \cot \frac{\sigma - \psi}{4} \right], \tag{7.165}$$

$$U_t^{hp}(\sigma, \psi) = \frac{1}{4 \sin^2 \gamma_0} \left[\cot \frac{\sigma + \psi}{4} - \cot \frac{\sigma - \psi}{4} \right]. \tag{7.166}$$

For the directions of the diffraction cone ($\vartheta = \pi - \gamma_0$), functions (7.163) and (7.164) can be represented in the form

$$\mathcal{F}_h^{(1)}(\pi - \gamma_0, \varphi) = g(\varphi, \varphi_0, \alpha) - g^{hp}(\varphi, \varphi_0) \tag{7.167}$$

and

$$\mathcal{F}_s^{(1)}(\pi - \gamma_0, \varphi) = f(\varphi, \varphi_0, \alpha) - f^{hp}(\varphi, \varphi_0), \tag{7.168}$$

where f and g are the Sommerfeld functions (2.62) and (2.64), and

$$g^{hp}(\varphi, \varphi_0) = -\frac{1}{4} \left[\cot \frac{\pi - \varphi - \varphi_0}{4} + \cot \frac{\pi - \varphi + \varphi_0}{4} \right]$$

$$- \varepsilon(\alpha - \varphi_0) \frac{1}{4} \left[\cot \frac{\pi + \varphi - \varphi_0}{4} + \cot \frac{\pi + \varphi + \varphi_0 - 2\alpha}{4} \right], \tag{7.169}$$

$$f^{hp}(\varphi, \varphi_0) = \frac{1}{4} \left[\cot \frac{\pi - \varphi - \varphi_0}{4} - \cot \frac{\pi - \varphi + \varphi_0}{4} \right]$$

$$- \varepsilon(\alpha - \varphi_0) \frac{1}{4} \left[\cot \frac{\pi + \varphi - \varphi_0}{4} - \cot \frac{\pi + \varphi + \varphi_0 - 2\alpha}{4} \right] \tag{7.170}$$

are the functions associated with the field scattered by the tangential half-planes.

Notice also that for analytic analysis and numeric calculation, the cotangent forms of functions f and g are more convenient than the Sommerfeld expressions (2.62) and (2.64). These forms are determined as

$$g(\varphi, \varphi_0, \alpha) = -\frac{1}{2n} \left[\cot \frac{\pi - \varphi - \varphi_0}{2n} + \cot \frac{\pi - \varphi + \varphi_0}{2n} \right.$$

$$\left. + \cot \frac{\pi + \varphi + \varphi_0}{2n} + \cot \frac{\pi + \varphi - \varphi_0}{2n} \right] \tag{7.171}$$

and

$$f(\varphi, \varphi_0, \alpha) = \frac{1}{2n} \left[\cot \frac{\pi - \varphi - \varphi_0}{2n} - \cot \frac{\pi - \varphi + \varphi_0}{2n} \right.$$

$$\left. + \cot \frac{\pi + \varphi + \varphi_0}{2n} - \cot \frac{\pi + \varphi - \varphi_0}{2n} \right], \tag{7.172}$$

with $n = \alpha/\pi$.

202 Chapter 7 Elementary Acoustic and Electromagnetic Edge Waves

Functions f, g and f^{hp}, g^{hp} are singular in the directions of the incident and reflected plane waves ($\varphi = \pi + \varphi_0$, $\varphi = \pi - \varphi_0$, $\varphi = 2\alpha - \pi - \varphi_0$), but their difference is finite. For instance,

$$\mathcal{F}_h^{(1)}(\pi - \gamma_0, \pi - \varphi_0) = -\frac{1}{2n}\cot\frac{\varphi_0}{n} + \frac{1}{4}\cot\frac{\varphi_0}{2}$$
$$-\frac{1}{2n}\cot\frac{\pi - \varphi_0}{n} + \varepsilon(\alpha - \varphi_0)\frac{1}{4}\cot\frac{\pi - \varphi_0}{2}$$
$$+\frac{1}{2n}\cot\frac{\alpha - \pi}{n} - \varepsilon(\alpha - \varphi_0)\frac{1}{4}\cot\frac{\alpha - \pi}{2}, \quad (7.173)$$

$$\mathcal{F}_s^{(1)}(\pi - \gamma_0, \pi - \varphi_0) = -\frac{1}{2n}\cot\frac{\varphi_0}{n} + \frac{1}{4}\cot\frac{\varphi_0}{2} - \frac{1}{2n}\cot\frac{\pi - \varphi_0}{n}$$
$$+\varepsilon(\alpha - \varphi_0)\frac{1}{4}\cot\frac{\pi - \varphi_0}{2} - \frac{1}{2n}\cot\frac{\alpha - \pi}{n}$$
$$+\varepsilon(\alpha - \varphi_0)\frac{1}{4}\cot\frac{\alpha - \pi}{2}, \quad (7.174)$$

$$\mathcal{F}_h^{(1)}(\pi - \gamma_0, 2\alpha - \pi - \varphi_0) = -\frac{1}{2n}\cot\frac{\varphi_0 - (\alpha - \pi)}{n} + \frac{1}{4}\cot\frac{\varphi_0 - (\alpha - \pi)}{2}$$
$$+\frac{1}{2n}\cot\frac{\alpha - \pi}{n} - \frac{1}{4}\cot\frac{\alpha - \pi}{2}$$
$$-\frac{1}{2n}\cot\frac{\alpha - \varphi_0}{n} + \frac{1}{4}\cot\frac{\alpha - \varphi_0}{2}, \quad (7.175)$$

and

$$\mathcal{F}_s^{(1)}(\pi - \gamma_0, 2\alpha - \pi - \gamma_0) = -\frac{1}{2n}\cot\frac{\varphi_0 - (\alpha - \pi)}{n} + \frac{1}{4}\cot\frac{\varphi_0 - (\alpha - \pi)}{2}$$
$$-\frac{1}{2n}\cot\frac{\alpha - \pi}{n} + \frac{1}{4}\cot\frac{\alpha - \pi}{2} - \frac{1}{2n}\cot\frac{\alpha - \varphi_0}{n}$$
$$+\frac{1}{4}\cot\frac{\alpha - \varphi_0}{2}. \quad (7.176)$$

These equations clearly show that functions $\mathcal{F}_{h,s}^{(1)}$ are free from the grazing singularity. Indeed, for the grazing directions of the incident wave ($\varphi_0 = \pi, \varphi_0 = \alpha - \pi$), these functions have the finite values

$$\mathcal{F}_h^{(1)}(\pi - \gamma_0, 0) = \mathcal{F}_h^{(1)}(\pi - \gamma_0, \alpha) = -\frac{1}{n}\cot\frac{\pi}{n} + \frac{1}{4}\tan\frac{\alpha}{2} \quad (7.177)$$

and

$$\mathcal{F}_s^{(1)}(\pi - \gamma_0, 0) = \mathcal{F}_s^{(1)}(\pi - \gamma_0, \alpha) = -\frac{1}{4}\tan\frac{\alpha}{2}. \quad (7.178)$$

It also follows from these equations that functions $\mathcal{F}_{h,s}^{(1)}$ are equal to zero when $\alpha = 2\pi$ and the wedge transforms into the half-plane. This result is the obvious consequence of the general expressions (7.163) and (7.164) and the identities $V_t = V_t^{hp}, U_t = U_t^{hp}$, which are valid in the case $\alpha = 2\pi$.

To find the total field scattered by finite objects, one should also calculate the contribution generated by the uniform component distributed over the *finite* elementary strips $(0 \leq \xi_{1,2} \leq l)$. In the far zone $(R \gg kl^2)$, it is determined by the integrals

$$du_{h1}^{(0)} = \frac{d\zeta}{4\pi} ik \sin\gamma_0 \sin\vartheta \sin\varphi \frac{e^{ikR}}{R} \int_0^l j_{h1}^{(0)}(\xi_1, \varphi_0) e^{-ik\xi_1 \cos\beta_1} d\xi_1 \qquad (7.179)$$

and

$$du_{s1}^{(0)} = -\frac{d\zeta}{4\pi} \sin\gamma_0 \frac{e^{ikR}}{R} \int_0^l j_{s1}^{(0)}(\xi_1, \varphi_0) e^{-ik\xi_1 \cos\beta_1} d\xi_1. \qquad (7.180)$$

Replacing $\xi_1, \beta_1, \varphi, \varphi_0$ here with $\xi_2, \beta_2, \alpha - \varphi, \alpha - \varphi_0$, one obtains the equations associated with the field from strip 2 $(0 \leq \xi_2 \leq l)$. These integrals are easily calculated in closed form. We show only those results that relate to the grazing incidence $(\varphi_0 = \pi)$:

$$du_{h1}^{(0)} = u_0 e^{ik\zeta \cos\gamma_0} \frac{d\zeta}{4\pi} \frac{e^{ikR}}{R} \frac{\sin\gamma_0 \sin\vartheta \sin\varphi}{1 - \cos\beta_1} \left[e^{ikl(1-\cos\beta_1)} - 1 \right], \qquad (7.181)$$

$$du_{s1}^{(0)} = u_0 e^{-ik\zeta \cos\gamma_0} \frac{d\zeta}{2\pi} \frac{e^{ikR}}{R} \sin\gamma_0 \sqrt{\frac{2}{1-\cos\beta_1}} \frac{e^{i3\pi/4}}{\sqrt{\pi}} \int_0^{\sqrt{kl(1-\cos\beta_1)}} e^{it^2} dt. \qquad (7.182)$$

For the grazing scattering direction $(\beta_1 = 0, \varphi = 0)$, it follows from these equations that

$$du_{h1}^{(0)} = 0 \qquad (7.183)$$

and

$$du_{s1}^{(0)} = u_0 e^{-ik\zeta \cos\gamma_0} \frac{d\zeta}{2\pi} \frac{e^{ikR}}{R} \sin\gamma_0 \sqrt{\frac{2kl}{\pi}} e^{i3\pi/4}. \qquad (7.184)$$

As was expected, the field $du_{h,s}^{(0)}$ is also free from the grazing singularity.

7.9.2 Electromagnetic EEWs

This theory is based on the paper by Ufimtsev (2006b).

Asymptotic expressions for electromagnetic EEWs established in Section 7.8 possess the grazing singularity. Careful analysis reveals the reason for this singularity. As in the case of acoustic waves, it turns out that the original definitions of quantities $\vec{j}^{(0)}$ and $\vec{j}^{(1)}$ are not adequate for actual surface currents induced under the grazing incidence ($\varphi_0 = \pi$ or $\varphi_0 = \alpha - \pi$). In this case, for instance, the component $j_n^{(1)}$, normal to the edge, does not vanish away from the edge, but instead it transforms into a plane wave there. Also, the component $j_n^{(0)}$ includes the *absent* reflected wave. Therefore, one should modify the original definitions of $\vec{j}^{(0,1)}$ to avoid the grazing singularity.

With this purpose we introduce a *new uniform component* $\vec{j}^{(0)} \equiv \vec{j}^{\,\mathrm{hp}}$ identical to the electric surface current induced on the *illuminated side* of the tangential perfectly conducting half-plane. A *new nonuniform component* of the surface current $\vec{j}^{(1)}$ is the difference $\vec{j}^{(1)} = \vec{j}^{(t)} - \vec{j}^{\,\mathrm{hp}}$, where $\vec{j}^{(t)}$ is the total current induced on the tangential perfectly conducting wedge.

EEWs Radiated by the Nonuniform Component $\vec{j}^{(1)}$

The incident electromagnetic wave (7.129) undergoes diffraction at a scattering perfectly conducting object (Fig. 8.1) and creates there a surface current $\vec{j}^{(1)}$. This current radiates diffracted EEWs. Far from the diffraction point ζ ($kR \gg 1$), these waves can be presented again in the form of Equations (7.130) and (7.131) with new functions $\vec{F}^{(1)}$ and $\vec{G}^{(1)}$. These new functions can be found in the same way as described in Section 7.8 and in the papers by Butorin et al. (1987) and Ufimtsev (1991), with a simple modification based on the replacement (7.159). We provide here only the final results.

The field generated by $\vec{j}^{(1)}$ is described as

$$\mathrm{d}\vec{E}^{(1)} = \frac{\mathrm{d}\zeta}{2\pi}\vec{\mathcal{E}}^{(1)}(\vartheta,\varphi)\frac{e^{ikR}}{R} \qquad \text{and} \qquad \mathrm{d}\vec{H}^{(1)} = [\nabla R \times \mathrm{d}\vec{E}^{(1)}]/Z_0 \qquad (7.185)$$

with

$$\vec{\mathcal{E}}^{(1)}(\vartheta,\varphi) = [E_{0t}(\zeta)\vec{\mathcal{F}}^{(1)}(\vartheta,\varphi) + Z_0 H_{0t}(\zeta)\vec{\mathcal{G}}^{(1)}(\vartheta,\varphi)]e^{ik\phi^i(\zeta)}, \qquad (7.186)$$

where

$$\mathcal{F}_\vartheta^{(1)}(\vartheta,\varphi) = [U_t(\sigma_1,\varphi_0) - U_t^{\mathrm{hp}}(\sigma_1,\varphi_0)]\sin\vartheta$$

$$+ [U_t(\sigma_2,\alpha - \varphi_0) - \varepsilon(\alpha - \varphi_0)U_t^{\mathrm{hp}}(\sigma_2,\alpha - \varphi_0)]\sin\vartheta, \qquad (7.187)$$

$$\mathcal{F}_\varphi^{(1)}(\vartheta,\varphi) = 0, \qquad (7.188)$$

$$\mathcal{G}_\vartheta^{(1)}(\vartheta,\varphi) = [\sin\gamma_0\cos\vartheta\cos\varphi - \cos\gamma_0\sin\vartheta\cos\sigma_1][V_t(\sigma_1,\varphi_0) - V_t^{\mathrm{hp}}(\sigma_1,\varphi_0)]$$

$$- [\sin\gamma_0\cos\vartheta\cos(\alpha - \varphi) - \cos\gamma_0\sin\vartheta\cos\sigma_2]$$

$$\times [V_t(\sigma_2,\alpha - \varphi_0) - \varepsilon(\alpha - \varphi_0)V_t^{\mathrm{hp}}(\sigma_2,\alpha - \varphi_0)], \qquad (7.189)$$

and

$$\mathcal{G}_{\varphi}^{(1)}(\vartheta,\varphi) = -[V_t(\sigma_1,\varphi_0) - V_t^{\text{hp}}(\sigma_1,\varphi_0)]\sin\varphi\sin\gamma_0$$

$$- [V_t(\sigma_2,\alpha-\varphi_0) - \varepsilon(\alpha-\varphi_0)V_t^{\text{hp}}(\sigma_2,\alpha-\varphi_0)]\sin(\alpha-\varphi)\sin\gamma_0.$$
$$(7.190)$$

It is supposed here that $0 < \varphi_0 \le \pi$.

In the directions $\vartheta = \pi - \gamma_0$ associated with the diffraction cone, these expressions are simplified as

$$\mathcal{F}_{\vartheta}^{(1)}(\pi-\gamma_0,\varphi) = -\frac{1}{\sin\gamma_0}[f(\varphi,\varphi_0,\alpha)-f^{\text{hp}}(\varphi,\varphi_0)], \qquad \mathcal{F}_{\varphi}^{(1)}(\vartheta,\varphi) = 0,$$
$$(7.191)$$

$$\mathcal{G}_{\varphi}^{(1)}(\pi-\gamma_0,\varphi) = \frac{1}{\sin\gamma_0}[g(\varphi,\varphi_0,\alpha)-g^{\text{hp}}(\varphi,\varphi_0)], \qquad \mathcal{G}_{\vartheta}^{(1)}(\pi-\gamma_0,\varphi) = 0,$$
$$(7.192)$$

with functions f, g and $f^{\text{hp}}, g^{\text{hp}}$ defined above in Section 7.9.1. Comparison with Equations (7.167) and (7.168) reveals the following relationships between the electromagnetic and acoustic EEWs:

$$\mathcal{F}_{\vartheta}^{(1)}(\pi-\gamma_0,\varphi) = -\frac{1}{\sin\gamma_0}\mathcal{F}_{\text{s}}^{(1)}(\pi-\gamma_0,\varphi) \qquad (7.193)$$

and

$$\mathcal{G}_{\varphi}^{(1)}(\pi-\gamma_0,\varphi) = \frac{1}{\sin\gamma_0}F_{\text{h}}^{(1)}(\pi-\gamma_0,\varphi). \qquad (7.194)$$

According to Section 7.9.1 these functions are free from the grazing singularities as well as from the singularities in the directions of the incident and reflected rays.

Field Radiated by the Uniform Component $\vec{j}^{(0)} \equiv \vec{j}^{\text{hp}}$

Here we investigate the field radiated by the current $\vec{j}^{\text{hp}}$ induced on the finite elementary strips ($0 \le \xi_{1,2} \le l$) belonging to the finite plane faces of a scattering object (Fig. 7.3). In this investigation we apply Cartesian coordinates x, y, z and x', y', z associated with faces 1 and 2, respectively. Axes x and x' belong to faces 1 and 2, respectively, and they are parallel to tangents τ_1 and τ_2. Utilizing the known solution for the half-plane diffraction problem, one finds the current on strip 1,

$$
j_{x,1}^{(0)} \equiv j_{x,1}^{hp} = 2H_{0z}e^{-ik\zeta \cos \gamma_0}e^{ik\xi_1(\cos^2 \gamma_0 - \sin^2 \gamma_0 \cos \varphi_0)}
$$

$$
\times \left[1 - \frac{e^{-i\pi/4}}{\sqrt{\pi}} \cdot \int_{\sqrt{2k\xi_1} \sin \gamma_0 \cos \frac{\varphi_0}{2}}^{\infty} e^{it^2}\, dt \right], \tag{7.195}
$$

$$
j_{z,1}^{(0)} \equiv j_{z,1}^{hp} = -2e^{-ik\zeta \cos \gamma_0}e^{ik\xi_1 \cos^2 \gamma_0}\frac{1}{\sin \gamma_0}
$$

$$
\times \left\{ Y_0 E_{0z} \left[\sin \varphi_0 e^{-ik\xi_1 \sin^2 \gamma_0 \cos \varphi_0}\frac{e^{-i\pi/4}}{\sqrt{\pi}} \cdot \int_{\sqrt{2k\xi_1} \sin \gamma_0 \cos \frac{\varphi_0}{2}}^{\infty} e^{it^2}\, dt \right. \right.
$$

$$
\left. + \frac{\sin \frac{\varphi_0}{2}}{\sin \gamma_0}\frac{e^{-i3\pi/4}}{\sqrt{2\pi k\xi_1}}e^{ik\xi_1 \sin^2 \gamma_0} - \sin \varphi_0 e^{-ik\xi_1 \sin^2 \gamma_0 \cos \varphi_0} \right]
$$

$$
+ H_{0z} \cos \gamma_0 \left[\cos \varphi_0 e^{-ik\xi_1 \sin^2 \gamma_0 \cos \varphi_0}\frac{e^{-i\pi/4}}{\sqrt{\pi}} \cdot \int_{\sqrt{2k\xi_1} \sin \gamma_0 \cos \frac{\varphi_0}{2}}^{\infty} e^{it^2}\, dt \right.
$$

$$
\left. \left. + \frac{\cos \frac{\varphi_0}{2}}{\sin \gamma_0}\frac{e^{-i3\pi/4}}{\sqrt{2\pi k\xi_1}}e^{ik\xi_1 \sin^2 \gamma_0} - \cos \varphi_0 e^{-ik\xi_1 \sin^2 \gamma_0 \cos \varphi_0} \right] \right\} \tag{7.196}
$$

where $Y_0 = 1/Z_0$ is the admittance of free space. The current components $j_{x',2}^{(0)}$ and $j_{z,2}^{(0)}$ on strip 2 are determined by Equations (7.195) and (7.196) with the replacements

$$
H_{0z} \to -H_{0z}, \qquad \xi_1 \to \xi_2, \qquad \varphi_0 \to \alpha - \varphi_0. \tag{7.197}
$$

The field radiated by these currents is determined (in terms of the retarded vector-potential $d\vec{A}$) as

$$
dE_\varphi = -Z_0\, dH_\vartheta = ikZ_0[-dA_{x,1} \sin \varphi + \varepsilon(\alpha - \varphi_0)dA_{x',2} \sin(\alpha - \varphi)], \tag{7.198}
$$

and

$$
dE_\vartheta = Z_0\, dH_\varphi = ikZ_0\{-[dA_{z,1} + \varepsilon(\alpha - \varphi_0)dA_{z,2}] \sin \vartheta
$$

$$
+ [dA_{x,1} \cos \varphi + \varepsilon(\alpha - \varphi_0)dA_{x',2} \cos(\alpha - \varphi)] \cos \vartheta\}. \tag{7.199}
$$

It is supposed here that $0 < \varphi_0 \le \pi$. For the far zone $(R \gg kl^2)$, one can use the approximation

$$
d\vec{A}_{1,2} = \frac{d\zeta}{4\pi} \sin \gamma_0 \frac{e^{ikR}}{R} \int_0^l \vec{j}_{1,2}(\xi_{1,2})e^{-ik\xi_{1,2} \cos \beta_{1,2}}\, d\xi_{1,2}. \tag{7.200}
$$

After substitution of $\vec{j}_{1,2}^{(0)}$ into Equation (7.200), this leads to the following expressions:

$$dA_{x,1} = -\frac{1}{ik}H_{0z}e^{-ik\zeta\cos\gamma_0}\frac{d\zeta}{2\pi}[B_3(\varphi,\varphi_0) - B_1(\varphi,\varphi_0)]\sin\gamma_0\frac{e^{ikR}}{R}, \tag{7.201}$$

$$dA_{z,1} = -\frac{1}{ik}e^{-ik\zeta\cos\gamma_0}\frac{d\zeta}{2\pi}\frac{e^{ikR}}{R}$$

$$\times \left\{ Y_0 E_{0z} \left[B_3(\varphi,\varphi_0)\sin\varphi_0 + B_2(\varphi)\frac{\sin\frac{\varphi_0}{2}}{\sin\gamma_0} - B_1(\varphi,\varphi_0)\sin\varphi_0 \right] \right.$$

$$\left. + H_{0z}\cos\gamma_0 \left[B_3(\varphi,\varphi_0)\cos\varphi_0 + B_2(\varphi)\frac{\cos\frac{\varphi_0}{2}}{\sin\gamma_0} - B_1(\varphi,\varphi_0)\cos\varphi_0 \right] \right\}, \tag{7.202}$$

where

$$B_1(\varphi,\varphi_0) = \frac{e^{ikl\Phi(\varphi,\varphi_0)} - 1}{\Phi(\varphi,\varphi_0)}, \tag{7.203}$$

$$B_2(\varphi) = \frac{e^{-i\pi/4}}{\sqrt{\pi}}\sqrt{\frac{2}{\Psi(\varphi)}}\int_0^{\sqrt{kl\Psi(\varphi)}} e^{it^2}\,dt, \tag{7.204}$$

$$B_3(\varphi,\varphi_0) = \frac{1}{\Phi(\varphi,\varphi_0)}\left[e^{ikl\Phi(\varphi,\varphi_0)}\frac{e^{-i\pi/4}}{\sqrt{\pi}}\int_{\sqrt{2kl}\sin\gamma_0\cos\frac{\varphi_0}{2}}^{\infty} e^{it^2}\,dt - \frac{1}{2}\right.$$

$$\left. + B_2(\varphi)\sin\gamma_0\cos\frac{\varphi_0}{2}\right] \tag{7.205}$$

with

$$\Phi(\varphi,\varphi_0) = \cos^2\gamma_0 - \sin^2\gamma_0\cos\varphi_0 - \cos\beta_1, \tag{7.206}$$

$$\Psi(\varphi) = 1 - \cos\beta_1. \tag{7.207}$$

Components $dA_{x',2}$ and $dA_{z,2}$ are found from $dA_{x,1}$ and $dA_{z,1}$, respectively, with replacements $H_{0z} \to -H_{0z}$, $\varphi \to \alpha - \varphi$, $\varphi_0 \to \alpha - \varphi_0$, and $\beta_1 \to \beta_2$.

We then substitute the above equations for the vector $d\vec{A}$ into Equations (7.198) and (7.199) and obtain the field expressions in the form of Equations (7.130) and (7.131):

$$d\vec{E}^{(0)} = \frac{d\zeta}{2\pi}\vec{\mathcal{E}}^{(0)}(\vartheta,\varphi)\frac{e^{ikR}}{R}, \qquad d\vec{H}^{(0)} = [\nabla R \times d\vec{E}^{(0)}]/Z_0, \tag{7.208}$$

$$\vec{\mathcal{E}}^{(0)}(\vartheta,\varphi) = [E_{0t}(\zeta)\vec{\mathcal{F}}^{(0)}(\vartheta,\varphi) + Z_0 H_{0t}(\zeta)\vec{\mathcal{G}}^{(0)}(\vartheta,\varphi)]e^{ik\phi^i(\zeta)}, \tag{7.209}$$

where

$$\mathcal{F}_\vartheta^{(0)} = \mathcal{F}_{1\vartheta}(\varphi, \varphi_0, \vartheta) + \varepsilon(\alpha - \varphi_0)\mathcal{F}_{1\vartheta}(\alpha - \varphi, \alpha - \varphi_0, \vartheta), \qquad \mathcal{F}_\varphi^{(0)} = 0, \quad (7.210)$$

$$\mathcal{G}_\vartheta^{(0)} = \mathcal{G}_{1\vartheta}(\varphi, \varphi_0, \vartheta) - \varepsilon(\alpha - \varphi_0)\mathcal{G}_{1\vartheta}(\alpha - \varphi, \alpha - \varphi_0, \vartheta), \qquad (7.211)$$

$$\mathcal{G}_\varphi^{(0)} = \mathcal{G}_{1\varphi}(\varphi, \varphi_0, \vartheta) + \varepsilon(\alpha - \varphi_0)\mathcal{G}_{1\varphi}(\alpha - \varphi, \alpha - \varphi_0, \vartheta). \qquad (7.212)$$

Here,

$$\mathcal{F}_{1\vartheta}(\varphi, \varphi_0, \vartheta) = \left[B_3(\varphi, \varphi_0)\sin\varphi_0 + B_2(\varphi)\frac{\sin\dfrac{\varphi_0}{2}}{\sin\gamma_0} - B_1(\varphi, \varphi_0)\sin\varphi_0 \right]\sin\vartheta,$$

$$(7.213)$$

$$\mathcal{G}_{1\vartheta}(\varphi, \varphi_0, \vartheta) = \left[B_3(\varphi, \varphi_0)\cos\varphi_0 + B_2(\varphi)\frac{\cos\dfrac{\varphi_0}{2}}{\sin\gamma_0} - B_1(\varphi, \varphi_0)\cos\varphi_0 \right]\cos\gamma_0\sin\vartheta$$

$$- [B_3(\varphi, \varphi_0) - B_1(\varphi, \varphi_0)]\sin\gamma_0\cos\vartheta\cos\varphi, \qquad (7.214)$$

$$\mathcal{G}_{1\varphi}(\varphi, \varphi_0, \vartheta) = [B_3(\varphi, \varphi_0) - B_1(\varphi, \varphi_0)]\sin\gamma_0\sin\varphi. \qquad (7.215)$$

Notice that functions $B_{1,2,3}$ are finite when $\Phi = 0$ and $\Psi = 0$. In particular, for the grazing incidence ($\varphi_0 = \pi$) and for the grazing scattering ($\vartheta = \pi - \gamma_0$, $\varphi = 0$), they are equal to

$$B_1 = ikl, \qquad B_2 = \sqrt{2kl}\frac{e^{-i\pi/4}}{\sqrt{\pi}}, \qquad B_3 = \frac{1}{2}kl. \qquad (7.216)$$

As expected, the field generated by the uniform component of the surface current is also free from the grazing singularity.

Thus, the asymptotic theory developed in Sections 7.9.1 and 7.9.2 is valid for all directions of incidence and scattering. It is well suited for calculation of bistatic scattering in the case when both planar faces of the edge are illuminated by the incident wave ($\alpha - \pi \leq \varphi_0 \leq \pi$). For other incidence directions φ_0, one can apply the original theory presented in Sections 7.1 to 7.8.

Here it is pertinent to mention the alternative approach (Michaeli, 1987; Breinbjerg, 1992; Johansen, 1996) for elimination of the grazing singularity. The uniform and nonuniform components of the surface current are defined there according to the original PTD, and the grazing singularity is eliminated by truncation of elementary strips ($0 \leq x_{1,2} \leq l$). Compared to this approach, a distinctive feature of the present theory is as follows: It introduces a new *nonuniform* scattering source $\vec{j}^{(1)}$ that generates an elementary edge wave *regular* in all scattering directions. In other words, it allows the extraction of the *fringe* component from the total field in a *pure explicit form*.

7.10   SOME REFERENCES RELATED TO ELEMENTARY EDGE WAVES

The investigation of EEWs has a long history. In Kirchhoff's approach, the EEWs were first discovered by Maggi (Maggi, 1888; Bakker and Copson, 1950). The same result was rediscovered by Rubinowicz (1917). A similar approach to electromagnetic EEWs was developed by Kottler (1923).

Attempts to define EEWs more strictly (on the basis of the Sommerfeld (1896, 1935) exact solution of the wedge diffraction problem) were first undertaken by Bateman (1955) and Rubinowicz (1965). However, their expressions for EEWs satisfy the Dirichlet or Neuman boundary conditions *everywhere* at the faces of the canonical tangent wedge. For this reason, these EEWs predict incorrect values for the diffracted field at those parts of the *virtual* tangent wedge that are extended outside the real scattering object (as shown in Fig. 7.10 by dotted lines). Infact, according to this definition of EEWs, the *infinite* plane areas of *free space* (outside the scattering object) formally become perfectly reflecting.

The same drawback exists in another theory of EEWs suggested in by Tiberio et al. (1994, 1995, 2004). In PTD a similar shortcoming occurs only at the extensions of *infinitely narrow* elementary strips (Fig. 7.3). As PTD is a source-based theory, this shortcoming can be completely removed by the truncation of elementary strips (Johansen, 1996).

The directivity patterns of electromagnetic EEWs can be interpreted as *equivalent edge currents* (EECs). The EECs introduced in the work of Knott and Senior (1973) and Knott (1985) are based on GTD and are valid only for the directions of the diffracted rays. The EECs based on PTD are applicable for arbitrary scattering directions (Michaeli, 1986, 1987; Breinbjerg, 1992; Johansen, 1996). Notice that the *untruncated* EECs developed by Michaeli (1986) are in complete agreement with the EEWs derived in Section 7.8. However, his *truncated* EECs (Michaeli, 1987) are not free from some shortcomings, which were overcome by Johansen (1996). The paper by Johansen (1996) also contains additional references related to the EEC concept.

Another interpretation of the directivity pattern of EEWs is the so-called *incremental length diffraction coefficient* (ILDC). The term ILDC was introduced by Mitzner (1974), who determined the ILDCs on the basis of PTD. The ILDC concept was further developed in the work of Tiberio et al. (2004).

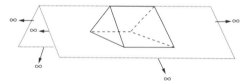

Figure 7.10   Perfectly reflecting solid prism of finite size (solid and dashed lines) and infinite faces of the virtual tangent wedge (dotted lines).

Notice also the asymptotic theory for plane screens (Wolf, 1967), which is similar to PTD and the method of matched asymptotic expansions (Tran Van Nhieu, 1995, 1996), which lead to the PTD ray asymptotics.

PROBLEMS

7.1 Start with Equations (7.3) and (7.4) and derive Equations (7.12) and (7.13) for the scattering surface sources j_s and j_h on strip 1.

7.2 Start with Equations (7.3) and (7.4) and derive Equations (7.16) and (7.17) for the scattering surface sources j_s and j_h on strip 2.

7.3 Start with Equations (7.19) and (7.20), use Equations (7.12) and (7.13), and verify the field expressions (7.25) and (7.26).

7.4 Start with Equations (7.19) and (7.20), use Equations (7.16) and (7.17), and verify the field expressions (7.29) and (7.30).

7.5 Start with Equation (7.37) for the pole $\sigma(p)$, verify its form (7.42), (7.43), and retrace the trajectory of this pole in the complex plane (η), when the argument p changes from $-\infty$ to $+\infty$.

7.6 Apply the Cauchy residue theorem to the integral (7.35) and verify its transformation into the form (7.64).

7.7 Apply the Cauchy residue theorem to the integral (7.36) and verify its transformation into the form (7.65).

7.8 Apply the stationary phase technique to the integrals (7.64) and (7.65) and prove the asymptotics (7.89), (7.90) for EEWs.

7.9 Verify that the directivity patterns $F_{s,h}^{(1)}$ of EEWs are always real functions, although their arguments $\sigma_{1,2}$ can be complex quantities.

7.10 Explain why the function $F_s^{(1)}$ equals zero for the grazing incidence ($\varphi_0 = 0$ or $\varphi_0 = \alpha$).

7.11 Functions $U_t(\sigma_1, \varphi_0)$, $V_t(\sigma_1, \varphi_0)$, as well as functions $U_0(\beta_1, \varphi_0)$, $V_0(\beta_1, \varphi_0)$, are singular at the point $\sigma_1 = \varphi_0$. Show that their differences, functions $U = U_t - U_0$ and $V = V_t - V_0$ remain finite there. Prove Equations (7.106) and (7.107).

7.12 Show that for the scattering directions $\vartheta = \pi - \gamma_0$, $0 \le \varphi \le \alpha$ (which belong to the diffraction cone, outside the wedge), functions $F_s^{(1)}$, $F_h^{(1)}$ transform into functions $f^{(1)}$, $g^{(1)}$, respectively. Prove Equations (7.115).

7.13 Show that for the scattering directions $\vartheta = \pi - \gamma_0$, $\alpha \le \varphi \le 2\pi$ (which belong to the diffraction cone, inside the wedge), the total field of EEWs equals zero. Prove Equations (7.116). Explain why this happens.

7.14 Prove that the incident wave

$$E_z^{inc} = E_{0z}e^{-ikz\cos\gamma_0}e^{-ikr\sin\gamma_0\cos(\varphi-\varphi_0)},$$

$$H_z^{inc} = H_{0z}e^{-ikz\cos\gamma_0}e^{-ikr\sin\gamma_0\cos(\varphi-\varphi_0)}$$

generates on elementary strip 1 (Fig. 7.3) the nonuniform currents

$$j_x^{(1)} = \frac{1}{2\alpha} H_{0z} e^{-ikz\cos\gamma_0} \int_D e^{-ik_1 x\cos\eta} [w^{-1}(\eta + \varphi_0) + w^{-1}(\eta - \varphi_0)] d\eta,$$

$$j_z^{(1)} = \frac{e^{-ikz\cos\gamma_0}}{2\alpha\sin\gamma_0} \times \left\{ H_{0z} \cos\gamma_0 e^{-ikz\cos\gamma_0} \right.$$

$$\times \int_D e^{-ik_1 x\cos\eta} [w^{-1}(\eta + \varphi_0) + w^{-1}(\eta - \varphi_0)] \cos\eta\, d\eta$$

$$\left. - Y_0 E_{0z} \int_D e^{-ik_1 x\cos\eta} [w^{-1}(\eta + \varphi_0) - w^{-1}(\eta - \varphi_0)] \sin\eta\, d\eta \right\}.$$

Hints:

- Represent the field excited by the incident wave around the wedge in the form

$$\vec{E} = \vec{E}(x, y)e^{-ikz\cos\gamma_0}, \qquad \vec{H} = \vec{H}(x, y)e^{-ikz\cos\gamma_0}.$$

- Use the Maxwell equations and express components $E_{r,\varphi}(x, y), H_{r,\varphi}$ as functions of components $E_z(x, y), H_z(x, y)$. See Equation (5.4) in Ufimtsev (1962).

- According to Chapter 4,

$$E_z = E_{0z} e^{-ikz\cos\gamma_0} [v(k_1 r, \varphi - \varphi_0) - v(k_1 r, \varphi + \varphi_0)],$$

$$H_z = H_{0z} e^{-ikz\cos\gamma_0} [v(k_1 r, \varphi - \varphi_0) + v(k_1 r, \varphi + \varphi_0)].$$

- Then define the current $\vec{j}^{(1)}$ as $\vec{j}^{(1)} = \hat{n} \times [\vec{H} - \vec{H}^{go}]$, where $\vec{H}$ is the total field around the wedge and $\vec{H}^{go}$ is its geometrical optics part. The x-axis is shown in Figure 7.3.

7.15 Use the same manipulations as those in Problem 7.14 and find the current $\vec{j}^{(1)}$ on elementary strip 2 (Fig. 7.3). Show that its components are determined by the equations:

$$j_{x'}^{(1)} = -\frac{1}{2\alpha} H_{0z} e^{-ikz\cos\gamma_0}$$

$$\times \int_D e^{-ik_1 x'\cos\eta} [w^{-1}(\eta + \alpha - \varphi_0) + w^{-1}(\eta - \alpha + \varphi_0)] d\eta,$$

$$j_z^{(1)} = -\frac{e^{-ikz\cos\gamma_0}}{2\alpha\sin\gamma_0}$$

$$\times \left\{ H_{0z} \cos\gamma_0 \int_D e^{-ik_1 x'\cos\eta} [w^{-1}(\eta + \alpha - \varphi_0) + w^{-1}(\eta - \alpha + \varphi_0)] \cos\eta\, d\eta \right.$$

$$\left. + Y_0 E_{0z} \int_D e^{-ik_1 x'\cos\eta} [w^{-1}(\eta + \alpha - \varphi_0) - w^{-1}(\eta - \alpha + \varphi_0)] \sin\eta\, d\eta \right\}.$$

The x'-axis is shown in Figure 7.3.

7.16   Find the vector-potential $dA_{x,z}^{(1)}$ generated by the nonuniform current $j_{x,z}^{(1)}$ induced on elementary strip 1 (Fig. 7.3). Follow the procedure shown below:

- Start with Equation (1.89). Substitute there the current $j_{x,z}^{(1)}$ found in Problem 7.14.
- Use Equation (7.21) for the Green function.
- Calculate the integral over variable ξ_1 (along the strip) in closed form.
- Apply the Cauchy theorem to the integral over variable η.
- Apply the asymptotic procedure (7.69) to integrals of the type of Equation (7.68).
- Represent the vector-potential in the form of a spherical wave diverging from the stationary point.

When you have this result, use Equations (1.92) and (1.93) and obtain the asymptotic expression for the wave generated by elementary strip 1. Having this, use the replacements

$$H_{0z} \rightarrow -H_{0z}, \qquad \beta_1 \rightarrow \beta_2, \qquad \sigma_1 \rightarrow \sigma_2, \qquad \varphi_0 \rightarrow \alpha - \varphi_0, \qquad \varphi \rightarrow \alpha - \varphi,$$

and obtain the wave generated by strip 2. The sum of these waves is the EEW shown in Equation (7.130).

7.17   Section 7.9.1 develops the asymptotics of acoustic EEWs free from the grazing singularity. Show that for the directions of the diffraction cone ($\vartheta = \pi - \gamma_0$), functions $\mathcal{F}_{h,s}^{(1)}$ take the form of Equations (7.167) and (7.168). Then prove Equations (7.173) and (7.174) for the specular direction $\varphi = \pi - \varphi_0$ and verify Equations (7.177) and (7.178) for the grazing incidence ($\varphi_0 = \pi$) and the grazing scattering ($\varphi = 0$).

7.18   Section 7.9.2 develops the asymptotics of electromagnetic EEWs free from the grazing singularity. Show that for the directions of the diffraction cone ($\vartheta = \pi - \gamma_0$), functions $\vec{\mathcal{F}}^{(1)}, \vec{\mathcal{G}}^{(1)}$ take the form of Equations (7.191) and (7.192). Then verify the relationships (7.193) and (7.194) between acoustic and electromagnetic waves.

Chapter **8**

Ray and Caustics Asymptotics for Edge Diffracted Waves

This chapter is based on the papers by Ufimtsev (1989, 1991).

8.1 RAY ASYMPTOTICS

The following relationships exist between the acoustic and electromagnetic diffracted rays:

$$u_s = E_t, \text{ if } u^{\text{inc}}(\zeta) = E_t^{\text{inc}}(\zeta); \qquad u_h = H_t, \text{ if } u^{\text{inc}}(\zeta) = H_t^{\text{inc}}(\zeta),$$

where $\hat{t}$ is the tangent to the edge at the diffraction point ζ.

8.1.1 Acoustic Waves

The theory of EEWs is applied here for calculation of scattering at a smoothly curved edge L with a slowly changing angle $\alpha(\zeta)$ between its faces (Fig. 8.1). In a small vicinity of the point ζ on the edge, an arbitrary incident field

$$u^{\text{inc}}(\zeta) = u_0(\zeta)e^{ik\phi^i(\zeta)} \tag{8.1}$$

can be locally considered as a plane wave propagating in the direction

$$\hat{k}^i = \nabla'\phi^i = \text{grad}'\phi^i. \tag{8.2}$$

Therefore, replacing the quantity $u^{\text{inc}}(\zeta)$ in Equations (7.89) and (7.90) and Equations (7.96) and (7.97) by Equation (8.1), we obtain the asymptotic expressions

Fundamentals of the Physical Theory of Diffraction. By Pyotr Ya. Ufimtsev
Copyright © 2007 John Wiley & Sons, Inc.

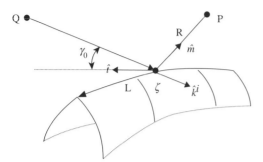

Figure 8.1   Element of a scattering edge L with a curvilinear coordinate ζ along the edge; $\hat{t}$ is the unit vector tangential to the edge at the point ζ.

for EEWs generated by an arbitrary incident wave. The resulting diffracted wave arising at the edge L and created by the nonuniform/fringe sources $j^{(1)}_{s,h}$ is a linear superposition of EEWs (7.89), (7.90),

$$u^{(1)}_{s,h} = \frac{1}{2\pi} \int_L u_0(\zeta) F^{(1)}_{s,h}(\zeta, \hat{m}) \frac{e^{ik\Phi(\zeta)}}{R(\zeta)} \, d\zeta, \tag{8.3}$$

and the edge wave generated by the total scattering sources $j^{(t)}_{s,h} = j^{(0)}_{s,h} + j^{(1)}_{s,h}$ is a linear superposition of EEWs (7.96), (7.97)

$$u^{(t)}_{s,h} = \frac{1}{2\pi} \int_L u_0(\zeta) F^{(t)}_{s,h}(\zeta, \hat{m}) \frac{e^{ik\Phi(\zeta)}}{R(\zeta)} \, d\zeta. \tag{8.4}$$

Here,

$$\hat{m} = \nabla R, \qquad \Phi = \phi^i + R, \tag{8.5}$$

and R is the distance between the edge point ζ and the observation point $P(x, y, z)$. Notice that the differential operator ∇' in Equation (8.2) acts on coordinates of the edge point ζ, but the operator ∇ in Equation (8.5) acts on coordinates x, y, z of the observation point P.

A high-frequency approximation (with $k \gg 1$) of the scattered field can be obtained by the stationary-phase technique (Copson, 1965; Murray, 1984), whose details have been already considered in Section 6.1.2. The stationary point ζ_{st} is determined by the equation

$$\frac{d\Phi}{d\zeta} = \nabla'\Phi \cdot \hat{t} = \nabla'(\phi^i + R) \cdot \hat{t} = (\hat{k}^i - \hat{m}) \cdot \hat{t} = 0. \tag{8.6}$$

Denote by $\hat{k}^s$ the unit vector $\hat{m}$ directed from the stationary point ζ_{st} to the observation point . Then Equation (8.6) can be rewritten as

$$\hat{k}^s \cdot \hat{t} = \hat{k}^i \cdot \hat{t} = -\cos\gamma_0. \tag{8.7}$$

Thus, the scattering directions $\hat{k}^s$ form a cone with its axis along the tangent $\hat{t}$ to the edge at the stationary point. Such a cone is shown in Figure 4.4.

The function Φ describes the distance between the points Q and P along the straight lines $Q\zeta$ and ζP (Fig. 8.1). Hence, Equation (8.6) indicates that this distance is extremal (minimal or maximal) when the point ζ is stationary. In other words, the location of the stationary point ζ_{st} on edge L satisfies the Fermat principle.

In accordance with the stationary-phase technique, the first term of the asymptotic expression for the field (8.3), (8.4) equals

$$u_{s,h}^{(1)} = \frac{1}{2\pi} u^{inc}(\zeta_{st}) F_{s,h}^{(1)}(\zeta_{st}, \hat{k}^s) \frac{e^{ikR}}{R} \int_{-\infty}^{\infty} e^{i\frac{k\Phi''(\zeta_{st})}{2}(\zeta - \zeta_{st})^2} \, d\zeta, \qquad (8.8)$$

where $\Phi''(\zeta_{st}) = d^2\Phi(\zeta_{st})/d\zeta^2$, and R is the distance between the stationary point ζ_{st} and the observation point P. Due to the equality

$$\int_{-\infty}^{\infty} e^{\pm ix^2} \, dx = \sqrt{\pi} e^{\pm i\frac{\pi}{4}}, \qquad (8.9)$$

Equation (8.8) can be written as

$$u_{s,h}^{(1)} = u^{inc}(\zeta_{st}) F_{s,h}^{(1)}(\zeta_{st}, \hat{k}^s) \frac{e^{i\pi/4}}{\sqrt{2\pi k \Phi''(\zeta_{st})}} \frac{e^{ikR}}{R}, \qquad (8.10)$$

where

$$\sqrt{\Phi''(\zeta_{st})} = \sqrt{|\Phi''(\zeta_{st})|} e^{i\frac{\pi}{2}}, \qquad \text{if } \Phi''(\zeta_{st}) < 0. \qquad (8.11)$$

In terms of the local spherical coordinates R, ϑ, φ (introduced in Fig. 7.3), the unit vectors $\hat{k}^s$ have the directions $\vartheta = \pi - \gamma_0$, $0 \leq \varphi \leq 2\pi$. For these directions, functions $F_{s,h}^{(1)}(\zeta_{st}, \hat{k}^s)$ are determined by Equations (7.115) and (7.116). Hence

$$u_s^{(1)} = u^{inc}(\zeta_{st}) f^{(1)}(\varphi, \varphi_0, \alpha) \frac{e^{i\pi/4}}{\sqrt{2\pi k \Phi''(\zeta_{st})}} \frac{e^{ikR}}{R} \qquad (8.12)$$

and

$$u_h^{(1)} = u^{inc}(\zeta_{st}) g^{(1)}(\varphi, \varphi_0, \alpha) \frac{e^{i\pi/4}}{\sqrt{2\pi k \Phi''(\zeta_{st})}} \frac{e^{ikR}}{R} \qquad (8.13)$$

in the directions $0 \leq \varphi \leq \alpha$, and

$$u_s^{(1)} = -u^{inc}(\zeta_{st}) f^{(0)}(\varphi, \varphi_0, \alpha) \frac{e^{i\pi/4}}{\sqrt{2\pi k \Phi''(\zeta_{st})}} \frac{e^{ikR}}{R} \qquad (8.14)$$

and

$$u_h^{(1)} = -u^{inc}(\zeta_{st})g^{(0)}(\varphi, \varphi_0, \alpha)\frac{e^{i\pi/4}}{\sqrt{2\pi k \Phi''(\zeta_{st})}}\frac{e^{ikR}}{R} \tag{8.15}$$

in the directions $\alpha < \varphi < 2\pi$, related to the region inside the tangential wedge.

The total diffracted field $u_{s,h}^{tot}$ radiated by the total sources $j_{s,h}^{tot} = j_{s,h}^{(1)} + j_{s,h}^{(0)}$ is described by Equations (8.12) and (8.13), where one should replace $f^{(1)}(\varphi, \varphi_0, \alpha)$, $g^{(1)}(\varphi, \varphi_0, \alpha)$ by functions $f(\varphi, \varphi_0, \alpha)$, $g(\varphi, \varphi_0, \alpha)$. In the region $\alpha < \varphi < 2\pi$ (inside the tangential wedge), the total diffracted field $u^{(1)} + u^{(0)}$ asymptotically equals zero, because $u^{(1)} = -u^{(0)}$ in accordance with Equations (8.14) and (8.15).

The above asymptotics for edge diffracted waves can be presented in another form that reveals their ray structure. To do this, we utilize the following differential operations:

$$\Phi' = \frac{d\Phi}{d\zeta} = \hat{t} \cdot \nabla'(\phi^{inc} + R) = \hat{t} \cdot \hat{k}^i + \hat{t} \cdot \nabla'R = -\cos\gamma_0 + \hat{t} \cdot \nabla'R, \tag{8.16}$$

$$\Phi'' = \frac{d\Phi'}{d\zeta} = \sin\gamma_0 \frac{d\gamma_0}{d\zeta} + \frac{d\nabla'R}{d\zeta} \cdot \hat{t} + \nabla'R \cdot \frac{d\hat{t}}{d\zeta}, \tag{8.17}$$

$$\frac{d\nabla'R}{d\zeta} \cdot \hat{t} = \frac{1}{R}[1 - (\hat{t} \cdot \nabla'R)^2], \tag{8.18}$$

$$\frac{d\hat{t}}{d\zeta} = \frac{\hat{v}}{a}. \tag{8.19}$$

Here $\hat{v}$ is the unit vector of the principal normal to the edge L, and a is the radius of curvature of the edge.

At the stationary point, $\nabla'R = -\hat{k}^s$ and $\hat{t} \cdot \nabla'R = -\hat{t} \cdot \hat{k}^s = \cos\gamma_0$. Therefore,

$$\frac{d\nabla'R}{d\zeta} \cdot \hat{t} = \frac{\sin^2\gamma_0}{R}, \tag{8.20}$$

$$\nabla'R \cdot \frac{d\hat{t}}{d\zeta} = -\frac{\hat{k}^s \cdot \hat{v}}{a}. \tag{8.21}$$

In view of relationships (8.16)–(8.19) and (8.20), (8.21),

$$\Phi''(\zeta_{st}) = \frac{1}{R}\left(1 + \frac{R}{\rho}\right)\sin^2\gamma_0, \tag{8.22}$$

where

$$\frac{1}{\rho} = \frac{1}{\sin\gamma_0}\left(\frac{d\gamma_0}{d\zeta} - \frac{\hat{k}^s \cdot \hat{v}}{a\sin\gamma_0}\right). \tag{8.23}$$

The quantity ρ is a caustic parameter; it determines the distance $(R = -\rho)$ along the ray from the edge to the caustic.

Now the edge diffracted field can be written in the ray form

$$u_{s,h}^{(1)} = u^{inc}(\zeta_{st}) \cdot (DF) \cdot (DC) \cdot e^{ikR}, \tag{8.24}$$

where

$$DF = \frac{1}{\sqrt{R|1 + R/\rho|}} \tag{8.25}$$

is the rays' divergence factor, and

$$DC = \frac{e^{\pm i\frac{\pi}{4}}}{\sin \gamma_0 \sqrt{2\pi k}} \begin{Bmatrix} f^{(1)}(\varphi, \varphi_0, \alpha) \\ g^{(1)}(\varphi, \varphi_0, \alpha) \end{Bmatrix} \tag{8.26}$$

can be interpreted as the diffraction coefficient. The quantities R and $R + \rho$ are the two principal radii of curvature of the diffracted phase front. The upper sign in $e^{\pm i\pi/4}$ is taken if $\Phi''(\zeta_{st}) > 0$ and the lower one if $\Phi''(\zeta_{st}) < 0$. The last multiplier in Equation (8.24), e^{ikR}, is the phase factor.

The divergence factor shows how the edge waves, being cylindrical-like waves in the vicinity of the edge $(R \ll |\rho|)$,

$$(DF)e^{ikR} \approx \frac{e^{ikR}}{\sqrt{R}}, \tag{8.27}$$

transform into spherical waves,

$$(DF)e^{ikR} \approx \sqrt{|\rho|} \frac{e^{ikR}}{R}, \tag{8.28}$$

at a large distance from the edge $(R \gg \rho)$.

The total edge diffracted fields can be also represented in ray form (with $kR \gg 1$)

$$u_{s,h}^{(t)} = u^{inc}(\zeta_{st}) \frac{1}{\sqrt{R(1 + R/\rho)}} \frac{e^{i\frac{\pi}{4}}}{\sin \gamma_0 \sqrt{2\pi k}} \begin{Bmatrix} f(\varphi, \varphi_0, \alpha) \\ g(\varphi, \varphi_0, \alpha) \end{Bmatrix} e^{ikR}. \tag{8.29}$$

We note that all variable parameters and coordinates in Equation (8.29) relate to the stationary point ζ_{st}.

Notice that the PTD ray asymptotics (8.14), (8.15) with the second derivative $\Phi''(\zeta)$ is much easier for the calculation than the GTD form (8.29), which involves complicated calculations of the caustic parameter $\rho(\zeta)$.

8.1.2 Electromagnetic Waves

According to Equations (7.130) and (7.131), the EEWs diverging from a scattering edge L create the combined wave

$$\vec{E}^{(1,t)} = \frac{1}{2\pi} \int_L \vec{\mathcal{E}}^{(1,t)}(\zeta) \frac{e^{ikR(\zeta)}}{R(\zeta)} \, d\zeta \tag{8.30}$$

and

$$\vec{H}^{(1,t)} = \frac{1}{2\pi Z_0} \int_L [\nabla R(\zeta) \times \vec{\mathcal{E}}^{(1,t)}(\zeta)] \frac{e^{ikR(\zeta)}}{R(\zeta)} \, d\zeta, \tag{8.31}$$

with

$$\vec{\mathcal{E}}^{(1,t)}(\zeta) = [E_{0t}(\zeta)\vec{F}^{(1,t)}(\zeta) + Z_0 H_{0t}(\zeta)\vec{G}^{(1,t)}(\zeta)]e^{ik\phi^i(\zeta)}. \tag{8.32}$$

Here, the integrands contain the fast oscillating factor $\exp[ik\Phi(\zeta)]$, where $\Phi(\zeta) = R(\zeta) + \phi^i(\zeta)$. Therefore, the application of the stationary-phase technique to these integrals results in the following ray asymptotics:

$$\left.\begin{array}{l} E_\vartheta^{(1)} = Z_0 H_\varphi^{(1)} \\ E_\vartheta^{(t)} = Z_0 H_\varphi^{(t)} \end{array}\right\} = -\frac{E_t^{\text{inc}}(\zeta_{\text{st}})}{\sin\gamma_0} \left[\frac{f^{(1)}(\varphi,\varphi_0,\alpha)}{f(\varphi,\varphi_0,\alpha)}\right] \frac{e^{i\pi/4}}{\sqrt{2\pi k \Phi''(\zeta_{\text{st}})}} \frac{e^{ikR}}{R} \tag{8.33}$$

and

$$\left.\begin{array}{l} E_\varphi^{(1)} = -Z_0 H_\vartheta^{(1)} \\ E_\varphi^{(t)} = -Z_0 H_\vartheta^{(t)} \end{array}\right\} = \frac{Z_0 H_t^{\text{inc}}(\zeta_{\text{st}})}{\sin\gamma_0} \left[\frac{g^{(1)}(\varphi,\varphi_0,\alpha)}{g(\varphi,\varphi_0,\alpha)}\right] \frac{e^{i\pi/4}}{\sqrt{2\pi k \Phi''(\zeta_{\text{st}})}} \frac{e^{ikR}}{R}. \tag{8.34}$$

They can also be written in the form of Equation (8.29):

$$\left.\begin{array}{l} E_\vartheta^{(1)} = Z_0 H_\varphi^{(1)} \\ E_\vartheta^{(t)} = Z_0 H_\varphi^{(t)} \end{array}\right\} = -\frac{E_t^{\text{inc}}(\zeta_{\text{st}})}{\sin^2\gamma_0} \left[\frac{f^{(1)}(\varphi,\varphi_0,\alpha)}{f(\varphi,\varphi_0,\alpha)}\right] \frac{e^{i\pi/4}}{\sqrt{2\pi k}} \frac{e^{ikR}}{\sqrt{R(1+R/\rho)}}, \tag{8.35}$$

and

$$\left.\begin{array}{l} E_\varphi^{(1)} = -Z_0 H_\vartheta^{(1)} \\ E_\varphi^{(t)} = -Z_0 H_\vartheta^{(t)} \end{array}\right\} = \frac{Z_0 H_t^{\text{inc}}(\zeta_{\text{st}})}{\sin^2\gamma_0} \left[\frac{g^{(1)}(\varphi,\varphi_0,\alpha)}{g(\varphi,\varphi_0,\alpha)}\right] \frac{e^{i\pi/4}}{\sqrt{2\pi k}} \frac{e^{ikR}}{\sqrt{R(1+R/\rho)}}. \tag{8.36}$$

Taking into account Equations (7.149) and (7.150), one can represent these approximations in terms of the field components tangential to the scattering edge:

$$\left.\begin{array}{c} E_t^{(1)} \\ E_t^{(t)} \end{array}\right\} = E_t^{\text{inc}}(\zeta_{\text{st}}) \left[\begin{array}{c} f^{(1)}(\varphi,\varphi_0,\alpha) \\ f(\varphi,\varphi_0,\alpha) \end{array}\right] \frac{e^{i\pi/4}}{\sqrt{2\pi k \Phi''(\zeta_{\text{st}})}} \frac{e^{ikR}}{R}, \qquad (8.37)$$

$$\left.\begin{array}{c} H_t^{(1)} \\ H_t^{(t)} \end{array}\right\} = H_t^{\text{inc}}(\zeta_{\text{st}}) \left[\begin{array}{c} g^{(1)}(\varphi,\varphi_0,\alpha) \\ g(\varphi,\varphi_0,\alpha) \end{array}\right] \frac{e^{i\pi/4}}{\sqrt{2\pi k \Phi''(\zeta_{\text{st}})}} \frac{e^{ikR}}{R}. \qquad (8.38)$$

8.1.3 Comments on Ray Asymptotics

- A comparison of Equations (8.12), (8.13) and (8.37), (8.38) reveals the following relationships between acoustic and electromagnetic diffracted rays:

$$u_s = E_t, \quad \text{if } u^{\text{inc}}(\zeta) = E_t^{\text{inc}}(\zeta) \qquad (8.39)$$

$$u_h = H_t, \quad \text{if } u^{\text{inc}}(\zeta) = H_t^{\text{inc}}(\zeta), \qquad (8.40)$$

where ζ is the diffraction point on a scattering edge. These relationships (together with Equations (7.149), (7.150)) allow one to completely determine the field of electromagnetic rays diffracted at a perfectly conducting object, if one knows the acoustic rays diffracted at soft and rigid objects of the same shape and size. Notice that these relationships were established earlier, in the paper by Ufimtsev (1995).

- The ray asymptotics (8.29), (8.35), (8.36) (for the field generated by the total scattering sources $j^{(t)} = j^{(0)} + j^{(1)}$) were postulated in the Geometrical Theory of Diffraction (GTD) (Keller, 1962). Now it is seen that GTD can be interpreted as the ray asymptotic form of PTD for the total diffracted field. Notice that the ray asymptotics of Equation (8.29) type (but in the Kirchhoff approximation) were obtained first by Rubinowicz (1924). In the paper by Ufimtsev (1995), it is shown that the above ray asymptotics can be easily obtained by the direct extension of the Rubinowicz theory.

- In contrast to PTD, GTD is not applicable in the regions where the field does not have a ray structure and where the actual diffraction phenomena happen (GO boundaries, foci, caustics). Several ray-based techniques have been developed to overcome the deficiencies of GTD (Kouyoumjian and Pathak, 1974; James, 1980; Borovikov and Kinber, 1994). Among them, the most developed for practical applications is the Uniform Theory of Diffraction (Kouyoumjian and Pathak, 1974; McNamara et al. 1990).

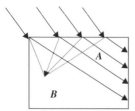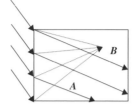

Figure 8.2 Rectangular facet of the scattering edge. The edge diffracted rays (solid arrows) exist only in region A and satisfy the boundary conditions there. Individual elementary edge rays (dotted arrows) in region B do not satisfy the boundary conditions, but they asymptotically cancel each other there.

- The ray asymptotics (8.29), as well as (8.35), and (8.36) for the fields $\vec{E}^{(t)}, \vec{H}^{(t)}$ are invariant with respect to the permutations $\vartheta \leftrightarrow \gamma_0, \varphi \leftrightarrow \varphi_0$, and therefore they satisfy the reciprocity principle. We note that these expressions are valid only in the directions of the diffraction cone ($\vartheta = \pi - \gamma_0$). Away from this cone, the total field $u_{s,h}^{tot}$ is asympotically (with $k \to \infty$) equal to zero, due to the absence of the stationary point on the edge. In this region, the individual elementary edge waves asymptotically cancel each other.

- The asymptotic expressions (8.29) (and $\vec{E}^{(t)}, \vec{H}^{(t)}$ in Equations (8.35), (8.36)) satisfy the boundary conditions on the planar faces of the edge. A situation with these conditions is illustrated in Figure 8.2.

- The ray asymptotics are not valid at caustics ($R = 0$ and $R = -\rho$), where they predict an infinitely large field intensity. The caustic $R = 0$ is located at the edge itself. The caustic $R = -\rho$ can be real or imaginary. A real caustic occurs outside the scattering object in the positive direction of the vector $\hat{k}^s$. An imaginary caustic is located in the direction contrary to $\hat{k}^s$. In particular, imaginary caustics may be inside the scattering body. The value $\Phi''(\zeta_{st}) > 0$ relates to the case when the edge diffracted ray has not yet reached a caustic, and the value $\Phi''(\zeta_{st}) < 0$ corresponds to the ray that has already passed a caustic and acquired there the additional phase shift equal to $-\pi/2$ (according to Equation (8.11)).

- The theory of EEWs developed in Chapter 7 allows one to calculate the edge diffraction field in the vicinity of any caustics away from the scattering edge. An example of such a calculation is considered in the following section.

8.2 CAUSTIC ASYMPTOTICS

Caustic asymptotics are presented here for both acoustic and electromagnetic waves. These asymptotics have the same structure and differ only in coefficients.

8.2.1 Acoustic Waves

Suppose that the edge diffracted rays form a smooth caustic C (Fig. 8.3). It is the envelope of diffracted rays, where a high-intensity field concentrates. According to Section 8.1 the diffracted field away from the caustic (and in front of the caustic) is the sum of two rays coming from the stationary points ζ_1 and ζ_2 on the scattering edge L:

$$u = u_0(\zeta_1)\frac{e^{i\pi/4}}{\sqrt{2\pi k \Phi''(\zeta_1)}}F(\zeta_1)\frac{e^{ik\Phi(\zeta_1)}}{R_1} + u_0(\zeta_2)\frac{e^{-i\pi/4}}{\sqrt{2\pi k |\Phi''(\zeta_2)|}}F(\zeta_2)\frac{e^{ik\Phi(\zeta_2)}}{R_2},$$

$$(8.41)$$

where $\Phi(\zeta) = \phi^i(\zeta) + R(\zeta)$ and $\phi^i(\zeta)$ is the phase of the incident wave at the point ζ on the scattering edge. Depending on the type of the function $F(\zeta_{1,2})$, Equation (8.41) represents either the field $u_{s,h}^{(1)}$ generated by the nonuniform/fringe sources $j_{s,h}^{(1)}$ or the field $u_{s,h}^{tot}$ generated by the total scattering sources $j_{s,h}^{(t)} = j_{s,h}^{(1)} + j_{s,h}^{(0)}$. Specifically, the functions

$$F_s^{(1)}(\zeta_{1,2}) = f^{(1)}(\varphi_{1,2}, \varphi_{01,02}, \alpha_{1,2}), \qquad F_h^{(1)}(\zeta_{1,2}) = g^{(1)}(\varphi_{1,2}, \varphi_{01,02}, \alpha_{1,2})$$

$$(8.42)$$

relate to the field $u_{s,h}^{(1)}$, and the functions

$$F_s^{(t)}(\zeta_{1,2}) = f(\varphi_{1,2}, \varphi_{01,02}, \alpha_{1,2}), \qquad F_h^{(t)}(\zeta_{1,2}) = g(\varphi_{1,2}, \varphi_{01,02}, \alpha_{1,2}) \qquad (8.43)$$

correspond to the field $u_{s,h}^{(t)}$. We note that functions f, g, $f^{(1)}$, $g^{(1)}$ are defined in Equations (2.62), (2.64), (4.14), (4.15), and (3.55) to (3.57). The function $\Phi''(\zeta_{1,2})$ can be represented in the form of Equation (8.22).

The first term in Equation (8.41) describes the ray that did not yet reach the caustic and the second term relates to the ray that has already touched the caustic.

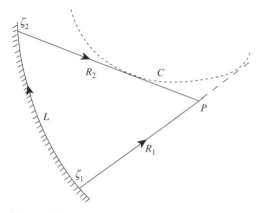

Figure 8.3   Edge diffracted rays at the point P in the vicinity of caustic C.

For the observation points behind the caustic, no stationary points exist on the edge and therefore no diffracted rays come here. This is a shadow region for diffracted rays. Below we develop the asymptotic approximation for the field in the illuminated region, in front of the caustic, including the points on the caustic itself.

We proceed with the general integral expression for the edge diffracted field of the Equations (8.3), (8.4) type,

$$u = \frac{1}{2\pi} \int_L u_0(\zeta) F(\zeta, \hat{m}) \frac{e^{ik\Phi}}{R} d\zeta, \qquad \hat{m} = \text{grad } R. \qquad (8.44)$$

In front of the caustic, the integrand in Equation (8.44) has two stationary points ζ_1 and ζ_2. As the observation point P approaches the caustic, the stationary points move toward each other, and merge when P reaches the caustic. In this case, $\Phi''(\zeta_{1,2}) \rightarrow 0$ and the ray asymptotics (8.41) become invalid. One can define a smooth caustic as a surface where both $\Phi'(\zeta) = 0$ and $\Phi''(\zeta) = 0$.

A main contribution to the integral (8.44) is provided by the stationary points. Real edges usually have ends. To extract the caustic effect in a pure form, we extend the integration limits in Equation (8.44) to infinity ($-\infty \leq \zeta \leq \infty$) and in this way remove the contributions of the ends to the field, which are usually less in magnitude than those of Equation (8.41).

Uniform asymptotics valid in the whole illuminated region, including the caustic, can be found by using the stationary-phase method, extended for the case of two merging stationary points (Chester, et al. 1957). In the following, we provide the basic details of this technique and derive the caustic asymptotics:

- As the first and second derivatives of the phase function $\Phi(\zeta)$ equal zero at the caustic, it is expedient to represent this function as a cubic polynomial

$$\Phi(\zeta) = \frac{1}{3}\tau^3 - \mu\tau + \psi \qquad (8.45)$$

  under the condition $\tau^2 = \mu = 0$ for the observation point on the caustic.

- The function $\tau(\zeta)$ shapes a three-sheeted Riemann surface. The regular branch of this function is selected by setting

$$\tau(\zeta_1) = \tau_1 = \mu^{1/2}, \qquad \tau(\zeta_2) = -\mu^{1/2}. \qquad (8.46)$$

  The quantities $\tau(\zeta_{1,2})$ are the stationary points of the function $\Phi[\zeta(\tau)]$. Parameters μ and ψ are found from Equation (8.45), setting $\zeta = \zeta_1$ and $\zeta = \zeta_2$:

$$\mu^{3/2} = \frac{3}{4}[\Phi(\zeta_2) - \Phi(\zeta_1)], \qquad \psi = \frac{1}{2}[\Phi(\zeta_1) + \Phi(\zeta_2)]. \qquad (8.47)$$

- According to Equation (8.41) and in agreement with Figure 8.3, the function $\Phi(\zeta)$ possesses the following properties:
  - $\Phi''(\zeta_1) \geq 0$ and $\Phi''(\zeta_2) \leq 0$. This means that $\Phi(\zeta_2) \geq \Phi(\zeta_1)$ and $\mu \geq 0$, $\mu^{1/2} \geq 0$, $\mu^{3/2} \geq 0$.

○ $\Phi'(\zeta) > 0$ for $\zeta_1 < \zeta < \zeta_2$ and $\Phi'(\zeta) < 0$ for $\zeta < \zeta_1$ and $\zeta > \zeta_2$.

○ $\Phi'''(\zeta) < 0$ in the vicinity of the merging point $\zeta_0 = \lim_{P \to C} \zeta_{1,2}$.

- It follows from Equation (8.45) that

$$\frac{d\zeta}{d\tau} = \frac{\tau^2 - \mu}{\Phi'(\zeta)}, \qquad \text{if } \Phi'(\zeta) \neq 0. \tag{8.48}$$

Due to the properties of function $\Phi(\zeta)$ and in view of relationships (8.46), this derivative is negative everywhere ($d\zeta/d\tau < 0$). This observation is helpful to choose the correct sign of $d\zeta/d\tau$ at the stationary points $\zeta_{1,2}$ where $\Phi'(\zeta_{1,2}) = 0$ and $\Phi''(\zeta_0) = 0$. The corresponding expressions for $d\zeta/d\tau$ at these points are found by the subsequent differentiation of (8.45):

$$\frac{d\zeta}{d\tau} = -\sqrt{\frac{2\mu^{1/2}}{|\Phi''(\zeta_{1,2})|}}, \qquad \text{if } \Phi'(\zeta_{1,2}) = 0, \tag{8.49}$$

$$\frac{d\zeta}{d\tau} = -\left[\frac{2}{|\Phi'''(\zeta_0)|}\right]^{3/2}, \qquad \text{if } \Phi'(\zeta_0) = \Phi''(\zeta_0) = 0, \tag{8.50}$$

where $\sqrt{|x|} > 0$ and $(|x|)^{3/2} > 0$.

- After transition to the new variable $\tau = \tau(\zeta)$ in the integral (8.44), we again avoid the contributions from the end points by extending the integration limits to infinity. To be consistent with Equation (8.46), we choose the integration limit $\tau = -\infty$ ($\tau = \infty$) when $\zeta = \infty$ ($\zeta = -\infty$). In addition we take into account that $d\zeta/d\tau = -|d\zeta/d\tau|$. Finally, the integral (8.44) can be represented as

$$u = \frac{1}{2\pi} e^{ik\psi} \int_{-\infty}^{\infty} G(\tau) e^{ik(\frac{\tau^3}{3} - \mu\tau)} \, d\tau \tag{8.51}$$

where

$$G(\tau) = \frac{u_0(\zeta)}{R(\zeta)} F(\zeta, \hat{m}) \left|\frac{d\zeta}{d\tau}\right|. \tag{8.52}$$

- Then, the function $G(\tau)$ is expanded into the series

$$G(\tau) = \sum_{n=0}^{\infty} p_n (\tau - \mu)^n + \sum_{n=0}^{\infty} q_n \tau (\tau - \mu)^n. \tag{8.53}$$

By integration of this series in Equation (8.51) one can obtain the asymptotic expansion valid for $k \to \infty$. We retain here only the two leading terms in Equation (8.53),

$$G(\tau) = p + q\tau \tag{8.54}$$

where $p = p_0$ and $q = q_0$. Setting here $\tau = \tau_1$ and $\tau = \tau_2$, one finds

$$p = \frac{1}{2}[G(\mu^{1/2}) + G(-\mu^{1/2})], \qquad q = \frac{1}{2\mu^{1/2}}[G(\mu^{1/2}) - G(-\mu^{1/2})].$$
(8.55)

• The further substitution of Equation (8.54) into (8.51) leads to the asymptotic expression

$$u \sim k^{-1/3}e^{ik\psi}[p\,\mathrm{Ai}(-k^{2/3}\mu) - ik^{-1/3}q\,\mathrm{Ai}'(-k^{2/3}\mu)], \qquad \text{with } k \to \infty, \quad (8.56)$$

where

$$\mathrm{Ai}(t) = \frac{1}{2\pi}\int_{-\infty}^{\infty} e^{i(\frac{x^3}{3}+tx)}\,\mathrm{d}x = \frac{1}{\pi}\int_{0}^{\infty}\cos\left(\frac{x^3}{3}+tx\right)\mathrm{d}x \qquad (8.57)$$

is the Airy function (Abramowitz and Stegun, 1972) and

$$\mathrm{Ai}'(t) = \frac{\mathrm{d}}{\mathrm{d}t}\mathrm{Ai}(t) = -\frac{1}{\pi}\int_{0}^{\infty} x\sin\left(\frac{x^3}{3}+tx\right)\mathrm{d}x. \qquad (8.58)$$

Notice also that

$$\mathrm{Ai}'(-t) = \frac{\mathrm{d}}{\mathrm{d}(-t)}\mathrm{Ai}(-t) = -\frac{\mathrm{d}}{\mathrm{d}t}\mathrm{Ai}(-t). \qquad (8.59)$$

The asymptotic approximation (8.56) is uniformly valid in the whole illuminated region, including the caustic. In particular, on the caustic itself,

$$u = k^{-1/3}p(\zeta_0)\mathrm{Ai}(0)e^{ik\Phi(\zeta_0)} + O(k^{-2/3}) \qquad (8.60)$$

where, according to Equation (10.4.4) in (Abramowitz and Stegun, 1972),

$$\mathrm{Ai}(0) = 3^{-2/3}/\Gamma(2/3) \approx 0.35502. \qquad (8.61)$$

In order to specify the final asymptotics (8.56) for the fields $u_{s,h}^{(1)}$, we obtain the explicit expressions for coefficients p and q:

$$\left.\begin{matrix} p_s^{(1)} \\ p_h^{(1)} \end{matrix}\right\} = \frac{1}{2}\left\{ \frac{u_0(\zeta_1)}{R(\zeta_1)}\begin{bmatrix} f^{(1)}(\varphi_1,\varphi_{01},\alpha_1) \\ g^{(1)}(\varphi_1,\varphi_{01},\alpha_1) \end{bmatrix}\left|\frac{\mathrm{d}\zeta(\tau_1)}{\mathrm{d}\tau}\right| \right.$$

$$\left. + \frac{u_0(\zeta_2)}{R(\zeta_2)}\begin{bmatrix} f^{(1)}(\varphi_2,\varphi_{02},\alpha_2) \\ g^{(1)}(\varphi_2,\varphi_{02},\alpha_2) \end{bmatrix}\left|\frac{\mathrm{d}\zeta(\tau_2)}{\mathrm{d}\tau}\right| \right\}$$
(8.62)

and

$$
\left.\begin{array}{c} q_s^{(1)} \\ q_h^{(1)} \end{array}\right\} = \frac{1}{2\mu^{1/2}} \left\{ \frac{u_0(\zeta_1)}{R(\zeta_1)} \left[\begin{array}{c} f^{(1)}(\varphi_1,\varphi_{01},\alpha_1) \\ g^{(1)}(\varphi_1,\varphi_{01},\alpha_1) \end{array} \right] \left| \frac{d\zeta(\tau_1)}{d\tau} \right| \right.
$$
$$
\left. - \frac{u_0(\zeta_2)}{R(\zeta_2)} \left[\begin{array}{c} f^{(1)}(\varphi_2,\varphi_{02},\alpha_2) \\ g^{(1)}(\varphi_2,\varphi_{02},\alpha_2) \end{array} \right] \left| \frac{d\zeta(\tau_2)}{d\tau} \right| \right\}, \tag{8.63}
$$

where the quantity μ is defined in Equation (8.47). After replacement of functions $f^{(1)}$, $g^{(1)}$ by f, g (or by $f^{(0)}$, $g^{(0)}$), the expressions (8.62) and (8.63) determine the coefficients $p_{s,h}^{(t)}$, $q_{s,h}^{(t)}$ (or $p_{s,h}^{(0)}$, $q_{s,h}^{(0)}$) related to the fields $u_{s,h}^{(t)}$ (or to $u_{s,h}^{(0)}$).

For large real arguments ($t \gg 1$), the Airy function and its derivative are determined by the asymptotic expressions (Abramowitz and Stegun, 1972):

$$
\text{Ai}(-t) \sim \pi^{-1/2}t^{-1/4} \sin\left(\frac{2}{3}t^{3/2} + \frac{\pi}{4}\right) \tag{8.64}
$$

and

$$
\text{Ai}'(-t) \sim -\pi^{-1/2}t^{1/4} \cos\left(\frac{2}{3}t^{3/2} + \frac{\pi}{4}\right). \tag{8.65}
$$

Utilizing these approximations one can show that, far from the caustic ($k^{2/3}\mu \gg 1$), the general asymptotics (8.56) transform into the ray asymptotics (8.41).

8.2.2 Electromagnetic Waves

For electromagnetic waves the caustic asymptotics are derived in the same way (Ufimtsev, 1991) and can be written as

$$
\left[\begin{array}{c} \vec{E}^{(1)} \\ \vec{H}^{(1)} \end{array} \right] \sim k^{-1/3} e^{ik\psi} \left\{ \left[\begin{array}{c} \vec{p}_e^{(1)} \\ \vec{p}_h^{(1)} \end{array} \right] \text{Ai}(-k^{2/3}\mu) - ik^{-1/3} \left[\begin{array}{c} \vec{q}_e^{(1)} \\ \vec{q}_h^{(1)} \end{array} \right] \text{Ai}'(-k^{2/3}\mu) \right\}
$$

with $k \to \infty$ \hfill (8.66)

Here,

$$
\vec{p}_e^{(1)} = \frac{1}{2} \left[\vec{\mathcal{E}}^{(1)}(\zeta_1) \frac{1}{R(\zeta_1)} \left| \frac{d\zeta(\tau_1)}{d\tau} \right| + \vec{\mathcal{E}}^{(1)}(\zeta_2) \frac{1}{R(\zeta_2)} \left| \frac{d\zeta(\tau_2)}{d\tau} \right| \right] \tag{8.67}
$$

and

$$\vec{q}_e^{(1)} = \frac{1}{2\sqrt{\mu}} \left[\vec{\mathcal{E}}^{(1)}(\zeta_1) \frac{1}{R(\zeta_1)} \left| \frac{d\zeta(\tau_1)}{d\tau} \right| - \vec{\mathcal{E}}^{(1)}(\zeta_2) \frac{1}{R(\zeta_2)} \left| \frac{d\zeta(\tau_2)}{d\tau} \right| \right], \qquad (8.68)$$

and, according to Equations (7.136), (7.141), and (7.143)

$$\mathcal{E}_{\vartheta_m}^{(1)}(\zeta_m) = -E_{t_m}^{\text{inc}}(\zeta_m) \frac{1}{\sin \gamma_{0m}} f^{(1)}(\varphi_m, \varphi_{0m}, \alpha_m) \qquad (8.69)$$

and

$$\mathcal{E}_{\varphi_m}^{(1)}(\zeta_m) = Z_0 H_{t_m}^{\text{inc}} \frac{1}{\sin \gamma_{0m}} g^{(1)}(\varphi_m, \varphi_{0m}, \alpha_m). \qquad (8.70)$$

The subscript $m = 1, 2$ indicates that a quantity with this subscript relates to the stationary point ζ_1 or ζ_2 at the edge.

Formulas like Equations (8.67) and (8.68) define the vectors $\vec{p}_h^{(1)}$ and $\vec{q}_h^{(1)}$. It is only necessary to replace the vector $\vec{\mathcal{E}}^{(1)}$ by $\vec{\mathcal{H}}^{(1)} = [\nabla R \times \vec{\mathcal{E}}^{(1)}]/Z_0$.

After replacement of functions $f^{(1)}$, $g^{(1)}$ by f, g (or by $f^{(0)}$, $g^{(0)}$), the expressions (8.62), (8.63) determine the coefficients $\vec{p}_{e,h}^{(t)}$, $\vec{q}_{e,h}^{(t)}$ (or $\vec{p}_{e,h}^{(0)}$, $\vec{q}_{e,h}^{(0)}$) related to the fields $\vec{E}^{(t)}$, $\vec{H}^{(t)}$ (or to $\vec{E}^{(0)}$, $\vec{H}^{(0)}$).

The derived asymptotics for the field $u_{s,h}^{(1)}$, $\vec{E}^{(1)}$, $\vec{H}^{(1)}$ (with functions $f^{(1)}$, $g^{(1)}$) have an important advantage compared to the asymptotics for the total field $u_{s,h}^{(t)}$, $\vec{E}^{(t)}$, $\vec{H}^{(t)}$. They remain finite at the boundaries of ordinary incident and reflected rays ($\varphi = \pi \pm \varphi_0$, $\varphi = 2\alpha - \pi - \varphi_0$), where functions f, g become singular.

It follows from Equations (8.56) and (8.66) that the structure of the caustic field is the same for acoustic and electromagnetic waves. The difference is only in the coefficients p and q, which are scalar quantities for acoustic waves and vectors for electromagnetic waves.

Notice also that asymptotics (8.56) and (8.66) are valid for the field calculation only in the illuminated region, in front of the caustic. When the observation point moves across the caustic into the shadow region, the diffracted field continuously changes and exponentially attenuates, because the elementary edge waves asymptotically cancel each other there. We do not consider this topic here. Details regarding the wave field in the vicinity of arbitrary caustics can be found in the review paper by Kravtsov and Orlov (1983).

PROBLEMS

8.1   Use the theory of EEWs from Section 7.5 and derive the edge wave scattered by an infinite straight edge of a wedge. The incident acoustic wave is given by

$$u^{\text{inc}} = u_0 \, e^{-ikz \cos \gamma_0} e^{-ikr \sin \gamma_0 \cos(\varphi - \varphi_0)}.$$

Consider the scattering by both a soft and a hard wedge. Integrate the EEWs over the whole edge. Apply the stationary-phase method and obtain the asymptotics (4.48) and (4.49).

8.2   Use the theory of EEWs from Section 7.8 and derive the edge wave scattered by an infinite straight edge of a perfectly conducting wedge. The incident electromagnetic wave is given by

$$E_z^{inc} = E_{0z}\, e^{-ikz \cos\gamma_0} e^{-ikr \sin\gamma_0 \cos(\varphi-\varphi_0)}$$

and

$$H_z^{inc} = H_{0z}\, e^{-ikz \cos\gamma_0} e^{-ikr \sin\gamma_0 \cos(\varphi-\varphi_0)}.$$

Integrate the EEWs over the whole edge. Apply the stationary-phase method and obtain the electromagnetic version of Equations (4.48) and (4.49).

8.3   Use the theory of EEWs from Section 7.5 and derive the edge wave $u_{s,h}^{(1)}$ scattered by a circular disk. The incident acoustic wave is given by

$$u^{inc} = u_0\, e^{ikz}.$$

Consider the scattering by both a soft and a hard disk. Integrate the EEWs over the whole edge.

- Find the focal asymptotics for the fields $u_{s,h}^{(1)}$. Compare with Equations (6.81) and (6.82).
- Apply the stationary-phase method and obtain the ray asymptotics. Compare with Equations (6.84) and (6.85).

8.4   Use the theory of EEWs from Section 7.8 and derive the edge wave $E_x^{(1)}$ scattered by a circular perfectly conducting disk. The incident electromagnetic wave is given by

$$E_x^{inc} = E_{0x}\, e^{ikz}.$$

Integrate the EEWs over the whole edge.

- Find the focal asymptotics for the fields $E_x^{(1)}$.
- Apply the stationary-phase method and obtain the ray asymptotics. Compare with Equations (6.84) and (6.85).

8.5   Use the theory of EEWs from Section 7.5 and derive the edge wave $u_{s,h}^{(1)}$ scattered by an elliptic disk ($y^2/a^2 + x^2/b^2 = 1$). The incident acoustic wave is given by

$$u^{inc} = u_0\, e^{ikz}.$$

Consider the scattering by both a soft and a hard disk. Integrate the EEWs over the whole edge.

- Find the focal asymptotics for the fields $u_{s,h}^{(1)}$. Compare with Equations (6.81) and (6.82).
- Apply the stationary-phase method and obtain the ray asymptotics. Compare with Equations (6.84) and (6.85).

8.6   Use the theory of EEWs from Section 7.8 and derive the edge wave $E_x^{(1)}$ scattered by an elliptic perfectly conducting disk ($y^2/a^2 + x^2/b^2 = 1$). The incident electromagnetic wave is given by

$$E_x^{inc} = E_{0x}\, e^{ikz}.$$

Integrate the EEWs over the whole edge.

- Find the focal asymptotics for the fields $E_x^{(1)}$.

- Apply the stationary-phase method and obtain the ray asymptotics. Compare with Equations (6.84) and (6.85).

8.7 Show that the caustic asymptotics (8.56) for acoustic waves transform into the ray asymptotics (8.41) when $k^{2/3}\mu \gg 1$.

8.8 Apply approximations (8.64), (8.65) for the Airy functions, use the caustic asymptotics (8.66) for electromagnetic waves, and obtain the ray asymptotics of the type of Equations (8.41) when when $k^{2/3}\mu \gg 1$.

Chapter 9

Multiple Diffraction of Edge Waves: Grazing Incidence and Slope Diffraction

9.1 STATEMENT OF THE PROBLEM AND RELATED REFERENCES

Clearly, the theory developed in the previous chapter can be applied to the investigation of multiple diffraction at edges that are spaced apart. Only two special cases need a individual investigation.

The first case is a grazing incidence of edge waves on acoustically hard *planar* plates. In the above asymptotic theory, the incident wave is approximated by an equivalent plane wave. However, a plane wave does not undergo diffraction at an infinitely thin plate under grazing incidence for the following reason. When this wave propagates in the direction parallel to the plate, its wave and amplitude fronts are perpendicular to the plate. As this incident field is constant in the direction normal to the plate, it automatically satisfies the boundary condition $du/dn = 0$ on the plate. Such a wave does not "see" the plate and propagates as if a free space is in its path. Because of this, the above theory predicts a zero diffracted field in this case. However, in the process of multiple diffraction, every diffracted wave is not plane. If its normal derivative du/dn is not zero on the plate, it undergoes diffraction. Such grazing diffraction is studied in Section 9.2.

The second case that also needs special treatment occurs when the scattering edge is located in the zero of the incident wave. This is the case of so-called *slope diffraction*. One distinguishes a slope diffraction of different orders, depending on the zero orders. Here we consider the most important one to be the slope diffraction of the first order, when the first derivative of the incident wave is not equal to zero. Such a situation occurs, for example, in reflector antennas, when one tries to decrease side lobes, and in the process of multiple diffraction between several scatterers, or between different parts of the same scatterer.

Fundamentals of the Physical Theory of Diffraction. By Pyotr Ya. Ufimtsev
Copyright © 2007 John Wiley & Sons, Inc.

Many authors have studied the phenomenon of slope diffraction. Ufimtsev (1958a,b,c, 1962) suggested uniform asymptotics for the secondary edge waves arising due to the slope diffraction on plane screens (strip, disk). Karp and Keller (1961) derived nonuniform ray asymptotics for the same edge waves. Mentzer et al. (1975) published uniform asymptotics similar to those found by Ufimtsev (1958a,b,c). The spectral theory of diffraction (Rahmat-Samii and Mittra, 1978) also enables one to investigate the slope diffraction. In the particular case of the half-plane diffraction problem, Boersma and Rahmat-Samii (1980) analyzed this phenomenon in the framework of the ray-based theories (uniform asymptotic theory (UAT) and uniform theory of diffraction (UTD)). Pathak (1988) constructed the general UTD for the slope diffraction at the wedge. The general PTD for the slope diffraction of electromagnetic waves (based on the concept of elementary edge waves) was elaborated in the papers by Ufimtsev (1991) and Ufimtsev and Rahmat-Samii (1995). A similar theory for both the grazing diffraction and the slope diffraction of acoustic waves was developed earlier in the work of Ufimtsev (1989, 1991). The theory presented below is based on the papers by Ufimtsev (1989, 1991) and Ufimtsev and Rahmat-Samii (1995).

9.2   GRAZING DIFFRACTION

The following relationship exists between the acoustic and electromagnetic diffracted rays arising due to the grazing diffraction at the plate S_1 (Fig. 9.1):

$$u_{\rm h} = H_t, \qquad \text{if } \frac{\partial u^{\rm inc}(\zeta)}{\partial n} = \frac{\partial H_t^{\rm inc}(\zeta)}{\partial n}$$

at the scattering edge L_1. Here, $\hat{t}$ is the tangent to the edge L_1 and $\hat{n}$ is the normal to the plate S_1 at the diffraction point ζ.

9.2.1   Acoustic Waves

Figure 9.1 shows the configuration appropriate to studying both the grazing diffraction and the slope diffraction. There are two acoustically hard scattering objects with edges L_1 and L_2. One of them is a planar plate S_1. The boundary conditions $du/dn = 0$ are imposed on the surfaces S_1 and S_2. Edge L_2 is located in the plane containing the plate S_1. No more than one edge diffracted ray comes to every point on edge L_1 (L_2) from edge L_2 (L_1). The wave initially diffracted at edge L_2 propagates to edge L_1 and undergoes grazing diffraction at plate S_1. This problem is investigated in the present section. The wave field diffracted at edge L_1 equals zero in the direction to edge L_2, where it undergoes slope diffraction. This problem is considered in the next section.

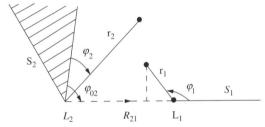

Figure 9.1 The problem of multiple diffraction. The edge L_2 is in the plane containing plate S_1 with edge L_1. Edge L_1 is perpendicular to the figure plane at the intersection point. Edge L_2 intersects the figure plane at an oblique angle. The line R_{21} shows the diffracted ray coming from L_2 to L_1. (Reprinted from Ufimtsev (1989) with the permission of the *Journal of Acoustical Society of America*.)

Suppose that the wave

$$u_2^{\text{inc}} = v_2(R_2, \varphi_2, \varphi_{02}) e^{ikR_2} \tag{9.1}$$

propagates from edge L_2 and undergoes grazing diffraction at plate S_1. This is a wave with a ray structure of the type of Equation (8.29). Consider the two first terms of its Taylor expansion in the vicinity of the grazing direction $\varphi_2 = \varphi_{02}$:

$$u_2^{\text{inc}} = v_2(R_2, \varphi_{02}, \varphi_{02}) e^{ikR_2} + \frac{\partial v_2(R_2, \varphi_{02}, \varphi_{02})}{\partial \varphi_2}(\varphi_2 - \varphi_{02}) e^{ikR_2} + \cdots \tag{9.2}$$

Here, the first term represents the wave that does not undergo diffraction at plate S_1, because its normal derivative on the plate equals zero [$\partial v_2(R_2, \varphi_{02}, \varphi_{02})/\partial \varphi_2 = \partial(\text{const})/\partial \varphi_2 = 0$]. Therefore, that part of the incident wave that experiences diffraction at the plate can be approximated by the wave

$$\frac{\partial v_2(R_2, \varphi_{02}, \varphi_{02})}{\partial \varphi_2}(\varphi_2 - \varphi_{02}) e^{ikR_2}, \tag{9.3}$$

which has the zero field in the grazing direction. Thus we see that the grazing diffraction of the wave (9.1) actually represents itself a particular case of the slope diffraction.

We approximate the wave (9.3) by the equivalent canonical wave

$$u_2^{\text{eq}} = u_{02} \frac{\partial}{\partial \varphi_{01}} e^{-ikz_1 \cos \gamma_{01}} e^{-ikr_1 \sin \gamma_{01} \cos(\varphi_1 - \varphi_{01})} \big|_{\varphi_{01} = \pi}$$

$$= u_{02} ik r_1 \sin \gamma_{01} \sin \varphi_1 e^{-ikz_1 \cos \gamma_{01}} e^{ikr_1 \sin \gamma_{01} \cos \varphi_1}, \tag{9.4}$$

obtained by the differentiation of the plane wave. The quantities r_1, φ_1, z_1 are local polar coordinates with the origin at the diffraction point on edge L_1 (Fig. 9.1), and the angle γ_{01} is shown in Figure 9.2, where $\hat{k}_1^i = \nabla R_{21}$.

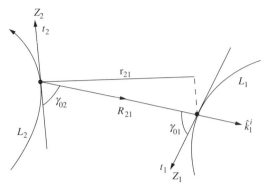

Figure 9.2   An edge wave arising at edge L_2 propagates in the direction $\hat{k}_1^i = \nabla R_{21}$ and undergoes the next diffraction at edge L_1. The unit vectors $\hat{t}_1$ and $\hat{t}_2$ are tangents to the edges L_1 and L_2. (Reprinted from Ufimtsev (1989) with the permission of the *Journal of Acoustical Society of America*).

The amplitude u_{02} of the equivalent canonical wave is defined by equating the normal derivatives of the real and equivalent incident waves at the diffraction point $z_1 = r_1 = 0$:

$$\frac{1}{R_{21}\sin\gamma_{02}}\frac{\partial v_2(R_2,\varphi_2,\varphi_{02})}{\partial\varphi_2}e^{ikR_2}\bigg|_{R_2=R_{21},\,\varphi_2=\varphi_{02}} = \frac{1}{r_1}\frac{\partial u_2^{eq}}{\partial\varphi_1}\bigg|_{z_1=r_1=0,\,\varphi_1=\pi}. \tag{9.5}$$

According to this equation,

$$u_{02} = -\frac{1}{ikR_{21}\sin\gamma_{01}\sin\gamma_{02}}\frac{\partial v_2(R_{21},\varphi_2,\varphi_{02})}{\partial\varphi_2}e^{ikR_{21}}, \qquad \text{with } \varphi_2 = \varphi_{02}, \tag{9.6}$$

or

$$u_{02} = w_{02}e^{ikR_{21}} \tag{9.7}$$

with

$$w_{02} = -\frac{1}{ikR_{21}\sin\gamma_{01}\sin\gamma_{02}}\frac{\partial v_2(R_{21},\varphi_2,\varphi_{02})}{\partial\varphi_2}\bigg|_{\varphi_2=\varphi_{02}}. \tag{9.8}$$

Now, we notice that the Helmholtz equation governing the wave field and its diffraction can be differentiated with respect to the parameter φ_{01} without changing the type of this equation. This means that the derivative (with respect to a free parameter) of any solution of the Helmholtz equation is also a solution of the same equation. The boundary conditions also admit differentiation with respect to φ_{01}. In Chapter 7, we found the edge diffracted field generated by the incident plane wave (7.2). The incident wave (9.4) is the derivative of the wave (7.2). Therefore, the diffracted field generated by the wave (9.4) can be found by the differentiation of the edge diffracted fields found in Chapter 7, if we replace $u^{inc}(\zeta)$ by $u_{02} = w_{02}\exp(ikR_{21})$. Before completing this procedure, however, we make another observation.

The incident wave propagating in the grazing direction to the plate creates the identical scattering sources $j_h^{(0)} = 2u^{inc}$ on both sides of the plate. According to Equation (1.10), the field generated by the sources induced on one side of the plate is completely cancelled by the field generated by the identical sources induced on the opposite side of the same plate. Therefore, in this particular case of the grazing incidence, the field $u_h^{(0)}$ equals zero and $u_h^{(1)} \equiv u_h^{(t)}$.

In view of this observation and the one made in the previous paragraph, the elementary edge wave generated at edge L_1 is found by the differentiation of Equation (7.97) with the simultaneous replacement of $u^{inc}(\zeta)$ by $w_{02}(\zeta)\exp(ikR_{21})$:

$$du_h^{(t)} = \frac{d\zeta}{2\pi} w_{02}(\zeta) \frac{\partial}{\partial\varphi_{01}} F_h^{(t)}(\zeta,\hat{m})\Big|_{\varphi_{01}=\pi} \cdot \frac{e^{ik[R_1(\zeta)+R_{21}(\zeta)]}}{R_1(\zeta)}. \tag{9.9}$$

The diffracted wave diverging from the whole edge L_1 is determined respectively by the integral

$$u_h^{(t)} = \frac{1}{2\pi} \int_{L_1} w_{02}(\zeta) \frac{\partial}{\partial\varphi_{01}} F_h^{(t)}(\zeta,\hat{m})\Big|_{\varphi_{01}=\pi} \cdot \frac{e^{ik[R_1(\zeta)+R_{21}(\zeta)]}}{R_1(\zeta)} d\zeta. \tag{9.10}$$

The ray asymptotic of this field is found by application of the stationary-phase technique described in Section 8.1. We omit all intermediate details and present the final expression

$$u_h^{(t)} = w_{02}(\zeta_{st}) \frac{1}{\sqrt{R_1(1+R_1/\rho_1)}} \frac{e^{i\pi/4}}{\sin\gamma_{01}\sqrt{2\pi k}} \frac{\partial g(\varphi_1,\varphi_{01},\alpha_1)}{\partial\varphi_{01}} e^{ik(R_1+R_{21})}\Big|_{\varphi_{01}=\pi}, \tag{9.11}$$

where $\alpha_1 = 2\pi$, $\sqrt{R_1(1+R_1/\rho_1)} > 0$ if $(1+R_1/\rho_1) > 0$, and $\sqrt{R_1(1+R_1/\rho_1)} = i\left|\sqrt{R_1(1+R_1/\rho_1)}\right|$ if $(1+R_1/\rho_1) < 0$. Here, all variable parameters and coordinates are the functions of the stationary point ζ_{st}, for example, $\gamma_{01}(\zeta_{st})$, $\varphi_1(\zeta_{st})$. The caustic parameter $\rho_1(\zeta_{st})$ is determined according to Equation (8.23) as

$$\frac{1}{\rho_1} = \frac{1}{\sin\gamma_{01}}\left(\frac{d\gamma_{01}}{d\zeta} - \frac{\hat{k}_1^s \cdot \hat{v}_1}{a_1\sin\gamma_{01}}\right), \tag{9.12}$$

where the unit vector

$$\hat{v}_1 = a_1 \frac{d\hat{t}_1}{d\zeta} \tag{9.13}$$

is the principal normal to edge L_1 and a_1 is the radius of curvature of this edge at the stationary point ζ_{st}. The unit vector $\hat{k}_1^s$ shows the directions of the diffracted rays (9.11) that form the diffraction cone. In accordance with Equation (8.7), this vector is defined by the equation

$$\hat{k}^s \cdot \hat{t}_1 = \hat{k}_1^i \cdot \hat{t}_1 = -\cos \gamma_{01}. \tag{9.14}$$

Asymptotic expression (9.11) can also be written in the form of Equation (8.13):

$$u_h^{(t)} \approx w_{02}(\zeta_{st}) \frac{\partial g(\varphi_1, \varphi_{01}, \alpha)}{\partial \varphi_{01}} \frac{e^{i\pi/4}}{\sqrt{2\pi k \Phi''(\zeta_{st})}} \frac{e^{ik(R_1 + R_{21})}}{R_1}, \tag{9.15}$$

where $\Phi(\zeta_{st}) = R_{21}(\zeta_{st}) + R_1(\zeta_{st})$. Note that the quantity $\Phi''(\zeta) = d^2\Phi(\zeta)/d\zeta^2$ is easier to calculate than the caustic parameter $\rho_1(\zeta)$ in Equation (9.11).

The above approximations (9.11) and (9.15) are the nonuniform asymptotics. They are singular in the directions $\varphi_1 = 0, 2\pi$, because

$$\frac{\partial}{\partial \varphi_{01}} g(\varphi_1, \varphi_{01}, \alpha_1)\big|_{\varphi_{01}=\pi} = -\frac{1}{2} \frac{\cos \frac{\varphi_1}{2}}{\sin^2 \frac{\varphi_1}{2}} \to \infty, \qquad \text{with } \varphi_1 \to 0, 2\pi. \tag{9.16}$$

This singularity is a consequence of the singularity of function $g(\varphi_1, \varphi_{01}, \alpha_1)$ in the directions $\varphi_1 = \pi \pm \varphi_0$. It can be treated as shown in Section 7.9.

It is worth noting that, in view of Equation (9.8), the wave (9.11), (9.15) arising due to the grazing/slope diffraction is less in magnitude by a small factor of $1/kR_{21}$ compared to the wave (8.29) generated by the ordinary edge diffraction.

9.2.2 Electromagnetic Waves

A similar problem for the grazing diffraction of electromagnetic waves was investigated in the paper by Ufimtsev (1991). Its solution is found in the same way as for acoustic waves. Here, it is supposed that the edge wave traveling from edge L_2 to edge L_1 is polarized perpendicularly to the plate S_1 ($\vec{E}_2^{tot} \perp S_1$) and its magnetic vector is parallel to this plate,

$$H_{2z_1}^{tot} = v_2(R_2, \varphi_2, \varphi_{02})e^{ikR_2}. \tag{9.17}$$

In the vicinity of edge L_1, this wave is approximated by the equivalent wave (9.4), where one should set $u_2^{eq} = H_{2z_1}^{eq}$. The quantity u_{02} in Equation (9.4) is defined by Equation (9.6), and the equivalent wave is written as

$$H_{2z_1}^{eq} = w_{02}e^{ikR_{21}}, \tag{9.18}$$

with the quantity w_{02} defined by Equation (9.8). The elementary edge wave arising at edge L_1 is calculated by the differentiation of the EEW given in Equations (7.135) and (7.136), assuming that $E_{0t} = 0$ and $H_{0t} = w_{02}\exp(ikR_{21})$.

By the integration of EEWs over edge L_1, one finds the wave diffracted at the edge L_1:

$$\vec{E}_{21}^{(t)} = \frac{1}{2\pi}Z_0\int_{L_1} w_{02}(\zeta)e^{ikR_{21}}\left[\left.\frac{\partial\vec{G}(\zeta)}{\partial\varphi_{01}}\right|_{\varphi_{01}=\pi}\right]\frac{e^{ikR_1}}{R_1}\,d\zeta_1 \qquad (9.19)$$

and

$$\vec{H}_{21}^{(t)} = \frac{1}{2\pi}\int_{L_1} w_{02}(\zeta)e^{ikR_{21}}\left[\nabla R_1\times\left.\frac{\partial\vec{G}(\zeta)}{\partial\varphi_{01}}\right|_{\varphi_{01}=\pi}\right]\frac{e^{ikR_1}}{R_1}\,d\zeta_1. \qquad (9.20)$$

Here are its ray asymptotics derived by the stationary-phase technique:

$$\vec{E}_{21}^{(t)} = \hat{e}_{\varphi_1}Z_0\frac{w_{02}(\zeta_{st})}{\sin^2\gamma_{01}}\frac{\partial g(\varphi_1,\pi,\alpha_1)}{\partial\varphi_{01}}\frac{e^{ik(R_1+R_{21})+i\pi/4}}{\sqrt{2\pi kR_1(1+R_1/\rho_1)}}, \qquad (9.21)$$

$$\vec{H}_{21}^{(t)} = [\nabla R_1\times\vec{E}^{(t)}]/Z_0, \qquad (9.22)$$

where $\hat{e}_{\varphi_1}$ is the unit vector associated with the polar angle φ_1 (Fig. 9.1). Therefore, in the first asymptotic approximation,

$$E_{\varphi_1}^{(t)} = -Z_0 H_{\vartheta_1} = Z_0\frac{w_{02}(\zeta_{st})}{\sin^2\gamma_{01}}\frac{\partial g(\varphi_1,\pi,\alpha_1)}{\partial\varphi_{01}}\frac{e^{ik(R_1+R_{21})+i\pi/4}}{\sqrt{2\pi kR_1(1+R_1/\rho_1)}}. \qquad (9.23)$$

The caustic parameter ρ_1 is defined by Equation (9.12). According to Equation (7.149) and (7.150),

$$H_{\vartheta_1}^{(t)} = -H_{t_1}^{(t)}/\sin\gamma_{01} \qquad (9.24)$$

and hence

$$H_{t_1}^{(t)} = w_{02}(\zeta_{st})\frac{1}{\sin\gamma_{01}}\frac{\partial g(\varphi_1,\pi,\alpha_1)}{\partial\varphi_{01}}\frac{e^{ik(R_1+R_{21})+i\pi/4}}{\sqrt{2\pi kR_1(1+R_1/\rho_1)}}. \qquad (9.25)$$

This is exactly the same ray asymptotic (9.11) found above for the acoustic waves. Thus, the relationship

$$H_t = u_h, \qquad \text{if } \frac{\partial H_t^{\text{inc}}}{\partial n} = \frac{\partial u^{\text{inc}}}{\partial n} \qquad (9.26)$$

at the scattering edge L_1

exists between the acoustic and electromagnetic waves arising due to the grazing diffraction.

9.3 SLOPE DIFFRACTION IN THE CONFIGURATION OF FIGURE 9.1

The following relationship exists between acoustic and electromagnetic waves arising due to the slope diffraction at edge L_2 (Fig. 9.1):

$$u_h = H_t, \qquad \text{if } \frac{\partial u^{\text{inc}}}{\partial n} = \frac{\partial H_t^{\text{inc}}}{\partial n}$$

at the diffraction point on edge L_2.

Here, $\hat{t}$ is the tangent to the edge L_2 and $\hat{n}$ is the normal to the plate S_1 (Fig. 9.1).

9.3.1 Acoustic Waves

Suppose that an external wave

$$u^{\text{ext}} = u_0 e^{ik\phi_0} \qquad (9.27)$$

generates at edge L_1 the diffracted wave (of the type of Equation (8.29)):

$$u_{1h}^{\text{inc}} = v_1(R_1, \varphi_1, \varphi_{01}) e^{ikR_1}$$

$$= u_0 \frac{e^{i\pi/4}}{\sin \gamma_0 \sqrt{2\pi k}} \frac{1}{\sqrt{R_1(1 + R_1/\rho_1)}} g(\varphi_1, \varphi_{01}, \alpha_1) e^{ikR_1}, \qquad (9.28)$$

with $\gamma_0 = \pi - \gamma_{01}$, $\cos \gamma_{01} = \hat{t}_1 \cdot \nabla \phi_0$, and the angle γ_{01} is shown in Figure 9.2. This wave hits the edge L_2 where it undergoes diffraction. That is why we denote it as u_{1h}^{inc}. The function $g(\varphi_1, \varphi_{01}, \alpha_1)$ is defined by Equation (2.64) and is equal to zero in the direction $\varphi_1 = \pi$ to edge L_2.

To calculate the slope diffraction of wave (9.28) at edge L_2, we approximate it by the equivalent wave

$$u_1^{eq} = u_{01} \frac{\partial}{\partial \varphi_{02}} e^{ikz_2 \cos \gamma_{02}} e^{-ikr_2 \sin \gamma_{02} \cos(\varphi_2 - \varphi_{02})}$$

$$= -u_{01} ikr_2 \sin \gamma_{02} \sin(\varphi_2 - \varphi_{02}) e^{ikz_2 \cos \gamma_{02}} e^{-ikr_2 \sin \gamma_{02} \cos(\varphi_2 - \varphi_{02})} \quad (9.29)$$

The angle γ_{02} is shown in Figure 9.2. The amplitude of this wave is determined by the requirement

$$\frac{1}{r_2} \frac{\partial u_1^{eq}}{\partial \varphi_2} \Bigg|_{z_2 = r_2 = 0, \; \varphi_2 = \varphi_{02}} = \frac{1}{R_1 \sin \gamma_{01}} \frac{\partial u_{1h}}{\partial \varphi_1} \Bigg|_{R_1 = R_{21}, \; \varphi_1 = \pi} \quad (9.30)$$

and equals

$$u_{01} = -\frac{1}{ikR_{21} \sin \gamma_{01} \sin \gamma_{02}} \frac{\partial v_1(R_1, \varphi_1, \varphi_{01})}{\partial \varphi_1} e^{ikR_{21}} \Bigg|_{R_1 = R_{21}, \; \varphi_1 = \pi}, \quad (9.31)$$

or

$$u_{01} = w_{01} e^{ikR_{21}} \quad (9.32)$$

where

$$w_{01} = -\frac{1}{ikR_{21} \sin \gamma_{01} \sin \gamma_{02}} \frac{\partial v_1(R_{21}, \varphi_1, \varphi_{01})}{\partial \varphi_1} \Bigg|_{\varphi_1 = \pi}. \quad (9.33)$$

According to the idea demonstrated in Section 9.2, the elementary edge waves generated at edge L_2 due to the slope diffraction are found by the differentiation of Equations (7.90) and (7.97) with respect to φ_{02}, and with the simultaneous replacement of $u^{inc}(\zeta)$ by (9.32):

$$du_h^{(1)} = \frac{d\zeta}{2\pi} w_{01}(\zeta) \frac{\partial}{\partial \varphi_{02}} F_h^{(1)}(\zeta) \frac{e^{ik[R_2(\zeta) + R_{21}(\zeta)]}}{R_2(\zeta)}, \quad (9.34)$$

$$du_h^{(t)} = \frac{d\zeta}{2\pi} w_{01}(\zeta) \frac{\partial}{\partial \varphi_{02}} F_h^{(t)}(\zeta) \frac{e^{ik[R_2(\zeta) + R_{21}(\zeta)]}}{R_2(\zeta)}. \quad (9.35)$$

We recall that the quantities $du_h^{(1)}$ and $du_h^{(t)}$ are the EEWs generated by the nonuniform $(j_h^{(1)})$ and the total $(j_h^{(t)} = j_h^{(0)} + j_h^{(1)})$ scattering sources, respectively.

The diffracted waves diverging from the whole edge are determined respectively by the integrals:

$$u_h^{(1)} = \frac{1}{2\pi} \int_{L_2} w_{01}(\zeta) \frac{\partial}{\partial \varphi_{02}} F_h^{(1)}(\zeta) \frac{e^{ik[R_2(\zeta) + R_{21}(\zeta)]}}{R_2(\zeta)} d\zeta \quad (9.36)$$

and

$$u_{\mathrm{h}}^{(\mathrm{t})} = \frac{1}{2\pi} \int_{L_2} w_{01}(\zeta) \frac{\partial}{\partial \varphi_{02}} F_{\mathrm{h}}^{(\mathrm{t})}(\zeta) \frac{e^{ik[R_2(\zeta)+R_{21}(\zeta)]}}{R_2(\zeta)} \, d\zeta. \tag{9.37}$$

The ray asymptotics of these waves are found by the stationary-phase technique:

$$u_{\mathrm{h}}^{(1)} = w_{01}(\zeta_{\mathrm{st}}) \frac{e^{i\pi/4}}{\sin \gamma_{02} \sqrt{2\pi k}} \frac{\partial g^{(1)}(\varphi_2, \varphi_{02}, \alpha_2)}{\partial \varphi_{02}} \frac{e^{ik(R_2+R_{21})}}{\sqrt{R_2(1+R_2/\rho_2)}}, \tag{9.38}$$

$$u_{\mathrm{h}}^{(\mathrm{t})} = w_{01}(\zeta_{\mathrm{st}}) \frac{e^{i\pi/4}}{\sin \gamma_{02} \sqrt{2\pi k}} \frac{\partial g(\varphi_2, \varphi_{02}, \alpha_2)}{\partial \varphi_{02}} \frac{e^{ik(R_2+R_{21})}}{\sqrt{R_1(1+R_2/\rho_2)}}. \tag{9.39}$$

Here, the caustic parameters ρ_2 is defined according to Equation (8.23) as

$$\frac{1}{\rho_2} = \frac{1}{\sin \gamma_{02}} \left(-\frac{d\gamma_{02}}{d\zeta} - \frac{\hat{k}_2^s \cdot \hat{v}_2}{a_2 \sin \gamma_{02}} \right) \tag{9.40}$$

where $\hat{v}_2 = a_2 d\hat{t}_2/d\zeta$ is the principal normal to the edge L_2 with radius a_2 at the stationary point ζ_{st}. These rays propagate in the direction of the vector $\hat{k}_2^s$ determined by the equation $\hat{k}_2^s \cdot \hat{t}_2 = \hat{k}_2^i \cdot \hat{t}_2 = \cos \gamma_{02}$. Function $g^{(1)}(\varphi_2, \varphi_{02}, \alpha_2)$ is defined according to Equation (4.15).

Notice also that Equation (9.40) differs from Equation (9.12) by the sign for the first term in parentheses. This difference is a consequence of Equations (8.16) and (8.17), which were used in the derivations of both Equations (9.12) and (9.40). According to Equations (8.16) and (8.17), these terms appear due to the differentiation of the dot products $\hat{k}_1^i \cdot \hat{t}_1 = -\cos \gamma_{0_1}$ and $\hat{k}_2^i \cdot \hat{t}_2 = \cos \gamma_{02}$. Here, the vector $\hat{k}_1^i$ ($\hat{k}_2^i$) shows the direction of the ray coming to edge L_1 (L_2) from edge L_2 (L_1), and the angles γ_{01} and γ_{02} are shown in Figure 9.2.

As well as in the case of grazing diffraction as considered in Section 9.2, the waves (9.38) and (9.39) arising due to the slope diffraction are also less in magnitude by a factor of $1/kR_{21}$ compared to the waves (8.13) and (8.29) generated by the ordinary edge diffraction. This result is not surprising, because the intensity of the incident field hitting the edge in the case of the slope diffraction is significantly less.

9.3.2 Electromagnetic Waves

Here, the above theory is extended for electromagnetic waves (Ufimtsev, 1991). The edge wave traveling from edge L_1 to edge L_2 can be considered as the sum of two waves with orthogonal polarization. One contains the component

$$H_{t_1}^{(t)} = v_1(R_1, \varphi_1, \varphi_{01})e^{ikR_1}, \tag{9.41}$$

with the function v_1 shown in Equation (9.28) and equal to zero in direction to edge L_2. It is clear that the diffraction of this wave at edge L_2 can be investigated in the same way as the diffraction of the acoustic wave (9.27). One approximates the incident wave by the equivalent wave

$$H_{1t_2}^{\mathrm{eq}} = w_{01}e^{ikR_{21}}\frac{\partial}{\partial\varphi_{02}}e^{ikz_2\cos\gamma_{02}}e^{-ikr_2\sin\gamma_{02}\cos(\varphi_2-\cos\varphi_{02})} \tag{9.42}$$

with w_{01} as defined in Equation (9.33). The EEWs are found by the differentiation of Equations (7.135) and (7.136) (with respect to the angle φ_{02}), where one should set $E_{0t} = 0$ and

$$H_{0t}\exp(ik\phi^i) = w_{01}\exp(ikR_{21}).$$

Then, for the total edge wave arising at wedge L_2, one obtains

$$\vec{E}^{(t)} = \frac{Z_0}{2\pi}\int_{L_2}w_{01}(\zeta)\frac{\partial\vec{G}^{(t)}(\zeta)}{\partial\varphi_{02}}\frac{e^{ik[R_2(\zeta)+R_{21}(\zeta)]}}{R_2(\zeta)}\,d\zeta, \tag{9.43}$$

$$\vec{H}^{(t)} = \frac{1}{2\pi}\int_{L_2}w_{01}(\zeta)\left[\nabla R_2 \times \frac{\partial\vec{G}^{(t)}(\zeta)}{\partial\varphi_{02}}\right]\frac{e^{ik[R_2(\zeta)+R_{21}(\zeta)]}}{R_2(\zeta)}\,d\zeta. \tag{9.44}$$

Asymptotic evaluation of these integrals leads to the ray asymptotics:

$$E_{\varphi_2}^{(t)} \approx Z_0 w_{01}(\zeta_{\mathrm{st}})\frac{e^{i\pi/4}}{\sin^2\gamma_{02}\sqrt{2\pi k}}\frac{\partial g(\varphi_2, \varphi_{02}, \alpha_2)}{\partial\varphi_{02}}\frac{e^{ik(R_2+R_{21})}}{\sqrt{R_1(1+R_2/\rho_2)}}, \tag{9.45}$$

$$H_{\vartheta_2}^{(t)} \approx - w_{01}(\zeta_{\mathrm{st}})\frac{e^{i\pi/4}}{\sin^2\gamma_{02}\sqrt{2\pi k}}\frac{\partial g(\varphi_2, \varphi_{02}, \alpha_2)}{\partial\varphi_{02}}\frac{e^{ik(R_2+R_{21})}}{\sqrt{R_1(1+R_2/\rho_2)}}. \tag{9.46}$$

As $H_{\vartheta_2}^{(t)} = -H_{t_2}^{(t)}/\sin\gamma_{02}$, one obtains the expression

$$H_{t_2}^{(t)} \approx w_{01}(\zeta_{\mathrm{st}})\frac{e^{i\pi/4}}{\sin\gamma_{02}\sqrt{2\pi k}}\frac{\partial g(\varphi_2, \varphi_{02}, \alpha_2)}{\partial\varphi_{02}}\frac{e^{ik(R_2+R_{21})}}{\sqrt{R_1(1+R_2/\rho_2)}}, \tag{9.47}$$

which completely agrees with Equation (9.39) and allows the formulation of the relationship

$$u_h = H_t, \quad \text{if } \frac{\partial}{\partial n} u_h^{inc} = \frac{\partial}{\partial n} H_t^{inc} \qquad (9.48)$$

on the scattering edge at $\zeta = \zeta_{st}$.

between the acoustic and electromagnetic rays arising due to the slope diffraction.

9.4 SLOPE DIFFRACTION: GENERAL CASE

The following relationships exist between the acoustic and electromagnetic diffracted rays arising due to the slope diffraction:

$$u_s = E_t, \quad \text{if } \frac{\partial}{\partial n} u^{inc}(\zeta_{st}) = \frac{\partial}{\partial n} E_t^{inc}(\zeta_{st}),$$

$$u_h = H_t, \quad \text{if } \frac{\partial}{\partial n} u^{inc}(\zeta_{st}) = \frac{\partial}{\partial n} H_t^{inc}(\zeta_{st}),$$

where ζ_{st} is the diffraction point on the scattering edge and $\hat{t}$ is the tangent to the edge.

9.4.1 Acoustic Waves

Suppose that the wave

$$u^{inc} = v_0 e^{ikR_Q} \qquad (9.49)$$

(with a ray structure of the type of Equation (8.29)) undergoes diffraction at the scattering object with edge L. The object can be acoustically soft or hard (with the boundary conditions $u = 0$ or $du/dn = 0$). The geometry of the problem is illustrated in Figures 9.3 and 9.4. The point Q belongs to the caustic of the incident wave.

It is assumed that $u^{inc} = 0$ and $\partial u^{inc}/\partial n \neq 0$ at the point ζ on the scattering edge L. The diffracted wave is calculated in the same manner as in Sections 9.2 and 9.3. The incident wave (9.49) is approximated by the equivalent wave

$$u^{eq} = u_0 \frac{\partial}{\partial \varphi_0} e^{-ikz \cos \gamma_0} e^{-ikr \sin \gamma_0 \cos(\varphi - \varphi_0)}$$

$$= -u_0 ikr \sin \gamma_0 \sin(\varphi - \varphi_0) e^{-ikz \cos \gamma_0} e^{-ikr \sin \gamma_0 \cos(\varphi - \varphi_0)}. \qquad (9.50)$$

The local polar coordinates r, φ, z are shown in Figure 9.4. The angle φ is measured from the illuminated face of the edge ($0 \leq \varphi \leq \alpha, 0 \leq \varphi_0 < \pi$).

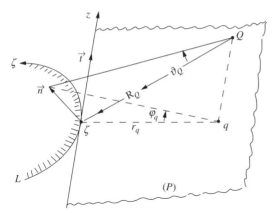

Figure 9.3   The plane P contains the tangent $\hat{t}$ to the edge L, the incident ray $Q\zeta$, and the point q (which is the projection of the point Q on the perpendicular r_q to the tangent $\hat{t}$). The vector $\hat{n}$ is the unit normal to the plane P. (Reprinted from Ufimtsev and Rahmat-Samii (1995) with the permission of *Annales des Telecommunications.*)

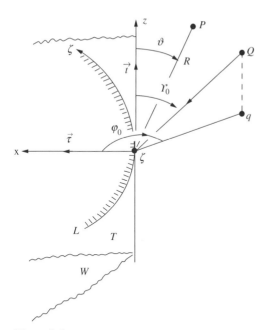

Figure 9.4   Here W is the tangential wedge to the scattering edge L, the plane T is the face of wedge W, the vector $\vec{\tau}$ is perpendicular to the tangent $\vec{t}$ and belongs to the plane T. The angle γ_0 indicates the direction of the incident ray. The point $P(R, \vartheta, \varphi)$ is the observation point. (Reprinted from Ufimtsev and Rahmat-Samii (1995) with the permission of *Annales des Telecommunications.*)

The amplitude u_0 of the equivalent wave is determined by the equation

$$\frac{\partial u^{\text{eq}}}{\partial n} = -\frac{1}{r}\frac{\partial u^{\text{eq}}}{\partial \varphi}\bigg|_{z=r=0,\ \varphi=\varphi_0} = \frac{\partial u^{\text{inc}}}{\partial n} = \frac{1}{R_Q}\frac{\partial u^{\text{inc}}}{\partial \vartheta_Q}\bigg|_{\vartheta_Q=0} = \frac{1}{r_q}\frac{\partial u^{\text{inc}}}{\partial \varphi_q}\bigg|_{\varphi_q=0} \tag{9.51}$$

and equals

$$u_0 = \frac{e^{ikR_Q}}{ikR_Q \sin \gamma_0}\frac{\partial v_0}{\partial \vartheta_Q}\bigg|_{\vartheta_Q=0}. \tag{9.52}$$

It can be written in the form

$$u_0 = w_0 e^{ik\phi^i}, \tag{9.53}$$

where $\phi^i = R_Q$ and

$$w_0 = \frac{1}{ikR_Q \sin \gamma_0}\frac{\partial v_0}{\partial \vartheta_Q}\bigg|_{\vartheta_Q=0}. \tag{9.54}$$

According to the idea introduced in Section 9.2, the waves diffracted at edge L are found by the differentiation of Equations (8.3), (8.4) with the simultaneous replacement of $u^{\text{inc}}(\zeta) = u_0(\zeta)\exp[ik\phi^i(\zeta)]$ by the quantity (9.53). As a result, the edge waves generated by the nonuniform/fringe scattering sources $(j_{s,h}^{(1)})$ are determined as

$$u_s^{(1)} = \frac{1}{2\pi}\int_L w_0(\zeta)\frac{\partial F_s^{(1)}(\zeta)}{\partial \varphi_0}\frac{e^{ik[R(\zeta)+R_Q(\zeta)]}}{R}\,d\zeta \tag{9.55}$$

and

$$u_h^{(1)} = \frac{1}{2\pi}\int_L w_0(\zeta)\frac{\partial F_h^{(1)}(\zeta)}{\partial \varphi_0}\frac{e^{ik[R(\zeta)+R_Q(\zeta)]}}{R}\,d\zeta, \tag{9.56}$$

and the edge waves radiated by the total scattering sources $(j_{s,h}^{(t)} = j_{s,h}^{(1)} + j_{s,h}^{(0)})$ are described by

$$u_s^{(t)} = \frac{1}{2\pi}\int_L w_0(\zeta)\frac{\partial F_s^{(t)}(\zeta)}{\partial \varphi_0}\frac{e^{ik[R(\zeta)+R_Q(\zeta)]}}{R}\,d\zeta \tag{9.57}$$

and

$$u_h^{(t)} = \frac{1}{2\pi}\int_L w_0(\zeta)\frac{\partial F_h^{(t)}(\zeta)}{\partial \varphi_0}\frac{e^{ik[R(\zeta)+R_Q(\zeta)]}}{R}\,d\zeta. \tag{9.58}$$

Their ray asymptotics can be derived by the stationary-phase technique demonstrated in Section 8.1. However, we can obtain them much faster by the differentiation of the ray asymptotics (8.12), (8.13) and (8.29) with the simultaneous replacement of $u^i(\zeta_{st})$ by (9.53):

$$
\begin{bmatrix} u_s^{(1)} \\ u_h^{(1)} \end{bmatrix} = w_0(\zeta_{st}) \frac{1}{\sqrt{R(1+R/\rho)}} \frac{e^{i\pi/4}}{\sin\gamma_0\sqrt{2\pi k}} \begin{bmatrix} \dfrac{\partial f^{(1)}(\varphi,\varphi_0,\alpha)}{\partial\varphi_0} \\ \dfrac{\partial g^{(1)}(\varphi,\varphi_0,\alpha)}{\partial\varphi_0} \end{bmatrix} e^{ik(R+R_Q)}, \quad (9.59)
$$

$$
\begin{bmatrix} u_s^{(t)} \\ u_h^{(t)} \end{bmatrix} = w_0(\zeta_{st}) \frac{1}{\sqrt{R(1+R/\rho)}} \frac{e^{i\pi/4}}{\sin\gamma_0\sqrt{2\pi k}} \begin{bmatrix} \dfrac{\partial f(\varphi,\varphi_0,\alpha)}{\partial\varphi_0} \\ \dfrac{\partial g(\varphi,\varphi_0,\alpha)}{\partial\varphi_0} \end{bmatrix} e^{ik(R+R_Q)}, \quad (9.60)
$$

where the stationary point ζ_{st} is calculated according to Equation (8.7) and the caustic parameter $\rho = \rho(\zeta_{st})$ is defined in Equation (8.23).

In view of Equation (9.54), the fields arising due to the slope diffraction are less in magnitude by a small factor $1/kR_Q$ compared to the ordinary diffracted fields (8.12), (8.13) and (8.29).

Notice that the integral representations (9.55), (9.56), (9.57), and (9.58) allow the calculation of the slope diffracted field in the vicinity of caustics and foci.

9.4.2 Electromagnetic Waves

Let us define an incident electromagnetic wave as

$$
\vec{E}^{inc} = \vec{E}_0^i e^{ikR_Q}, \qquad \vec{H}^{inc} = \vec{H}_0^i e^{ikR_Q}, \quad (9.61)
$$

where

$$
\vec{E}_0^i = Z_0[\vec{H}_0^i \times \nabla R_Q]. \quad (9.62)
$$

Suppose that in the direction to the scattering edge L, the quantities $\vec{E}_0^i$, $\vec{H}_0$ have zeros, but their first normal derivatives are not equal to zero:

$$
\vec{E}_0 = 0, \ \vec{H}_0^i = 0 \qquad \text{on } L \quad (9.63)
$$

$$
\frac{\partial \vec{E}_0^i}{\partial n} = \frac{1}{R_Q}\frac{\partial \vec{E}_0^i}{\partial \vartheta_Q} \neq 0, \qquad \frac{\partial \vec{H}_0^i}{\partial n} = \frac{1}{R_Q}\frac{\partial \vec{H}_0^i}{\partial \vartheta_Q} \neq 0 \qquad \text{on } L. \quad (9.64)
$$

As in the previous section, one approximates the actual incident wave (in the vicinity of the edge L) by the equivalent wave with components

$$
E_t^{eq} = -ikrE_{0t}^{eq}\sin\gamma_0\sin(\varphi-\varphi_0)e^{-ikz\cos\gamma_0}e^{-ikr\sin\gamma_0\cos(\varphi-\varphi_0)}, \quad (9.65)
$$

$$
H_t^{eq} = -ikrH_{0t}^{eq}\sin\gamma_0\sin(\varphi-\varphi_0)e^{-ikz\cos\gamma_0}e^{-ikr\sin\gamma_0\cos(\varphi-\varphi_0)}, \quad (9.66)
$$

which are the derivatives of the ordinary plane wave with respect to the incidence angle φ_0. The amplitudes of the equivalent wave are found with the requirement that the normal derivatives of this wave (on the scattering edge) are equal to those of the actual incident wave:

$$\frac{\partial E_t^{\text{eq}}}{\partial n} = -\left.\frac{\partial E_t^{\text{eq}}}{r\partial\varphi}\right|_{z=r=0,\ \varphi=\varphi_0} = ik\sin\gamma_0 E_{0t}^{\text{eq}} = \left.\frac{\partial E_{0t}^i}{R_Q\partial\vartheta_Q}e^{ikR_Q}\right|_{\vartheta_Q=0}, \qquad (9.67)$$

$$\frac{\partial H_t^{\text{eq}}}{\partial n} = -\left.\frac{\partial H_t^{\text{eq}}}{r\partial\varphi}\right|_{z=r=0,\ \varphi=\varphi_0} = ik\sin\gamma_0 H_{0t}^{\text{eq}} = \left.\frac{\partial H_{0t}^i}{R_Q\partial\vartheta_Q}e^{R_Q}\right|_{\vartheta_Q=0}. \qquad (9.68)$$

Hence

$$E_{0t}^{\text{eq}} = \frac{1}{ikR_Q\sin\gamma_0}e^{ikR_Q}\left.\frac{\partial E_{0t}^i}{\partial\vartheta_Q}\right|_{\vartheta_Q=0} \qquad (9.69)$$

and

$$H_{0t}^{\text{eq}} = \frac{1}{ikR_Q\sin\gamma_0}e^{ikR_Q}\left.\frac{\partial H_{0t}^i}{\partial\vartheta_Q}\right|_{\vartheta_Q=0}. \qquad (9.70)$$

As in the previous section, the elementary edge waves diffracted at edge L are found by the differentiation of Equations (7.135) and (7.136), with respect to the angle φ_0 and with the simultaneous replacement of $E_{0t}(\zeta)$, $H_{0t}(\zeta)$ by the quantities (9.69), (9.70):

$$d\vec{E}^{(t)} = \frac{d\zeta}{2\pi}\vec{\mathcal{E}}^{(t)}(\zeta)e^{ikR_Q(\zeta)}\frac{e^{ikR(\zeta)}}{R(\zeta)}, \qquad (9.71)$$

$$d\vec{H}^{(t)} = [\nabla R \times d\vec{E}^{(t)}]/Z_0. \qquad (9.72)$$

Here,

$$\vec{\mathcal{E}}^{(t)}(\zeta) = E_{0t}^{\text{eq}}(\zeta)\frac{\partial}{\partial\varphi_0}\vec{F}^{(t)}(\zeta,\vartheta,\varphi) + Z_0 H_{0t}^{\text{eq}}(\zeta)\frac{\partial}{\partial\varphi_0}\vec{G}^{(t)}(\zeta,\vartheta,\varphi). \qquad (9.73)$$

The total edge wave created by all EEWs is determined by the integrals

$$\vec{E}^{(t)} = \frac{1}{2\pi}\int_L \vec{\mathcal{E}}^{(t)}(\zeta)\frac{e^{ik[R(\zeta)+R_Q(\zeta)]}}{R(\zeta)}\,d\zeta \qquad (9.74)$$

and

$$\vec{H}^{(t)} = \frac{1}{2\pi Z_0}\int_L [\nabla R \times \vec{\mathcal{E}}^{(t)}(\zeta)]\frac{e^{ik[R(\zeta)+R_Q(\zeta)]}}{R(\zeta)}\,d\zeta. \qquad (9.75)$$

The ray asymptotics of this wave are found by the stationary-phase technique:

$$E_\varphi^{(t)} = -Z_0 H_\vartheta^{(t)} = Z_0 H_{0t}^{eq} \frac{e^{i\pi/4}}{\sin^2 \gamma_0 \sqrt{2\pi k}} \frac{\partial g(\varphi, \varphi_0, \alpha)}{\partial \varphi_0} \frac{e^{ik(R+R_Q)}}{\sqrt{R(1+R/\rho)}},$$ (9.76)

$$E_\vartheta^{(t)} = Z_0 H_\varphi^{(t)} = -E_{0t}^{eq} \frac{e^{i\pi/4}}{\sin^2 \gamma_0 \sqrt{2\pi k}} \frac{\partial f(\varphi, \varphi_0, \alpha)}{\partial \varphi_0} \frac{e^{ik(R+R_Q)}}{\sqrt{R(1+R/\rho)}},$$ (9.77)

where, for all functions of the variable ζ, one should take their values at the stationary point ζ_{st}. In view of Equations (7.149) and (7.150), these expressions can be written in terms of the components parallel to the tangent $\hat{t}(\zeta_{st})$,

$$E_t^{(t)} = E_{0t}^{eq} \frac{e^{i\pi/4}}{\sin \gamma_0 \sqrt{2\pi k}} \frac{\partial f(\varphi, \varphi_0, \alpha)}{\partial \varphi_0} \frac{e^{ik(R+R_Q)}}{\sqrt{R(1+R/\rho)}},$$ (9.78)

$$H_t^{(t)} = H_{0t}^{eq} \frac{e^{i\pi/4}}{\sin \gamma_0 \sqrt{2\pi k}} \frac{\partial g(\varphi, \varphi_0, \alpha)}{\partial \varphi_0} \frac{e^{ik(R+R_Q)}}{\sqrt{R(1+R/\rho)}}.$$ (9.79)

Their comparison with Equation (9.60) leads to the equivalence relationships between the acoustic and electromagnetic diffracted rays arising due to the slope diffraction:

$$u_s = E_t, \qquad \text{if } \frac{\partial}{\partial n} u^{inc}(\zeta_{st}) = \frac{\partial}{\partial n} E_t^{inc}(\zeta_{st}),$$ (9.80)

$$u_h = H_t, \qquad \text{if } \frac{\partial}{\partial n} u^{inc}(\zeta_{st}) = \frac{\partial}{\partial n} H_t^{inc}(\zeta_{st}),$$ (9.81)

where ζ_{st} is the diffraction point on the scattering edge.

PROBLEMS

9.1 Use the asymptotic expression (9.10) for the grazing diffraction of acoustic waves, apply the stationary-phase technique, and confirm the ray approximation (9.11).

9.2 Use the asymptotic expressions (9.19), (9.20) for the grazing diffraction of electromagnetic waves, apply the stationary-phase technique, and confirm the ray approximations (9.23), (9.25). Compare Equation (9.25) with Equation (9.11) and establish the equivalence relationship between acoustic and electromagnetic diffracted rays.

9.3 Use the asymptotic expressions (9.36), (9.37) for the slope diffraction of acoustic waves, apply the stationary-phase technique, and confirm the ray approximations (9.38), (9.39).

9.4 Use the asymptotic expressions (9.43), (9.44) for the slope diffraction of electromagnetic waves, apply the stationary-phase technique, and confirm the ray approximations (9.45), (9.46) and (9.47). Compare Equation (9.47) with Equation (9.39), and establish the equivalence relationship between acoustic and electromagnetic diffracted rays.

Chapter 10

Diffraction Interaction of Neighboring Edges on a Ruled Surface

The following relationships exist between acoustic and electromagnetic diffracted waves in the directions belonging to the diffraction cone:

$$u_h = H_t, \quad \text{if } u_h^{\text{inc}}(\zeta) = H_t^{\text{inc}}(\zeta), \quad u_s = E_t, \quad \text{if } \frac{\partial u_s^{\text{inc}}(\zeta)}{\partial n} = \frac{\partial E_t^{\text{inc}}(\zeta)}{\partial n}.$$

Here, $\hat{t}(\hat{n})$ is the tangent (normal) to the edge at the diffraction point ζ. These relationships, together with Equations (7.149) and (7.150), allow one to determine all components of the electromagnetic diffracted wave.

Consider a diffraction interaction of two edges with a common face. If this face is bent, the edge wave already undergoes diffraction on its way along the face to another edge. This problem is not amenable to theoretical treatment in a general case. However, the disturbing effect of the face can be neglected in the particular case illustrated in Figure 10.1. The common face S of edges L_1 and L is a ruled surface whose generatrices coincide with the edge-diffracted rays arising at the edge L_1 and propagating to the edge L. It is assumed that a plane tangential to the face does not change its orientation along the generatrix. Notice that a planar facet can be considered as a limiting case of a ruled surface. Therefore, the theory developed in the following is applicable in this case as well.

The present section is based on the papers by Ufimtsev (1989, 1991).

Fundamentals of the Physical Theory of Diffraction. By Pyotr Ya. Ufimtsev
Copyright © 2007 John Wiley & Sons, Inc.

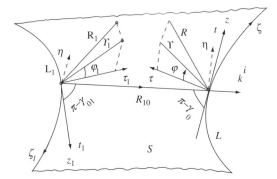

Figure 10.1   Element of a ruled surface S with two edges (L_1 and L). The unit vectors $\hat{t}_1, \hat{t}$ are tangential to the edges. The unit vectors $\hat{\tau}_1, \hat{\tau}$ are tangential to the surface S and perpendicular to $\hat{t}_1, \hat{t}$, respectively. The quantities r_1, φ_1, z_1 and r, φ, z are local polar coordinates. The unit vector $\hat{n}$ is normal to the plane tangential to S and containing the generatrix R_{10} as well as the tangents $\hat{t}_1, \hat{t}$ and $\hat{\tau}_1, \hat{\tau}$. (Reprinted from Ufimtsev (1989) with the permission of the *Journal of Acoustical Society of America*.)

10.1 DIFFRACTION AT AN ACOUSTICALLY HARD SURFACE

Suppose that the edge wave propagating from the edge L_1 has a ray structure and can be represented in the form of Equation (8.29) as

$$u_{1\text{h}}(R_1, \varphi_1) = u_{01} g(\varphi_1, \varphi_{01}, \alpha_1) \frac{e^{ikR_1 + i\pi/4}}{\sin \gamma_{01} \sqrt{2\pi kR_1 (1 + R_1/\rho_1)}}, \tag{10.1}$$

where the function $g(\varphi_1, \varphi_{01}, \alpha_1)$ is defined in Equation (2.64). In the vicinity of the edge L, this wave can be approximated by the two merging plane waves

$$\lim_{\varphi_0 \to 0} \frac{1}{2} u_{1\text{h}}(R_{10}, 0) e^{-ikz \cos \gamma_0} \left(e^{-ikr \sin \gamma_0 \cos(\varphi - \varphi_0)} + e^{-ikr \sin \gamma_0 \cos(\varphi + \varphi_0)} \right). \tag{10.2}$$

The first term here plays the role of the incident wave in the canonical wedge diffraction problem utilized in Chapter 7 to derive the asymptotic approximation (8.3), (8.4). Therefore, replacing the quantity $u_0(\zeta) \exp(ik\phi^i)$ in Equation (8.4) by $(1/2)u_{1\text{h}}(R_{10}, 0)$, we obtain the asymptotic expression

$$u_{\text{h}}^{(\text{t})} = \frac{1}{4\pi} \int_L u_{1\text{h}}(\zeta) F_{\text{h}}^{(\text{t})}(\zeta, \hat{m}) \frac{e^{ikR(\zeta)}}{R(\zeta)} d\zeta, \qquad \text{with } u_{1\text{h}}(\zeta) = u_{1\text{h}}(R_{10}, 0), \tag{10.3}$$

for the edge wave generated by the *total* scattering source $j_{\text{h}}^{(\text{t})} = j_{\text{h}}^{(1)} + j_{\text{h}}^{(0)}$. One should note that in this particular case, $j_{\text{h}}^{(0)} = u_{1\text{h}}(R_1, 0)$.

The ray asymptotic of this wave can be found by the stationary-phase technique. However, it can be obtained directly from Equation (8.29) if we replace there $u^{\text{inc}}(\zeta_{\text{st}})$ by $(1/2)u_{1\text{h}}(\zeta_{\text{st}})$ and set $\varphi_0 = 0$:

$$u_{\text{h}}^{(\text{t})} = \frac{1}{2} u_{1\text{h}}(\zeta_{\text{st}}) g(\varphi, 0, \alpha) \frac{e^{ikR + i\pi/4}}{\sin \gamma_0 \sqrt{2\pi k R (1 + R/\rho)}}, \qquad (10.4)$$

where

$$u_{1\text{h}}(\zeta_{\text{st}}) = u_{01} g(0, \varphi_{01}, \alpha_1) \frac{e^{ikR_{10} + i\pi/4}}{\sin \gamma_{01} \sqrt{2\pi k R_{10}(1 + R_{10}/\rho_1)}}. \qquad (10.5)$$

The letters α_1, α denote the external angles of edges L_1, L ($\pi \leq \alpha_1 \leq 2\pi$, $\pi \leq \alpha \leq 2\pi$), and the caustic parameters ρ_1, ρ are defined according to Equation (8.23).

The ray asymptotic (10.4) is not applicable near the shadow boundary $\varphi = \pi$, where it becomes singular. Instead, one can suggest the following heuristic approximation of Equation (10.3) valid in all directions along the diffraction cone ($\vartheta = \pi - \gamma_0$, $0 \leq \varphi \leq \alpha$):

$$u_{\text{h}}^{(\text{t})} = u_{1\text{h}}(R_{10}, 0) D_{\text{h}}(\chi, \varphi, \gamma_0) \frac{e^{ikR}}{\sqrt{R(1 + R/\rho)}}, \qquad (10.6)$$

where

$$D_{\text{h}}(\chi, \varphi, \gamma_0) = \sqrt{\frac{\chi}{\sin \gamma_0}} v(k\chi \sin \gamma_0, \varphi) e^{-ik\chi \sin \gamma_0}, \qquad (10.7)$$

$$v(s, \varphi) = \frac{\frac{2}{N} \sin \frac{\pi}{N} \cos \frac{\varphi}{2}}{\cos \frac{\pi}{N} - \cos \frac{\varphi}{N}} e^{-is \cos \varphi} \frac{e^{-i\pi/4}}{\sqrt{\pi}} \int_{\sqrt{2s} \cos \frac{\varphi}{2}}^{\text{sgn}(\cos \frac{\varphi}{2})\infty} e^{it^2} \, dt, \qquad (10.8)$$

with $N = \alpha/\pi$, and

$$\chi = \frac{R_{10} R}{R_{10} + R} \sin \gamma_0. \qquad (10.9)$$

According to the relationships

$$v(s, \pi \mp 0) = \mp \frac{1}{2} e^{is} \qquad (10.10)$$

and

$$\rho = \rho_1 + R_{10}, \qquad \text{for } \varphi = \pi, \qquad (10.11)$$

the diffracted field $u^{(t)}$ is discontinuous at the shadow boundary ($\varphi_1 = 0, \varphi = \pi$):

$$u_h^{(t)} = \mp \frac{1}{2} u_{1h}(R_{10} + R, 0). \tag{10.12}$$

However, the sum of the diffracted and incident fields ($u_h^{(t)} + u_{1h}$) is continuous there. One can also show that the normal derivative of this sum is also continuous at the shadow boundary.

Utilizing the asymptotic approximation

$$\int_x^{[\mathrm{sgn}(x)]\infty} e^{it^2} \, dt \sim -\frac{e^{ix^2}}{2ix}, \qquad \text{with } |x| \gg 1, \tag{10.13}$$

it is easy to verify that the uniform approximation (10.6) reduces to the ray asymptotic (10.4) in the directions away from the shadow boundary, where $\sqrt{2k\chi \sin \gamma_0} \left| \cos \frac{\varphi}{2} \right| \gg 1$.

10.2 DIFFRACTION AT AN ACOUSTICALLY SOFT SURFACE

The geometry of the problem is shown in Figure 10.1. Suppose that the edge wave

$$u_{1s}(R_1, \varphi_1) = u_{01} f(\varphi_1, \varphi_{01}, \alpha_1) \frac{e^{ikR_1 + i\pi/4}}{\sin \gamma_{01} \sqrt{2\pi k R_1 (1 + R_1/\rho_1)}} \tag{10.14}$$

propagates from the edge L_1 along the face S and, due to the boundary condition, equals zero in the direction $\varphi_1 = 0$ to the edge L. We note that $f(0, \varphi_{01}, \alpha_1) = 0$ in accordance with the definition given by Equation (2.62). Thus, the diffraction of the wave (10.14) at the edge L is a particular case of the slope diffraction, and it can be treated, as shown below, with a little modification of the technique developed in Chapter 9.

First, we notice that an appropriate wave (equivalent to the incident wave (10.14) in the vicinity of edge L) can be constructed from the combination of the incident and reflected plane waves

$$e^{-ikz \cos \gamma_0} (e^{-ikr \sin \gamma_0 \cos(\varphi - \varphi_0)} - e^{-ikr \sin \gamma_0 \cos(\varphi + \varphi_0)}) \tag{10.15}$$

running along the face to the edge. Namely,

$$u_s^{eq} = u_0^{eq} e^{-ikz \cos \gamma_0} \frac{\partial}{\partial \varphi_0} (e^{-ikr \sin \gamma_0 \cos(\varphi - \varphi_0)} - e^{-ikr \sin \gamma_0 \cos(\varphi + \varphi_0)})|_{\varphi_0 = 0}$$

$$= -u_0^{eq} 2ikr \sin \gamma_0 \sin \varphi \, e^{-ikz \cos \gamma_0} e^{-ikr \sin \gamma_0 \cos \varphi}. \tag{10.16}$$

The amplitude factor u_0^{eq} of the equivalent wave is determined by the requirement

$$\frac{1}{R_{10} \sin \gamma_{01}} \frac{\partial u_{1s}(R_1, \varphi_1)}{\partial \varphi_1} \bigg|_{R_1 = R_{10}, \varphi_1 = 0} = \frac{1}{r} \frac{\partial u_s^{eq}}{\partial \varphi} \bigg|_{z = r = 0, \varphi = 0} \tag{10.17}$$

and equals

$$u_0^{\text{eq}} = -\frac{1}{i2kR_{10}\sin\gamma_{01}\sin\gamma_0}\left.\frac{\partial u_{1s}(R_1,\varphi_1)}{\partial\varphi_1}\right|_{R_1=R_{10},\varphi_1=0}. \tag{10.18}$$

According to the idea introduced in Chapter 9, the edge wave arising due to the slope diffraction of the equivalent wave (10.16) is determined by the derivative of (8.4) with the simultaneous replacement of $u_0(\zeta)\exp(ik\phi^i)$ by u_0^{eq}:

$$u_s^{(t)} = \frac{1}{2\pi}\int_L u_0^{\text{eq}}(\zeta)\left.\frac{\partial}{\partial\varphi_0}F_s^{(t)}(\zeta,\hat{m})\right|_{\varphi_0=0}\cdot\frac{e^{ikR(\zeta)}}{R(\zeta)}\,d\zeta. \tag{10.19}$$

The ray asymptotic of the diffracted field (10.19) can be found by the stationary-phase technique or directly by the differentiation of Equation (8.29) with the replacement of $u_0(\zeta)\exp(ik\phi^i)$ by u_0^{eq}:

$$u_s^{(t)} = u_0^{\text{eq}}(\zeta_{\text{st}})\frac{\partial f(\varphi,0,\alpha)}{\partial\varphi_0}\frac{e^{ikR+i\pi/4}}{\sin\gamma_0\sqrt{2\pi kR(1+R/\rho)}}. \tag{10.20}$$

This function is singular in the direction of the shadow boundary ($\varphi = \pi$). Instead of it one can suggest the following heuristic approximation of Equation (10.19) valid in all directions along the diffraction cone ($\vartheta = \pi - \gamma_0, 0 \le \varphi \le \alpha$):

$$u_s^{(t)} = \frac{i}{k}\frac{\partial u_{1s}(\zeta_{\text{st}})}{\partial n}D_s(\chi,\varphi,\gamma_0)\frac{e^{ikR}}{\sqrt{R(1+R/\rho)}}, \tag{10.21}$$

where

$$\frac{\partial u_{1s}(\zeta_{\text{st}})}{\partial n} = \frac{1}{R_{10}\sin\gamma_{01}}\left.\frac{\partial u_{1s}(R_1,\varphi_1)}{\partial\varphi_1}\right|_{R_1=R_{10},\varphi_1=0}, \tag{10.22}$$

$$D_s(\chi,\varphi,\alpha) = -k\chi\sqrt{\frac{\chi}{\sin\gamma_0}}w(k\chi\sin\gamma_0,\varphi)e^{-ik\chi\sin\gamma_0}, \tag{10.23}$$

and

$$\chi = \frac{R_{10}R}{R_{10}+R}\sin\gamma_0. \tag{10.24}$$

Here,

$$w(s,\varphi) = -\frac{2\sqrt{2}}{N^2}\frac{\sin\frac{\pi}{N}\sin\frac{\varphi}{N}\cos^2\frac{\varphi}{2}}{\left(\cos\frac{\pi}{N}-\cos\frac{\varphi}{N}\right)^2}\mathbb{F}\left(\sqrt{2s}\cos\frac{\varphi}{2}\right)\frac{e^{i(s-\pi/4)}}{\sqrt{\pi s}}, \tag{10.25}$$

with $N = \alpha/\pi$ and

$$\mathbb{F}(x) = 1 + i2xe^{-ix^2} \cdot \int_x^{\mathrm{sgn}(x)\infty} e^{it^2} dt. \tag{10.26}$$

For large arguments $|x| \gg 1$,

$$\mathbb{F}(x) \approx \frac{i}{2x^2} \tag{10.27}$$

and

$$w(s, \varphi) \approx -\frac{2}{N^2} \frac{\sin \dfrac{\pi}{N} \sin \dfrac{\varphi}{N}}{\left(\cos \dfrac{\pi}{N} - \cos \dfrac{\varphi}{N}\right)^2} \frac{e^{i(s+\pi/4)}}{\sqrt{\pi}(2s)^{3/2}}. \tag{10.28}$$

One can show that the field $u_{1s}(R_1, \varphi_1) + u_s^{(t)}$ and its normal derivative are continuous at the shadow boundary ($\varphi_1 = 0, \varphi = \pi$). Away from this boundary (when $\sqrt{2k\chi \sin \gamma_0} \left|\cos \frac{\varphi}{2}\right| \gg 1$), the function (10.21) reduces to the ray asymptotic (10.20).

Here it is pertinent to mention the review paper by Molinet (2005) related to the excitation of two-dimensional edge waves by the creeping waves and whispering-gallery waves propagating over convex and concave scattering surfaces, respectively.

10.3 DIFFRACTION OF ELECTROMAGNETIC WAVES

In a general case, a wave diffracted at edge L_1 can be considered as the sum of two waves with orthogonal polarizations, that is, with the components H_t and E_t. Because they play the role of the waves incident on edge L, we denote them as H_t^{inc}, E_t^{inc}. The wave with component H_t^{inc} can be represented as Equation (10.1) and its diffraction at edge L is calculated in the same way as in Section 10.1. The wave with component E_t^{inc} can be represented as Equation (10.14) and its diffraction at edge L is calculated as shown in Section 10.2.

These calculations result in the following ray asymptotics:

$$E_\varphi^{(t)} = -Z_0 H_\vartheta^{(t)} = \frac{1}{2} Z_0 H_t^{\mathrm{inc}}(\zeta_{\mathrm{st}}) g(\varphi, 0, \alpha) \frac{e^{ikR+i\pi/4}}{\sin^2 \gamma_0 \sqrt{2\pi kR(1+R/\rho)}}, \tag{10.29}$$

$$E_\vartheta^{(t)} = Z_0 H_\varphi^{(t)} = -E_t^{\mathrm{eq}}(\zeta_{\mathrm{st}}) \frac{\partial f(\varphi, 0, \alpha)}{\partial \varphi_0} \frac{e^{ikR+i\pi/4}}{\sin^2 \gamma_0 \sqrt{2\pi kR(1+R/\rho)}}, \tag{10.30}$$

where

$$H_t^{\mathrm{inc}}(\zeta_{\mathrm{st}}) = u_{01} g(0, \varphi_{01}, \alpha_1) \frac{e^{ikR_{10}+i\pi/4}}{\sin \gamma_{01} \sqrt{2\pi kR_{10}(1+R_{10}/\rho_1)}}, \tag{10.31}$$

$$E_t^{\mathrm{eq}}(\zeta_{\mathrm{st}}) = -\frac{1}{i2kR_{10} \sin \gamma_{01} \sin \gamma_0} \frac{\partial E_t^{\mathrm{inc}}(R_1, \varphi_1)}{\partial \varphi_1}\Bigg|_{R_1=R_{10}, \varphi_1=0}, \tag{10.32}$$

and

$$E_t^{inc} = u_{01} f(\varphi_1, \varphi_{01}, \alpha_1) \frac{e^{ikR_{10}+i\pi/4}}{\sin \gamma_{01} \sqrt{2\pi k R_{10}(1 + R_{10}/\rho_1)}}.$$ (10.33)

In view of Equations (7.149) and (7.150), one can rewrite the ray asymptotics in terms of the components parallel to the tangent to the edge at the diffraction point ζ_{st}:

$$H_t^{(t)} = \frac{1}{2} H_t^{inc}(\zeta_{st}) g(\varphi, 0, \alpha) \frac{e^{ikR+i\pi/4}}{\sin \gamma_0 \sqrt{2\pi k R(1 + R/\rho)}},$$ (10.34)

and

$$E_t^{(t)} = E_t^{eq}(\zeta_{st}) \frac{\partial f(\varphi, 0, \alpha)}{\partial \varphi_0} \frac{e^{ikR+i\pi/4}}{\sin \gamma_0 \sqrt{2\pi k R(1 + R/\rho)}}.$$ (10.35)

Their comparison with Equations (10.4) and (10.20) reveals the same relationships between acoustic and electromagnetic diffracted rays as those established in the previous sections:

$$u_h = H_t, \quad \text{if } u_h^{inc} = H_t^{inc},$$ (10.36)

and

$$u_s = E_t, \quad \text{if } \frac{\partial u_s^{inc}}{\partial n} = \frac{\partial E_t^{inc}}{\partial n},$$ (10.37)

at the diffraction point on the scattering edge.

The uniform asymptotics for the diffracted wave (valid for the directions $\vartheta = \pi - \gamma_0, 0 \le \varphi \le \alpha$) are described by the electromagnetic versions of Equations (10.6) and (10.21):

$$E_\varphi^{(t)} = -Z_0 H_\vartheta^{(t)} = Z_0 H_t^{inc}(\zeta_{st}) D_h(\chi, \varphi, \gamma_0) \frac{e^{ikR}}{\sin \gamma_0 \sqrt{R(1 + R/\rho)}}$$ (10.38)

and

$$E_\vartheta^{(t)} = Z_0 H_\varphi^{(t)} = \frac{1}{ik \sin \gamma_0} \frac{\partial E_t^{inc}(\zeta_{st})}{\partial n} D_s(\chi, \varphi, \gamma_0) \frac{e^{ikR}}{\sqrt{R(1 + R/\rho)}}.$$ (10.39)

Here, the superscript "t" means that this wave (diffracted at edge L) is radiated by the surface current, $\vec{j}^{(t)} = \vec{j}^{(0)} + \vec{j}^{(1)}$. Together with the incident wave (diverging from edge L_1), they form the total field. One can show that the total field and its normal derivatives are continuous at the shadow boundary for the incident wave ($\varphi = \pi$). Away from this boundary ($\sqrt{k\chi \sin \gamma_0} \left| \cos \frac{\varphi}{2} \right| \gg 1$), the uniform asymptotics (10.38), (10.39) transform into the ray asymptotics (10.29), (10.30).

Asymptotics (10.38), (10.39) written in the terms of the components E_t and H_t completely agree with those of Equations (10.6) and (10.21), and confirm the relationships (10.36), (10.37) between the acoustic and electromagnetic waves.

PROBLEMS

10.1 Show that the field $u_{1h} + u_h^{(t)}$ of acoustic waves and its normal derivative are continuous at the shadow boundary ($\varphi = 0$). Functions u_{1h} and $u_h^{(t)}$ are defined by Equations (10.1) and (10.6), respectively.

10.2 Show that away from the shadow boundary, function (10.6) for acoustic waves transforms asymptotically into the ray approximation (10.4).

10.3 Show that the field $u_{1S} + u_S^{(t)}$ of acoustic waves and its normal derivative are continuous at the shadow boundary ($\varphi = 0$). Functions u_{1S} and $u_S^{(t)}$ are defined by Equations (10.14) and (10.21), respectively.

10.4 Show that away from the shadow boundary, function (10.21) for acoustic waves transforms asymptotically into the ray approximation (10.20).

10.5 Show that away from the shadow boundary ($\varphi = \pi$), function (10.38) for electromagnetic waves transforms asymptotically into the ray approximation (10.34).

10.6 Show that away from the shadow boundary ($\varphi = \pi$), function (10.39) for electromagnetic waves transforms asymptotically into the ray approximation (10.35).

Chapter 11

Focusing of Multiple Acoustic Edge Waves Diffracted at a Convex Body of Revolution with a Flat Base

The theory presented below is based on the papers by Ufimtsev (1989, 1991).

11.1 STATEMENT OF THE PROBLEM AND ITS CHARACTERISTIC FEATURES

This problem is illustrated in Figure 11.1, which shows a convex body of revolution excited by the axisymmetrical incident wave

$$u^{\mathrm{inc}} = u_0 \mathrm{e}^{ik\phi^i} \tag{11.1}$$

constant along the edge.

The axis of symmetry (z-axis) is a focal line for elementary edge waves/rays. With respect to the observation points P on this axis, each diffraction point at the edge is a point of the stationary phase. The elementary rays propagate in the directions of the edge-diffraction cones, which transform (in this particular case) into the meridian planes. Because of that the directivity patterns of elementary edge rays are expressed in terms of the Sommerfeld functions f and g, as shown in Sections 7.6 and 8.1. The functions f and g are defined in Equations (2.62) and (2.64), and they describe the field generated by the total scattering sources $j_{\mathrm{s,h}}^{\mathrm{tot}} = j_{\mathrm{s,h}}^{(0)} + j_{\mathrm{s,h}}^{(1)}$ induced near the edge. Analysis of this field is the main objective of the present chapter.

Here, we ignore the exponentially small multiple edge waves created by the creeping waves (running over the front part, $z < 0$, of the object) and take into account only the multiple diffraction of edge waves propagating over the flat base. The denotation $u_{\mathrm{s,h}}^{(m)}$ will be used for the field of multiple edge waves, where the index $m = 2, 3, \ldots$ indicates the order of diffraction.

Fundamentals of the Physical Theory of Diffraction. By Pyotr Ya. Ufimtsev

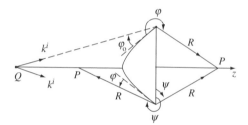

Figure 11.1 Body of revolution excited by the source Q. Focusing of diffracted edge waves occurs at points P on the z-axis. (Reprinted from Ufimtsev (1989) with the permission of *The Journal of Acoustical Society of America.*)

The first-order (primary) edge waves excited directly by the incident wave (11.1) are determined by the integral expression (8.4) applied to the circular edge. Due to the symmetry of the problem, the integrand in Equation (8.4) is constant and the integration over the edge results in the expression

$$\left.\begin{array}{c} u_s^{\mathrm{pr}} \\ u_h^{\mathrm{pr}} \end{array}\right\} = u_0 e^{ik\phi^i} \cdot a \left[\begin{array}{c} f(\varphi, \varphi_0, \alpha) \\ g(\varphi, \varphi_0, \alpha) \end{array}\right] \frac{e^{ikR}}{R}, \tag{11.2}$$

where a is the radius of the edge and α is the external angle between the faces of the edge. This asymptotic expression is valid for any point of observation on the focal line outside the scattering object, under the condition $kR \gg 1$.

Multiple edge waves $u_{s,h}^{(m)}$ with $m = 2, 3, \ldots$ (arising due to diffraction of waves running over the flat base) can be found with application of Equations (10.3) and (10.19), where functions $F_{s,h}^{\mathrm{tot}}$ transform into functions f and g. For calculation of the edge waves running over the flat base, one can utilize the ray asymptotics (10.4) and (10.20), where one should set $\gamma_0 = \pi/2, R = 2a$, and $\rho = -a$. On their way to the opposite point of the edge, these waves intersect the focal line and acquire the phase shift equal to $(-\pi/2)$, which is a direct consequence of the factor $1/\sqrt{1 + R/a} = 1/\sqrt{1 - 2a/a} = i$.

Now one can proceed to the calculation of multiple edge waves.

11.2 MULTIPLE HARD DIFFRACTION

According to Equation (10.3), the field created by the $(m + 1)$-order edge waves on the focal line can be represented in the form

$$u_h^{(m+1)}(P) = \frac{1}{4\pi} g(\psi, 0, \alpha) \frac{e^{ikR}}{R} \int_L \bar{u}_h^{(m)}(\zeta) d\zeta = \frac{a}{2} g(\psi, 0, \alpha) \bar{u}_h^{(m)} \frac{e^{ikR}}{R} \tag{11.3}$$

Here, $\bar{u}_h^{(m)}$ denotes the m-order edge wave propagating along the flat base to the opposite point ζ at the edge, where it undergoes diffraction and creates the elementary waves of the $(m + 1)$ order. More precisely, $\bar{u}_h^{(m)}$ is the field of the m-order wave at the diffraction point ζ. One can show that

$$\bar{u}_h^{(1)} \equiv u_h^{pr} = \frac{1}{2} u_0 e^{ik\phi^i} g(\alpha, \varphi_0, \alpha) \frac{e^{i(2ka - \pi/4)}}{\sqrt{\pi ka}} \tag{11.4}$$

and

$$\bar{u}_h^{(m)} = \bar{u}_h^{(m-1)} g(0, 0, \alpha) \frac{e^{i(2ka - \pi/4)}}{4\sqrt{\pi ka}}, \qquad m = 2, 3, 4, \ldots . \tag{11.5}$$

These relationships lead to

$$\bar{u}_h^{(m)} = u_0 e^{ik\phi^i} 2g(\alpha, \varphi_0, \alpha) \left[g(0, 0, \alpha) \right]^{m-1} \left[\frac{e^{i(2ka - \pi/4)}}{4\sqrt{\pi ka}} \right]^m, \qquad m = 1, 2, 3, \ldots . \tag{11.6}$$

Therefore, the total field of all edge waves on the focal line equals

$$u_h^{ew}(P) = u_h^{pr}(P) + \sum_{m=2}^{\infty} u_h^{(m)}(P)$$

$$= u_0 e^{ik\phi^i} \cdot a \frac{e^{ikR}}{R} \left\{ g(\varphi, \varphi_0, \alpha) + g(\psi, 0, \alpha) g(\alpha, \varphi_0, \alpha) \frac{e^{i(2ka - \pi/4)}}{4\sqrt{\pi ka}} \right.$$

$$\left. + g(\psi, 0, \alpha) g(\alpha, \varphi_0, \alpha) \sum_{m=3}^{\infty} [g(0, 0, \alpha)]^{m-2} \left[\frac{e^{i(2ka - \pi/4)}}{4\sqrt{\pi ka}} \right]^{m-1} \right\}. \tag{11.7}$$

Here, the series is the geometric progression that can be converted to its sum. The physical meaning of Equation (11.7) is clear. The first term in the braces relates to the primary edge waves, the second to the secondary waves, and the third term represents the sum of all multiple edge waves of order 3 and higher.

The total scattered field on the focal line also includes the reflected rays (6.187) in front of the object ($z < 0$) and the shadow radiation (6.228) behind the object ($z > 0$). This approximation for the scattered field actually represents the *incomplete* asymptotic expansion, because it includes only the *first term* in the individual asymptotic expansion for each multiple edge wave. Also, Equation (6.187) is only the first term in the asymptotic expansion for the reflected field.

Expression (11.7) can be used to calculate the total scattering cross-section. In the case of the incident plane wave, $u^{inc} = e^{ikz}$, this quantity is defined as

$$\sigma_{h,s} = \frac{4\pi}{k} \text{Im}(u_{h,s}^{tot} \cdot Re^{-ikR}), \tag{11.8}$$

where $u_{h,s}^{tot}$ is the total field scattered in the *forward* direction ($\psi = \pi/2$). This field consists of the following components:

- The shadow radiation, which is equivalent to the PO field (for the forward direction) and determined by Equation (6.228), where one should set $u_0 = 1$;
- The primary edge waves generated by the nonuniform/fringe scattering sources $j_{h,s}^{(1)}$ and determined by Equations (6.41), (6.42) or (11.2), where one should set $u_0 = 1$ and $u_0 \exp(ik\phi^i) = 1$, respectively;
- The sum of all multiple edge waves of order 2 and higher.

The substitution of this total field into Equation (11.8) results in the following asymptotic expression:

$$\sigma_h = 2\pi a^2 \left\{ 1 + \frac{2}{ka} g\left(\frac{\pi}{2}, 0, \alpha\right) g(\alpha, \varphi_0, \alpha) \right.$$
$$\left. \times \sum_{m=1}^{\infty} [g(0,0,\alpha)]^{m-1} \frac{\sin[m(2ka - \pi/4)]}{4^m (\pi ka)^{m/2}} \right\}, \qquad (11.9)$$

which is *incomplete* in the sense mentioned above. Notice that the series in Equation (11.9) equals zero, when $2ka - \pi/4 = l\pi$ ($l = 1, 2, 3, \ldots$). In this case, all corrections to the first term in Equation (11.9) are determined by the higher-order terms in the individual asymptotic expansions for each multiple edge wave.

11.3 MULTIPLE SOFT DIFFRACTION

The primary edge wave excited by the incident wave (11.1) is determined by Equation (11.2). The higher-order edge waves arise due to the *slope diffraction* of waves running along the flat base of the scattering object. These higher-order edge waves are calculated on the basis of Equation (10.19). For the $(m + 1)$-order edge wave arriving at point P on the focal line, it can be written in the form

$$u_s^{(m+1)}(P) = \frac{1}{2\pi} \frac{\partial f(\psi, 0, \alpha)}{\partial \varphi_0} \frac{e^{ikR}}{R} \int_L \bar{u}_s^{(m)} d\zeta = a\bar{u}_s^{(m)} \frac{\partial f(\psi, 0, \alpha)}{\partial \varphi_0} \frac{e^{ikR}}{R}. \qquad (11.10)$$

Here, $\bar{u}_s^{(m)}$ is the amplitude factor of the wave, which is equivalent to the m-order edge wave coming to the edge point ζ from its opposite point $\bar{\zeta} = \zeta - \pi a$. This quantity is calculated with application of Equations (10.18) and (10.20), where one should set $\gamma_0 = \gamma_{01} = \pi/2$, $R = 2a$, and $\rho = -a$. These calculations result in

$$\bar{u}_s^{(1)} = -u_0 e^{ik\phi^i} \frac{\partial f(\alpha, \varphi_0, \alpha)}{\partial \varphi} \frac{e^{i(2ka+\pi/4)}}{8ka\sqrt{\pi ka}}, \qquad (11.11)$$

$$\bar{u}_s^{(m)} = \bar{u}_s^{(m-1)} \frac{\partial^2 f(0, 0, \alpha)}{\partial \varphi \partial \varphi_0} \frac{e^{i(2ka+\pi/4)}}{8ka\sqrt{\pi ka}}, \qquad (11.12)$$

or

$$\bar{u}_s^{(m)} = -u_0 e^{ik\phi^i} \frac{\partial f(\alpha, \varphi_0, \alpha)}{\partial \varphi} \left[\frac{\partial^2 f(0,0,\alpha)}{\partial \varphi \partial \varphi_0} \right]^{m-1} \left[\frac{e^{i(2ka+\pi/4)}}{8ka\sqrt{\pi ka}} \right]^m. \quad (11.13)$$

After the substitution of (11.13) into Equation (11.10) we obtain

$$u_s^{(m+1)}(P) = -u_0 e^{ik\phi^i} \cdot a \frac{e^{ikR}}{R}$$

$$\times \frac{\partial f(\psi, 0, \alpha)}{\partial \varphi_0} \frac{\partial f(\alpha, \varphi_0, \alpha)}{\partial \varphi} \left[\frac{\partial^2 f(0,0,\alpha)}{\partial \varphi \, \partial \varphi_0} \right]^{m-1} \left[\frac{e^{i(2ka+\pi/4)}}{8ka\sqrt{\pi ka}} \right]^m. \quad (11.14)$$

The total field of all edge waves on the focal lines equals

$$u_s^{ew}(P) = u_s^{pr}(P) + \sum_{m=1}^{\infty} u_s^{(m)}(P)$$

$$= u_0 e^{ik\phi^i} \cdot a \frac{e^{ikR}}{R} \left\{ f^{(1)}(\varphi, \varphi_0, \alpha) - \frac{\partial f(\psi, 0, \alpha)}{\partial \varphi_0} \frac{\partial f(\alpha, \varphi_0, \alpha)}{\partial \varphi} \right.$$

$$\left. \times \sum_{m=2}^{\infty} \left[\frac{\partial^2 f(0,0,\alpha)}{\partial \varphi \, \partial \varphi_0} \right]^{m-2} \left[\frac{e^{i(2ka+\pi/4)}}{8ka\sqrt{\pi ka}} \right]^{m-1} \right\}. \quad (11.15)$$

The series in Equation (11.15) is a geometrical progression.

Now we apply Equation (11.15) for calculation of the total scattering cross-section. In the case of the incident plane wave $u^{inc} = e^{ikz}$, it is defined by Equation (11.8), where

$$u_s^{tot} = u^{sh} + u_s^{(1)} + \sum_{m=2}^{\infty} u_s^{(m)}. \quad (11.16)$$

Here, u^{sh} is the shadow radiation (6.228) (where one should set $u_0 = 1$), the quantity

$$u_s^{(1)} = a f^{(1)}(\varphi, \varphi_0, \alpha) \frac{e^{ikR}}{R} \quad (11.17)$$

is the primary edge wave generated by the nonuniform scattering sources $j_s^{(1)}$, and the series represents the sum of all edge waves of order 2 and higher. Thus, the total scattering cross-section of the acoustically soft object equals

$$\sigma_s = 2\pi a^2 \left\{ 1 - \frac{2}{ka} \frac{\partial f(\pi/2, 0, \alpha)}{\partial \varphi_0} \frac{\partial f(\alpha, \varphi_0, \alpha)}{\partial \varphi} \right.$$

$$\left. \times \sum_{m=1}^{\infty} \left[\frac{\partial^2 f(0,0,\alpha)}{\partial \varphi \, \partial \varphi_0} \right]^{m-1} \frac{\sin[m(2ka + \pi/4)]}{(8ka)^m (\pi ka)^{m/2}} \right\}. \quad (11.18)$$

We emphasize again that approximations (11.15) and (11.18) are the *incomplete* asymptotic expansions in the sense discussed above in Section 11.2. The series in Equation (11.18) equals zero when $2ka + \pi/4 = l\pi$ $(l = 1, 2, 3, \ldots)$. In this case, all corrections to the first term in (11.18) are determined by the higher-order terms in the individual asymptotic expansions for each multiple edge wave.

PROBLEMS

11.1 Prove Equations (11.4) to (11.6) for the primary and multiple acoustic edge waves on a hard scattering object. Explain all of the details, including caustic parameters, phase shifts, directivity factors, and fractional coefficients.

11.2 Prove Equations (11.11) to (11.13) for the primary and multiple acoustic edge waves on a soft scattering object. Explain all of the details, including caustic parameters, phase shifts, directivity factors, and fractional coefficients.

Chapter 12

Focusing of Multiple Edge Waves Diffracted at a Disk

The theory presented in this chapter is based on the papers by Ufimtsev (1989, 1991). It represents the extension of the previous chapter to the disk diffraction problem, where it is necessary to take into account the edge waves propagating along both faces of the disk (Fig. 12.1). This problem is complicated by the fact that the wave traveling along one face of the disk generates (due to diffraction at the edge) the higher–order waves not only on the same face but also on the other face. However, its solution can be lightened if we utilize the symmetry of the scattered field. Let us consider the scattering at an arbitrary plate located in the plane $z = 0$. It follows from Equation (1.10) that

$$u_s^{sc}(-z) = u_s^{sc}(z), \qquad u_h^{sc}(-z) = -u_h^{sc}(z). \tag{12.1}$$

Here, the first equality is obvious and the second is caused by the factor

$$\frac{\partial}{\partial n}\frac{e^{ikr}}{r} = \nabla'\frac{e^{ikr}}{r}\cdot\hat{n} = -\nabla\frac{e^{ikr}}{r}\cdot\hat{n} = -\frac{d}{dr}\frac{e^{ikr}}{r}(\nabla r\cdot\hat{n}) = -\frac{d}{dr}\frac{e^{ikr}}{r}(\hat{r}\cdot\hat{n}), \tag{12.2}$$

where $\hat{r}(-z)\cdot\hat{n} = -\hat{r}(z)\cdot\hat{n}$.

The geometry of the problem is shown in Figure 12.1. The incident wave is given by $u^{inc} = e^{ikz}$. The scattered field is investigated at the points P on the z-axis, which is the focal line of the edge-diffracted waves.

12.1 MULTIPLE HARD DIFFRACTION

The primary edge waves excited by the incident wave directly are given by Equation (11.2), where one should set $u_0 \exp(ik\phi^i) = 1$. The $(m+1)$-order waves are

Fundamentals of the Physical Theory of Diffraction. By Pyotr Ya. Ufimtsev
Copyright © 2007 John Wiley & Sons, Inc.

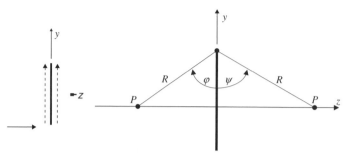

Figure 12.1   Disk projection on the plane $y0z$ (bold solid lines) and the edge waves (dashed lines). Angles φ and $\psi = 2\pi - \varphi$ (measured from the left and right faces of the disk) determine the directions to the observation point P.

determined by Equation (11.3) adjusted to the disk problem as

$$u_{\mathrm{h}}^{(m+1)}(P) = \frac{1}{4\pi} \int_{L} \left[\bar{u}_{\mathrm{l}}^{(m,t)}(\zeta) g(\varphi, 0, \alpha) + \bar{u}_{\mathrm{r}}^{(m,t)}(\zeta) g(\psi, 0, \alpha) \right] \frac{e^{ikR}}{R} \, d\zeta$$

$$= \frac{a}{2} \left[\bar{u}_{\mathrm{l}}^{(m,t)} g(\varphi, 0, \alpha) + \bar{u}_{\mathrm{r}}^{(m,t)} g(\psi, 0, \alpha) \right] \frac{e^{ikR}}{R}, \qquad m = 1, 2, 3, \ldots.$$

$$(12.3)$$

Here, a is the radius of the disk, $\psi = 2\pi - \varphi, \alpha = 2\pi$. The quantity $\bar{u}_{\mathrm{l}}^{(m,t)}$ is the *total* m-order edge wave arrived along the *left* face of the disk to the edge point ζ,

$$\bar{u}_{\mathrm{l}}^{(m,t)} = \bar{u}_{\mathrm{ll}}^{(m)} + \bar{u}_{\mathrm{rl}}^{(m)}.$$

$$(12.4)$$

The quantity $\bar{u}_{\mathrm{ll}}^{(m)}$ denotes the m-order wave on the *left* face (at the edge point ζ) generated (at the edge point $\zeta - \pi a$) by the $(m-1)$-order wave $\bar{u}_{\mathrm{l}}^{(m-1,t)}$ that arrived at the edge along the *left* face. Analogously, the quantity $\bar{u}_{\mathrm{rl}}^{(m)}$ is the m-order wave on the *left* face (at the point ζ of the edge) generated (at the point $\zeta - \pi a$) by the $(m-1)$-order wave $\bar{u}_{\mathrm{r}}^{(m-1,t)}$ that arrived there along the *right* face. With this type of denotations the sense of the following function becomes clear

$$\bar{u}_{\mathrm{r}}^{(m,t)} = \bar{u}_{\mathrm{lr}}^{(m)} + \bar{u}_{\mathrm{rr}}^{(m)}.$$

$$(12.5)$$

Due to Equation (12.1),

$$\bar{u}_{\mathrm{l}}^{(m,t)} = -\bar{u}_{\mathrm{r}}^{(m,t)}, \qquad \bar{u}_{\mathrm{ll}}^{(m)} = -\bar{u}_{\mathrm{lr}}^{(m)}, \qquad \bar{u}_{\mathrm{rr}}^{(m)} = -\bar{u}_{\mathrm{rl}}^{(m)}.$$

$$(12.6)$$

In addition, according to Equation (2.64),

$$g(\psi, 0, \alpha) = -g(\varphi, 0, \alpha) = \frac{1}{\cos \dfrac{\varphi}{2}}.$$

$$(12.7)$$

In view of these relationships, Equation (12.3) can be rewritten as

$$u_h^{(m+1)}(P) = a\bar{u}_l^{(m,t)} g(\varphi, 0, \alpha) \frac{e^{ikR}}{R}$$ (12.8)

or

$$u_h^{(m+1)}(P) = a\bar{u}_r^{(m,t)} g(\psi, 0, \alpha) \frac{e^{ikR}}{R}.$$ (12.9)

The quantities $\bar{u}_l^{(1,t)} \equiv \bar{u}_l^{pr}$ and $\bar{u}_r^{(1,t)} \equiv \bar{u}_r^{pr}$ are the primary edge waves at the disk face, which passed through the focal line and arrived at the edge point ζ. They can be found with application of Equation (8.29), where one should set $\gamma_0 = \pi/2$, $R = 2a$, $\rho = -a$, and $g = g(0, \pi/2, \alpha)$ for $\bar{u}_l^{pr}$, and $g = g(\alpha, \pi/2, \alpha)$ for $\bar{u}_r^{pr}$. In the same way one can find the quantities $\bar{u}_{ll}^{(m)}, \bar{u}_{lr}^{(m)}, \bar{u}_{rl}^{(m)}$ with $m = 2, 3, \ldots$; only the function g will be different, namely $g = g(0, 0, \alpha) = -1$. We omit all intermediate manipulations and obtains the results:

$$\bar{u}_l^{pr} = -\bar{u}_r^{pr} = g(0, \pi/2, \alpha)\lambda = -\sqrt{2}\lambda$$ (12.10)

and

$$\bar{u}_l^{(m,t)} = -\bar{u}_l^{(m-1,t)}\lambda = (-1)^m \sqrt{2}\lambda^m,$$ (12.11)

where

$$\lambda = \frac{e^{i(2ka - \pi/4)}}{2\sqrt{\pi ka}}.$$ (12.12)

Hence

$$u_h^{(m+1)}(P) = a\sqrt{2}g(\varphi, 0, \alpha)(-1)^m \lambda^m \frac{e^{ikR}}{R}$$ (12.13)

and the total focal field of all edge waves equals

$$u_h^{ew}(P) = a \left[g(\varphi, \pi/2, \alpha) + \sqrt{2}g(\varphi, 0, \alpha) \sum_{m=1}^{\infty} (-1)^m \lambda^m \right] \frac{e^{ikR}}{R}.$$ (12.14)

The function $g(\varphi, \pi/2, \alpha)$ is singular in the directions $\varphi = \pi/2$ and $\varphi = 3\pi/2$. Because of this, the total scattered field in the far zone should be represented in the traditional PTD form:

$$u_h^{sc} = u_h^{(0)} + u_h^{fr} + \sum_{m=2}^{\infty} u_h^{(m)}.$$ (12.15)

Here,

$$u_h^{(0)} = \pm \frac{ika^2}{2} \frac{e^{ikR}}{R} \text{ in the directions } \varphi = \begin{cases} 3\pi/2 \\ \pi/2 \end{cases}$$ (12.16)

is the field generated by the uniform scattering sources $j_h^{(0)}$ or, in other words, it is the PO approximation. The quantity

$$u_h^{\text{fr}} = a g^{(1)}(\varphi, \pi/2, \alpha) \frac{e^{ikR}}{R}, \tag{12.17}$$

with $g^{(1)}(\pi/2, \pi/2, \alpha) = -g(3\pi/2, \pi/2, \alpha) = -1/2$, is the field generated by the nonuniform/fringe scattering sources $j_h^{(1)}$ caused by the primary diffraction of the incident wave at the disk. The series in Equations (12.14) and (12.15) represent the contributions generated by that part of $j_h^{(1)}$ that is caused by the multiple diffraction.

After the substitution of Equation (12.15) into Equation (11.8) one finds the total scattering cross-section of the disk

$$\sigma_h = 2\pi a^2 \left\{ 1 + \frac{2}{ka} \sum_{m=1}^{\infty} (-1)^m \frac{\sin[m(2ka - \pi/4)]}{2^{m-1}(\pi ka)^{m/2}} \right\}. \tag{12.18}$$

This is the *incomplete* asymptotic approximation, which includes only the *first* term of the *total* asymptotic expansion (with $ka \to \infty$) for each multiple edge wave. Comparison with the exact asymptotic solution (Witte and Westpfahl, 1970), which contains six first terms for the total cross-section, confirms that Equation (12.18) is correct.

12.2 MULTIPLE SOFT DIFFRACTION

The focal field created by the primary edge waves (excited by the incident wave $u^{\text{inc}} = \exp(ikz)$) is determined according to Equation (11.2) as

$$u_s^{\text{pr}} = a f(\varphi, \pi/2, \alpha) \frac{e^{ikR}}{R}, \tag{12.19}$$

where $\alpha = 2\pi$ and

$$f(\varphi, \varphi_0, \alpha) = \frac{1}{2} \left(-\frac{1}{\cos \dfrac{\varphi - \varphi_0}{2}} + \frac{1}{\cos \dfrac{\varphi + \varphi_0}{2}} \right). \tag{12.20}$$

This is the edge wave generated by the total scattering sources $j_s^{\text{tot}} = j_s^{(0)} + j_s^{(1)}$. The focal field of the primary edge waves created only by the nonuniform component $j_s^{(1)}$ is also described by Equation (12.1), where one should replace the function f by

$$f^{(1)}(\varphi, \varphi_0, \alpha) = f(\varphi, \varphi_0, \alpha) - f^{(0)}(\varphi, \varphi_0) \tag{12.21}$$

with

$$f^{(0)}(\varphi, \varphi_0) = \frac{\sin \varphi_0}{\cos \varphi + \cos \varphi_0}. \tag{12.22}$$

The higher-order edge waves arise due to the *slope diffraction* of waves running along the flat faces of the disk. They are calculated on the basis of Equation (10.19). In the case of diffraction at a solid convex body of revolution, the related technique was developed in Section 11.2. In the present section, this technique is extended for the investigation of diffraction at an acoustically soft disk. According to Equation (10.19), the focal field generated by the $(m+1)$-order edge waves is determined by

$$u_s^{(m+1)}(P) = \frac{1}{2\pi} \int_L \left[\bar{u}_1^{(m,t)} \frac{\partial f(\varphi,0,\alpha)}{\partial \varphi_0} + \bar{u}_r^{(m,t)} \frac{\partial f(\psi,0,\alpha)}{\partial \varphi_0} \right] \frac{e^{ikR}}{R} \, d\zeta \qquad (12.23)$$

where $d\zeta = a\,d\theta$ is the differential arc length of the disk edge L. The geometry of the problem is shown in Figure 12.1. The angles φ and ψ are measured from different faces of the disk and $\psi = 2\pi - \varphi$. Due to the axial symmetry of the problem, Equation (12.23) is reduced to

$$u_s^{(m+1)}(P) = a \left[\bar{u}_1^{(m,t)} \frac{\partial f(\varphi,0,\alpha)}{\partial \varphi_0} + \bar{u}_r^{(m,t)} \frac{\partial f(\psi,0,\alpha)}{\partial \varphi_0} \right] \frac{e^{ikR}}{R}. \qquad (12.24)$$

Here, $\bar{u}_{1,r}^{(m,t)}$ is the amplitude factor of the wave, which is equivalent to the *total* m-order edge wave coming (to the edge point ζ from its opposite point $\bar{\zeta} = \zeta - \pi a$) along the left or right face, as indicated by the subscripts l, r. In accordance with Equation (12.1) the scattered field is also symmetric with respect to the disk plane, therefore

$$\bar{u}_1^{(m,t)} = \bar{u}_r^{(m,t)}. \qquad (12.25)$$

Besides,

$$\frac{\partial f(\psi,0,\alpha)}{\partial \varphi_0} = \frac{\partial f(\varphi,0,\alpha)}{\partial \varphi_0} = \frac{\sin\frac{\varphi}{2}}{2\cos^2\frac{\varphi}{2}}. \qquad (12.26)$$

Hence,

$$u_s^{(m+1)}(P) = 2a\bar{u}_1^{(m,t)} \frac{\partial f(\varphi,0,\alpha)}{\partial \varphi_0} \frac{e^{ikR}}{R}. \qquad (12.27)$$

The quantity

$$\bar{u}_1^{(m,t)} = \bar{u}_{ll}^{(m)} + \bar{u}_{rl}^{(m)} = 2\bar{u}_{ll}^{(m)} \qquad (12.28)$$

consists of two equal terms. The term $\bar{u}_{ll}^{(m)}$ relates to the m-order edge wave on the left face of the disk at the point ζ. This wave is generated by the $(m-1)$-order edge wave at the opposite point $\bar{\zeta} = \zeta - \pi a$, which arrived there along the left side of the disk. The term $\bar{u}_{rl}^{(m)}$ relates to the m-order edge wave (at the same point ζ on the left face of the disk) created by the $(m-1)$-order edge wave at the opposite point $\bar{\zeta} = \zeta - \pi a$

and arrived there along the *right* side of the disk. Because of the symmetry of the field, these terms are equal to each other.

The quantities $\bar{u}_{11}^{(m)}$ and $\bar{u}_1^{(m,t)}$ are calculated with application of Equations (10.18) and (10.20), where one should set $\gamma_0 = \gamma_{01} = \pi/2$, $R = 2a$, and $\rho = -a$. These calculations result in

$$\bar{u}_1^{(1,t)} \equiv \bar{u}_1^{(1)} = \mu \frac{\partial f(0, \pi/2, \alpha)}{\partial \varphi}, \qquad \text{with } \mu = \frac{e^{i(2ka+\pi/4)}}{8ka\sqrt{\pi ka}}, \tag{12.29}$$

$$\bar{u}_1^{(m,t)} = 2\mu \bar{u}_1^{(m-1,t)} \frac{\partial^2 f(0,0,\alpha)}{\partial \varphi \partial \varphi_0}, \qquad \text{with } m = 2,3,4\ldots \tag{12.30}$$

The denotation $\bar{u}_1^{(1,t)} = \bar{u}_1^{(1)}$ is used to emphasize that only *one* primary edge wave exists on each side of the disk, but *two* edge waves of any higher order are on every side. Thus,

$$\bar{u}_1^{(m,t)} = 2^{m-1} \mu^m \frac{\partial f(0, \pi/2, \alpha)}{\partial \varphi} \left[\frac{\partial^2 f(0,0,\alpha)}{\partial \varphi \partial \varphi_0} \right]^{m-1} \tag{12.31}$$

and

$$u_s^{(m+1)}(P) = a2^m \mu^m \frac{\partial f(0, \pi/2, \alpha)}{\partial \varphi} \left[\frac{\partial^2 f(0,0,\alpha)}{\partial \varphi \partial \varphi_0} \right]^{m-1} \frac{\partial f(\varphi, 0, \alpha)}{\partial \varphi_0} \frac{e^{ikR}}{R}, \tag{12.32}$$

with $m = 1, 2, 3, \ldots$ and

$$\frac{\partial f(0, \pi/2, \alpha)}{\partial \varphi} = \frac{1}{\sqrt{2}}, \qquad \frac{\partial^2 f(0,0,\alpha)}{\partial \varphi \partial \varphi_0} = \frac{1}{4}. \tag{12.33}$$

The total focal field created by all the edge waves together equals

$$u_s^{ew}(P) = a \frac{e^{ikR}}{R} \left\{ f(\varphi, \pi/2, \alpha) + \frac{\partial f(0, \pi/2, \alpha)}{\partial \varphi} \frac{\partial f(\varphi, 0, \alpha)}{\partial \varphi_0} \right.$$
$$\left. \times \sum_{m=1}^{\infty} 2^m \frac{e^{im(2ka+\pi/4)}}{(8ka\sqrt{\pi ka})^m} \left[\frac{\partial^2 f(0,0,\alpha)}{\partial \varphi \partial \varphi_0} \right]^{m-1} \right\}. \tag{12.34}$$

This expression can be used to calculate the total cross-section (11.8), where u_s^{tot} is the total field scattered in the forward direction $\varphi = 3\pi/2$. In the present case,

$$u_s^{(t)} = a \frac{e^{ikR}}{R} \left\{ i\frac{ka}{2} + f^{(1)}(3\pi/2, \pi/2, \alpha) + \frac{\partial f(0, \pi/2, \alpha)}{\partial \varphi} \frac{\partial f(3\pi/2, 0, \alpha)}{\partial \varphi_0} \right.$$
$$\left. \times \sum_{m=1}^{\infty} 2^m \frac{e^{im(2ka+\pi/4)}}{(8ka\sqrt{\pi ka})^m} \left[\frac{\partial^2 f(0,0,\alpha)}{\partial \varphi \partial \varphi_0} \right]^{m-1} \right\}. \tag{12.35}$$

with

$$f^{(1)}(3\pi/2, \pi/2, \alpha) = \frac{1}{2}, \qquad \frac{\partial f(3\pi/2, 0, \alpha)}{\partial \varphi_0} = \frac{1}{\sqrt{2}}. \tag{12.36}$$

The substitution of Equation (12.35) into Equation (11.8) determines the total cross-section

$$\sigma_s = 2\pi a^2 \left\{ 1 + \frac{2}{ka} \sum_{m=1}^{\infty} \frac{\sin[m(2ka + \pi/4)]}{2^{m-1}(8ka\sqrt{\pi ka})^m} \right\}. \tag{12.37}$$

This is the *incomplete* asymptotic expression, which includes only the first term of the *total* asymptotic expansion for every edge wave. It can be verified by comparison with the exact asymptotic expression (14.54) in Bowman et al. (1987), which contains the asymptotic terms up to the order of $(ka)^{-4}$. According to Equation (12.37),

$$\sigma_s = 2\pi a^2 \left\{ 1 + \frac{\cos(2ka - \pi/4)}{4\sqrt{\pi}(ka)^{5/2}} + \frac{\cos(4ka)}{64\pi(ka)^4} + O[(ka)^{-11/2}] \right\}. \tag{12.38}$$

All these three terms are identical to the exact ones. Thus, the comparison of asymptotics (12.18) and (12.37) with known exact results proves that PTD correctly predicts the first term in the total asymptotic expansion for every multiple edge wave.

12.3 MULTIPLE DIFFRACTION OF ELECTROMAGNETIC WAVES

Here, we investigate the diffraction of a plane wave

$$E_x^{\text{inc}} = Z_0 H_y^{\text{inc}} = e^{ikz} \tag{12.39}$$

at a perfectly conducting disk (Fig. 12.1). The basic features of this problem are essentially the same as those in the acoustical problems above. For this reason, we will not repeat them here and only briefly discuss a new specific feature caused by the vector nature of electromagnetic waves. Due to this nature and to the axial symmetry of the problem, one can separate the diffracted waves (of the second and higher orders) into two independent groups, with E_φ- and H_φ-polarizations. Multiple diffraction of the E_φ-waves (H_φ-waves) is calculated just like the diffraction of acoustic waves at a soft (hard) disk. This observation significantly facilitates the investigation, which results in the following approximations for the focal field on the z-axis ($z \gg kR^2$). The focal fields generated by all the multiple E_φ-waves and H_φ-waves are equal to

$$E_x = Z_0 H_y = \frac{a}{z} e^{ikz} \sum_{m=1}^{\infty} \frac{e^{im(2ka+\pi/4)}}{2^{4m}\pi^{m/2}(ka)^{3m/2}} \qquad (E_\varphi\text{-waves}) \tag{12.40}$$

and

$$E_x = Z_0 H_y = \frac{a}{z} e^{ikz} \sum_{m=1}^{\infty} (-1)^m \frac{e^{im(2ka-\pi/4)}}{2^m(\pi ka)^{m/2}} \qquad (H_\varphi\text{-waves}). \tag{12.41}$$

The focal field includes these fields plus the contributions generated by the current $\vec{j}^{(0)}$ (PO contribution) and the current $\vec{j}^{(1)}$ (related to the primary edge diffraction):

$$E_x^{(0)} = Z_0 H_y^{(0)} = \frac{ika^2}{2} \frac{e^{ikz}}{z} \tag{12.42}$$

and

$$E_x^{(1)} = Z_0 H_y^{(1)} = \frac{a}{2}[f^{(1)}(3\pi/2, \pi/2, 2\pi) + g^{(1)}(3\pi/2, \pi/2, 2\pi)]\frac{e^{ikz}}{z}. \tag{12.43}$$

However,

$$f^{(1)}(3\pi/2, \pi/2, 2\pi) = -g^{(1)}(3\pi/2, \pi/2, 2\pi) = -\frac{1}{2} \tag{12.44}$$

and so $E_x^{(1)} = H_y^{(1)} = 0$. Therefore, the total focal field equals

$$E_x^{(t)} = Z_0 H_y^{(t)} = a\frac{e^{ikz}}{z}\left[\frac{ika}{2} + \sum_{m=1}^{\infty}(-1)^m \frac{e^{im(2ka-\pi/4)}}{2^m(\pi ka)^{m/2}} + \sum_{m=1}^{\infty} \frac{e^{im(2ka+\pi/4)}}{2^{4m}\pi^{m/2}(ka)^{3m/2}}\right]. \tag{12.45}$$

Now, according to Equation (11.8), which is also valid for electromagnetic waves (with the replacement of $u^{(t)}$ by $E_x^{(t)}$), one obtains the total scattering cross-section:

$$\sigma = 2\pi a^2 \left\{1 + \frac{2}{ka}\sum_{m=1}^{\infty}(-1)^m \frac{\sin[m(2ka - \pi/4)]}{2^m(\pi ka)^{m/2}} + \frac{2}{ka}\sum_{m=1}^{\infty}\frac{\sin[m(2ka + \pi/4)]}{2^{4m}\pi^{m/2}(ka)^{3m/2}}\right\}. \tag{12.46}$$

It turns out that this quantity is connected by the relation

$$\sigma = \tfrac{1}{2}(\sigma_h + \sigma_s) \tag{12.47}$$

with the similar quantities (11.9) and (11.18) found for acoustic waves.

PROBLEMS

12.1 Prove Equation (12.14) for the focal field generated by all of the acoustic edge waves scattered at a hard disk. Explain all of the details, including caustic parameters, phase shifts, directivity factors, and fractional coefficients.

12.2 Prove Equation (12.34) for the focal field generated by all of the acoustic edge waves scattered at a soft disk. Explain all of the details, including caustic parameters, phase shifts, directivity factors, and fractional coefficients.

12.3 Prove Equation (12.40) for the focal field generated by the E_φ-group of the electromagnetic edge waves scattered at a perfectly conducting disk. Explain all of the details, including caustic parameters, phase shifts, directivity factors, and fractional coefficients.

12.4 Prove Equation (12.41) for the focal field generated by the H_φ-group of the electromagnetic edge waves scattered at a perfectly conducting disk. Explain all of the details, including caustic parameters, phase shifts, directivity factors, and fractional coefficients.

Chapter **13**

Backscattering at a Finite-Length Cylinder

13.1 ACOUSTIC WAVES

The geometry of the problem is shown in Figure 13.1. A solid circular cylinder with flat bases is illuminated by the incident plane wave

$$u^{\text{inc}} = u_0 e^{ik(y \sin \gamma + z \cos \gamma)}, \qquad \text{with } 0 \leq \gamma \leq \pi/2. \tag{13.1}$$

The total length of the cylinder and its diameter are denoted by $L = 2\ell$ and $d = 2a$, respectively. The scattered field is evaluated for the backscattering direction $\vartheta = \pi - \gamma, \varphi = 3\pi/2$.

13.1.1 PO Approximation

According to Equation (1.37), the PO fields backscattered by the acoustically hard and soft objects differ from each other only in sign. Hence, it is sufficient to exhibit the PO calculations only for the case of scattering at a hard cylinder.

First we calculate the far field scattered by the left base/disk of the hard cylinder. In this case, the application of Equation (1.37) leads to the expression

$$u_{\text{h}}^{(0)\text{disk}} = u_0 \frac{ik}{2\pi} \cos \vartheta \, e^{i2kl \cos \vartheta} \frac{e^{ikR}}{R} \int_0^a r' dr' \int_0^{2\pi} e^{i2kr' \sin \vartheta \sin \varphi'} d\varphi', \tag{13.2}$$

where $\vartheta = \pi - \gamma$. In view of Equation (6.55) we have

$$u_{\text{h}}^{(0)\text{disk}} = u_0 \frac{ia \cos \vartheta}{2 \sin \vartheta} J_1(2ka \sin \vartheta) e^{i2kl \cos \vartheta} \frac{e^{ikR}}{R}. \tag{13.3}$$

The application of Equation (1.37) to the field scattered by the cylindrical part of the object leads to the integral expression

$$u_{\text{h}}^{(0)\text{cyl}} = u_0 \frac{ika}{2\pi} \sin \vartheta \frac{e^{ikR}}{R} \int_{-l}^{l} e^{-i2kz' \cos \vartheta} dz' \int_{\pi}^{2\pi} e^{i2ka \sin \vartheta \sin \varphi'} \sin \varphi' d\varphi'. \tag{13.4}$$

Fundamentals of the Physical Theory of Diffraction. By Pyotr Ya. Ufimtsev
Copyright © 2007 John Wiley & Sons, Inc.

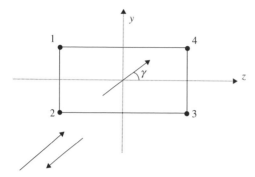

Figure 13.1   Cross-section of the cylinder by the $y0z$-plane. Dots 1, 2, and 3 are the stationary phase points visible in the sector $\pi/2 < \vartheta < \pi$.

Here, the integration is performed over the illuminated part of the scattering surface where $\pi \leq \varphi' \leq 2\pi$ under the condition $\pi/2 + 0 \leq \vartheta \leq \pi - 0$. In the limiting case when $\vartheta = \pi$, the integration encompasses the whole cylindrical surface ($0 \leq \varphi' \leq 2\pi$) and results in zero scattered field. However, Equation (13.4) is also valid in this case as it equals zero due to the factor $\sin \vartheta$. Thus,

$$u_{\mathrm{h}}^{(0)\mathrm{cyl}} = 0, \qquad \text{if } \vartheta = \pi. \tag{13.5}$$

The integral in Equation (13.4) over the variable z' is calculated in the closed form. The integral over the variable φ' is calculated (under the condition $2ka \sin \vartheta \gg 1$), by the stationary-phase technique (Copson, 1965; Murray, 1984). The details of this technique have already, been considered in Sections 6.1.2 and 8.1. The stationary point $\varphi'_{\mathrm{st}} = 3\pi/2$ is found from the equation

$$\frac{\mathrm{d}}{\mathrm{d}\varphi'} 2ka \sin \vartheta \sin \varphi' = 2ka \sin \vartheta \cos \varphi' = 0. \tag{13.6}$$

The asymptotic expression for the integral is given by

$$\int_{\pi}^{2\pi} \mathrm{e}^{i2ka \sin \vartheta \sin \varphi'} \sin \varphi' \, \mathrm{d}\varphi' \sim -\sqrt{\frac{\pi}{ka \sin \vartheta}} \mathrm{e}^{-i2ka \sin \vartheta + i\pi/4}. \tag{13.7}$$

Therefore, under the condition $2ka \sin \vartheta \gg 1$, the scattered field $u_{\mathrm{h}}^{(0)\mathrm{cyl}}$ is determined asymptotically as

$$u_{\mathrm{h}}^{(0)\mathrm{cyl}} \sim -u_0 \frac{ia \sin \vartheta}{2 \cos \vartheta} \sin(2kl \cos \vartheta) \frac{\mathrm{e}^{-i2ka \sin \vartheta + i\pi/4}}{\sqrt{\pi ka \sin \vartheta}} \frac{\mathrm{e}^{ikR}}{R}. \tag{13.8}$$

Having Equations (13.5) and (13.8), one can construct the approximation valid in the entire region $\pi/2 \leq \vartheta \leq \pi$. This can be done in a manner similar to that in Section 6.1.4. We use the asymptotic expressions for the Bessel functions with large arguments $(x \gg 1)$ and observe that

$$\frac{e^{-i2x+i\pi/4}}{\sqrt{\pi x}} \approx J_0(2x) - iJ_1(2x), \tag{13.9}$$

$$\frac{e^{-i2x+i\pi/4}}{\sqrt{\pi x}} \approx \frac{1}{i}[J_1(2x) - iJ_2(2x)], \tag{13.10}$$

and

$$\frac{e^{-i2x+i\pi/4}}{\sqrt{\pi x}} \approx e^{-in\pi/2}[J_n(2x) - iJ_{n+1}(2x)], \qquad n = 0, 1, 2, 3, \ldots. \tag{13.11}$$

Each of these combinations can be used to construct the approximation for the field in the region $\pi/2 \leq \vartheta \leq \pi$. We apply and analyze the simplest ones, (13.9) and (13.10). With these approximations the total PO field can be represented in the two following forms:

$$u_{h,1}^{(0)} = u_0 \frac{ia}{2} \frac{e^{ikR}}{R} \left\{ \frac{\cos \vartheta}{\sin \vartheta} J_1(2ka \sin \vartheta) e^{i2kl \cos \vartheta} \right.$$
$$\left. - \frac{\sin \vartheta}{\cos \vartheta} \sin(2kl \cos \vartheta) [J_0(2ka \sin \vartheta) - iJ_1(2ka \sin \vartheta)] \right\} \tag{13.12}$$

and

$$u_{h,2}^{(0)} = u_0 \frac{ia}{2} \frac{e^{ikR}}{R} \left\{ \frac{\cos \vartheta}{\sin \vartheta} J_1(2ka \sin \vartheta) e^{i2kl \cos \vartheta} \right.$$
$$\left. + i \frac{\sin \vartheta}{\cos \vartheta} \sin(2kl \cos \vartheta) [J_1(2ka \sin \vartheta) - iJ_2(2ka \sin \vartheta)] \right\}. \tag{13.13}$$

The scattering cross-section σ is defined by Equation (1.26). We have calculated the normalized scattering cross-section as

$$\sigma_{\text{norm}} = \sigma/\sigma_{\text{d}} \tag{13.14}$$

where the quantity

$$\sigma_{\text{d}} = \pi a^2 (ka)^2 \tag{13.15}$$

is the PO scattering cross-section of the disk under the normal incidence ($\vartheta = \pi$). The results are shown in Figures 13.2 and 13.3. The curves PO-1 and PO-2 relate to Equations (13.12) and (13.13), respectively.

The small discrepancy between the two curves in Figure 13.2 is caused by the different higher-order terms in the asymptotic expressions (13.9) and (13.10). It takes place when the cylinder diameter is not sufficiently large and equals only one wavelength ($d = 2a = \lambda$) when the argument of the Bessel functions does not exceed 2π. The discrepancy between the approximations PO-1 and PO-2 becomes practically

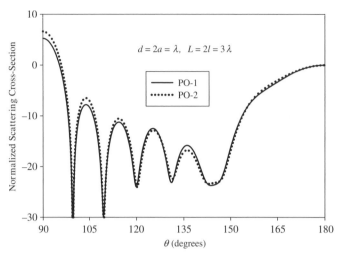

Figure 13.2 Backscattering at the finite cylinder according to the PO approximations PO-1 (Equation (13.12)) and PO-2 (Equation (13.13)).

negligible in the case when $d = 3\lambda$, as is clearly seen in Figure 13.3. Because of that we use in the following calculations the simplest approximation (13.12). There is another reason in favour of using Equation (13.12). It matches an analogous type of approximation for the field (13.22), (13.23) generated by the nonuniform/fringe source $j^{(1)}$. This field is investigated in the following section.

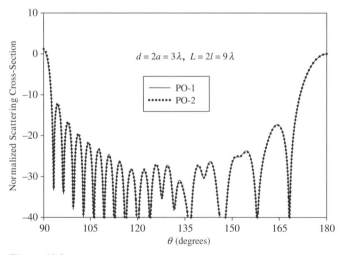

Figure 13.3 Backscattering at the finite cylinder according to the PO approximations PO-1 (Equation (13.12)) and PO-2 (Equation (13.13)).

13.1.2 Backscattering Produced by the Nonuniform Component $j^{(1)}$

There are three types of nonuniform components $j^{(1)}$ of the scattering sources on a finite cylinder. The edge/fringe component $j_{\text{fr}}^{(1)}$ concentrates near the edges. The component $j_{\text{cr}}^{(1)}$ associated with creeping waves concentrates near the shadow boundary on the cylindrical part of the object and exponentially attenuates away from this boundary. The third component $j_{\text{dif}}^{(1)}$ is caused by the transverse diffusion of the wave field between the adjacent rays reflected from the cylindrical surface. This component $j_{\text{dif}}^{(1)}$ exists on the illuminated part of the cylindrical surface away from the shadow boundary. Among these components of $j^{(1)}$, a main contribution to the backscattering is provided by the fringe component $j_{\text{fr}}^{(1)}$ and this contribution is investigated here. Notice also that the field generated by $j_{\text{dif}}^{(1)}$ is comparable with that produced by $j_{\text{fr}}^{(1)}$, but this situation occurs only in the direction of the specular rays reflected from the cylindrical surface. This topic is considered in the next chapter.

First we analyze the field generated by the fringe component located near the left edge ($z = -l$, $\sqrt{x^2 + y^2} = a$). According to Equation (8.3) it is determined as

$$u_{\text{s,h}}^{(1)\text{left}} = u_0 \frac{a}{2\pi} e^{i2kl\cos\vartheta} \frac{e^{ikR}}{R} \int_0^{2\pi} F_{\text{s,h}}^{(1)}(\varphi') e^{i2ka\sin\vartheta\sin\varphi'} d\varphi'. \tag{13.16}$$

In the direction $\vartheta = \pi$, functions $F_{\text{s,h}}^{(1)}$ transform into functions $f^{(1)}$ and $g^{(1)}$ as shown in Equations (7.120) and (7.121). Recall that these functions are defined in Section 4.1. Hence for this direction,

$$\left.\begin{array}{c} u_{\text{s}}^{(1)\text{left}} \\ u_{\text{h}}^{(1)\text{left}} \end{array}\right\} = u_0 a \begin{bmatrix} f^{(1)} \\ g^{(1)} \end{bmatrix} e^{-i2kl} \frac{e^{ikR}}{R}, \qquad (\vartheta = \pi). \tag{13.17}$$

For other directions ϑ, which satisfy the condition $2ka\sin\vartheta \gg 1$, the integral in Equation (13.16) is evaluated asymptotically by the stationary-phase technique. There are two stationary points ($\varphi_{\text{st},1} = \pi/2$ and $\varphi_{\text{st},2}' = 3\pi/2$) in the integrand of Equation (13.16). At these points, functions $F_{\text{s,h}}^{(1)}$ also transform into functions $f^{(1)}$ and $g^{(1)}$. The resulting asymptotic approximations for the field (13.16) are given as

$$u_{\text{s}}^{(1)\text{left}} = u_0 \frac{a}{2} e^{i2kl\cos\vartheta} \frac{e^{ikR}}{R} \frac{1}{\sqrt{\pi ka\sin\vartheta}}$$
$$\times [f^{(1)}(1)e^{i2ka\sin\vartheta - i\pi/4} + f^{(1)}(2)e^{-i2ka\sin\vartheta + i\pi/4}] \tag{13.18}$$

and

$$u_{\text{h}}^{(1)\text{left}} = u_0 \frac{a}{2} e^{i2kl\cos\vartheta} \frac{e^{ikR}}{R} \frac{1}{\sqrt{\pi ka\sin\vartheta}}$$
$$\times [g^{(1)}(1)e^{i2ka\sin\vartheta - i\pi/4} + g^{(1)}(2)e^{-i2ka\sin\vartheta + i\pi/4}]. \tag{13.19}$$

Here, the functions $f^{(1)}(1)$, $g^{(1)}(1)$ relate to the stationary point 1 ($\varphi'_{st,1} = \pi/2$) and the functions $f^{(1)}(2)$, $g^{(1)}(2)$ relate to the stationary point 2 ($\varphi'_{st,2} = 3\pi/2$). These points are shown in Figure 13.1.

In order to construct the field approximations in the entire region $\pi/2 \le \vartheta \le \pi$, we apply the idea suggested in Sections 6.1.4 and 13.1.1. In otherwords, we substitute the asymptotics

$$\frac{e^{i2ka \sin \vartheta - i\pi/4}}{\sqrt{\pi ka \sin \vartheta}} \approx J_0(2ka \sin \vartheta) + iJ_1(2ka \sin \vartheta), \tag{13.20}$$

$$\frac{e^{-i2ka \sin \vartheta + i\pi/4}}{\sqrt{\pi ka \sin \vartheta}} \approx J_0(2ka \sin \vartheta) - iJ_1(2ka \sin \vartheta) \tag{13.21}$$

into Equations (13.18) and (13.19) and then extend the obtained field expressions to the entire region $\pi/2 \le \vartheta \le \pi$. It turns out that the resulting expressions

$$u_s^{(1)\text{left}} = u_0 \frac{a}{2} e^{i2kl \cos \vartheta} \frac{e^{ikR}}{R} \left\{ f^{(1)}(1)[J_0(2ka \sin \vartheta) + i J_1(2ka \sin \vartheta)] \right.$$
$$\left. + f^{(1)}(2)[J_0(2ka \sin \vartheta) - i J_1(2ka \sin \vartheta)] \right\} \tag{13.22}$$

and

$$u_h^{(1)\text{left}} = u_0 \frac{a}{2} e^{i2kl \cos \vartheta} \frac{e^{ikR}}{R} \left\{ g^{(1)}(1)[J_0(2ka \sin \vartheta) + i J_1(2ka \sin \vartheta)] \right.$$
$$\left. + g^{(1)}(2)[J_0(2ka \sin \vartheta) - i J_1(2ka \sin \vartheta)] \right\} \tag{13.23}$$

exactly transform into Equation (13.17) when $\vartheta \to \pi$. Therefore, these expressions can be considered as appropriate approximations for the scattered field in all directions $\pi/2 \le \vartheta \le \pi$.

The contribution of the right edge ($z = +l$, $\sqrt{x^2 + y^2} = a$) to the field in the region $\pi/2 < \vartheta < \pi$ is described by the expression

$$u_{s,h}^{(1)\text{right}} = u_0 \frac{a}{2\pi} e^{-i2kl \cos \vartheta} \frac{e^{ikR}}{R} \int_\pi^{2\pi} F_{s,h}^{(1)}(\varphi') e^{i2ka \sin \vartheta \sin \varphi'} d\varphi', \tag{13.24}$$

analogous to Equation (13.16). Its asymptotic approximation found by the stationary-phase technique is determined as

$$\left. \begin{matrix} u_s^{(1)\text{right}} \\ u_h^{(1)\text{right}} \end{matrix} \right\} = u_0 \frac{a}{2} \begin{bmatrix} f^{(1)}(3) \\ g^{(1)}(3) \end{bmatrix} \frac{e^{-i2ka \sin \vartheta + i\pi/4}}{\sqrt{\pi ka \sin \vartheta}} e^{-i2kl \cos \vartheta} \frac{e^{ikR}}{R}, \tag{13.25}$$

where $2ka \sin \vartheta \gg 1$ and functions $f^{(1)}(3)$, $g^{(1)}(3)$ relate to the stationary point 3 ($\varphi'_{st,3} = 3\pi/2$) shown in Figure 13.1. With the application of Equation (13.21), this expression is extended to all directions $\pi/2 < \vartheta < \pi$ and provides the following approximations to the field (13.24):

$$u_s^{(1)\text{right}} = u_0 \frac{a}{2} f^{(1)}(3) \, [J_0(2ka \sin \vartheta) - iJ_1(2ka \sin \vartheta)] e^{-i2kl \cos \vartheta} \frac{e^{ikR}}{R} \qquad (13.26)$$

and

$$u_h^{(1)\text{right}} = u_0 \frac{a}{2} g^{(1)}(3) \, [J_0(2ka \sin \vartheta) - iJ_1(2ka \sin \vartheta)] e^{-i2kl \cos \vartheta} \frac{e^{ikR}}{R}. \qquad (13.27)$$

Thus, in the first approximation, the total field produced by the component $j_{\text{fr}}^{(1)}$ equals

$$u_s^{(1)} = u_0 \frac{a}{2} \frac{e^{ikR}}{R} \left\{ f^{(1)}(1) \, [J_0(2ka \sin \vartheta) + iJ_1(2ka \sin \vartheta)] e^{i2kl \cos \vartheta} \right.$$
$$\left. + [f^{(1)}(2)e^{i2kl \cos \vartheta} + f^{(1)}(3)e^{-i2kl \cos \vartheta}][J_0(2ka \sin \vartheta) - iJ_1(2ka \sin \vartheta)] \right\},$$
$$(13.28)$$

and

$$u_h^{(1)} = u_0 \frac{a}{2} \frac{e^{ikR}}{R} \left\{ g^{(1)}(1) \, [J_0(2ka \sin \vartheta) + iJ_1(2ka \sin \vartheta)] e^{i2kl \cos \vartheta} \right.$$
$$\left. + [g^{(1)}(2)e^{i2kl \cos \vartheta} + g^{(1)}(3)e^{-i2kl \cos \vartheta}][J_0(2ka \sin \vartheta) - iJ_1(2ka \sin \vartheta)] \right\}.$$
$$(13.29)$$

The functions $f^{(1)}$ and $g^{(1)}$ are determined according to Section 4.1 as

$$f^{(1)}(1) = \frac{\sin \frac{\pi}{n}}{n} \left(\frac{1}{\cos \frac{\pi}{n} - 1} - \frac{1}{\cos \frac{\pi}{n} - \cos \frac{\pi - 2\vartheta}{n}} \right) + \frac{1}{2} \frac{\cos \vartheta}{\sin \vartheta}, \qquad (13.30)$$

$$g^{(1)}(1) = \frac{\sin \frac{\pi}{n}}{n} \left(\frac{1}{\cos \frac{\pi}{n} - 1} + \frac{1}{\cos \frac{\pi}{n} - \cos \frac{\pi - 2\vartheta}{n}} \right) - \frac{1}{2} \frac{\cos \vartheta}{\sin \vartheta}, \qquad (13.31)$$

$$f^{(1)}(2) = \frac{\sin \frac{\pi}{n}}{n} \left(\frac{1}{\cos \frac{\pi}{n} - 1} - \frac{1}{\cos \frac{\pi}{n} - \cos \frac{2\vartheta}{n}} \right) - \frac{1}{2} \frac{\cos \vartheta}{\sin \vartheta} - \frac{1}{2} \frac{\sin \vartheta}{\cos \vartheta},$$
$$(13.32)$$

$$g^{(1)}(2) = \frac{\sin \frac{\pi}{n}}{n} \left(\frac{1}{\cos \frac{\pi}{n} - 1} + \frac{1}{\cos \frac{\pi}{n} - \cos \frac{2\vartheta}{n}} \right) + \frac{1}{2} \frac{\cos \vartheta}{\sin \vartheta} + \frac{1}{2} \frac{\sin \vartheta}{\cos \vartheta},$$
$$(13.33)$$

$$f^{(1)}(3) = \frac{\sin\frac{\pi}{n}}{n}\left(\frac{1}{\cos\frac{\pi}{n}-1} - \frac{1}{\cos\frac{\pi}{n}-\cos\frac{\pi+2\vartheta}{n}}\right) + \frac{1}{2}\frac{\sin\vartheta}{\cos\vartheta}, \qquad (13.34)$$

and

$$g^{(1)}(3) = \frac{\sin\frac{\pi}{n}}{n}\left(\frac{1}{\cos\frac{\pi}{n}-1} + \frac{1}{\cos\frac{\pi}{n}-\cos\frac{\pi+2\vartheta}{n}}\right) - \frac{1}{2}\frac{\sin\vartheta}{\cos\vartheta}, \qquad (13.35)$$

with $n = 3/2$.

Although certain terms in these functions are singular in the directions $\vartheta = \pi/2$ and $\vartheta = \pi$, these singularities always cancel each other, and the functions $f^{(1)}$, $g^{(1)}$ remain finite. In the direction $\vartheta = \pi/2$ they have the values

$$f^{(1)}(1) = 0, \qquad f^{(1)}(2) = f^{(1)}(3) = \frac{\frac{1}{n}\sin\frac{\pi}{n}}{\cos\frac{\pi}{n}-1} + \frac{1}{2n}\cot\frac{\pi}{n} \qquad (13.36)$$

and

$$g^{(1)}(1) = \frac{\frac{2}{n}\sin\frac{\pi}{n}}{\cos\frac{\pi}{n}-1}, \qquad g^{(1)}(2) = g^{(1)}(3) = \frac{\frac{1}{n}\sin\frac{\pi}{n}}{\cos\frac{\pi}{n}-1} - \frac{1}{2n}\cot\frac{\pi}{n}, \quad (13.37)$$

and in the direction $\vartheta = \pi$ they are determined as

$$f^{(1)}(1) = f^{(1)}(2) = \frac{\frac{1}{n}\sin\frac{\pi}{n}}{\cos\frac{\pi}{n}-1} + \frac{1}{2n}\cot\frac{\pi}{n}, \qquad f^{(1)}(3) = 0 \qquad (13.38)$$

and

$$g^{(1)}(1) = g^{(1)}(2) = \frac{\frac{1}{n}\sin\frac{\pi}{n}}{\cos\frac{\pi}{n}-1} - \frac{1}{2n}\cot\frac{\pi}{n}, \qquad g^{(1)}(3) = \frac{\frac{2}{n}\sin\frac{\pi}{n}}{\cos\frac{\pi}{n}-1}. \quad (13.39)$$

We also obtain the expressions for the functions $f^{(1)}(4)$ and $g^{(1)}(4)$ related to the stationary point 4 (Fig. 13.1), which becomes visible in the directions $\vartheta = \pi/2$ and $\vartheta = \pi$. For both directions, functions $f^{(1)}(4)$ and $g^{(1)}(4)$ have the same values,

$$f^{(1)}(4) = 0 \qquad \text{and} \qquad g^{(1)}(4) = \frac{\frac{2}{n}\sin\frac{\pi}{n}}{\cos\frac{\pi}{n}-1}. \qquad (13.40)$$

As the contribution from point 4 equals zero for the soft cylinder, the approximation (13.28) can be used in the entire region $\pi/2 \le \vartheta \le \pi$. In the case of expression (13.29) for the hard cylinder, it is valid (strictly speaking) for directions $\pi/2 + 0 \le \vartheta \le \pi - 0$, when point 4 is invisible. However, the contribution of point 4 for the hard cylinder is ka (or kl) times less than the PO field in the direction $\vartheta = \pi$ (or $\vartheta = \pi/2$) and it can be neglected in the case of large cylinders.

A quantitative influence of the field $u_{s,h}^{(1)}$ on the backscattering is illustrated graphically in the next section.

13.1.3 Total Backscattered Field

The total field is the sum

$$u_{s,h}^{(t)} = u_{s,h}^{(0)} + u_{s,h}^{(1)}, \tag{13.41}$$

where $u_{s}^{(0)} = -u_{h}^{(0)}$ and the terms $u_{h}^{(0)}$, $u_{s,h}^{(1)}$ are determined by Equations (13.12), (13.28), and (13.29). Utilizing these approximations, we have calculated the normalized scattering cross-section (13.14) and demonstrated the individual contribution of each term in Equation (13.41). Recall that the field $u_{s,h}^{(0)}$ is produced by the uniform scattering source $j_{s,h}^{(0)}$ and represents the PO approximation. The field $u_{s,h}^{(1)}$ is produced by the nonuniform source $j_{s,h}^{(1)}$ concentrated near the edges and is denoted below as the fringe component of the backscattering. The numerical results are presented in the following for two sets of geometrical parameters of the cylinder: (a) $d = 2a = \lambda$, $L = 2l = 3\lambda$; and (b) $d = 3\lambda$, $L = 9\lambda$. Here, d is the diameter and L is the length of the cylinder.

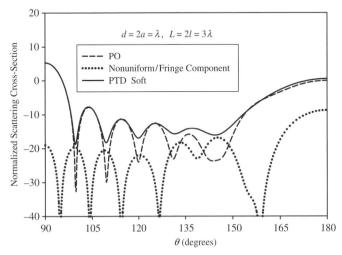

Figure 13.4 Backscattering at a soft cylinder. According to Equation (13.46), the PO curve here also displays the backscattering of electromagnetic waves (with E_x-polarization) from a perfectly conducting cylinder.

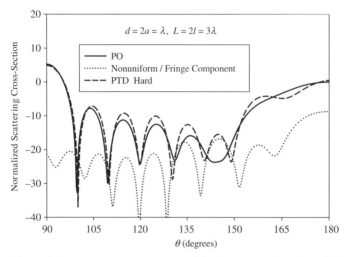

Figure 13.5   Backscattering at a hard cylinder. According to Equation (13.70), the PO curve here also displays the backscattering of electromagnetic waves (with H_x-polarization) from a perfectly conducting cylinder.

An interesting observation follows from Figures 13.4 to 13.7. Most of the maximums in the *soft fringe* field are located in the vicinity of the angular positions of the minimums of the PO field. The opposite situation is observed for the *hard fringe* field: its maximums are positioned near the maximums of the PO field. This observation explains why the minimums of the field scattered by soft cylinders are not as deep as those for the case of hard cylinders.

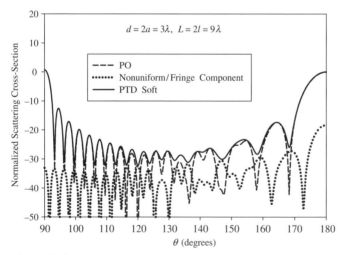

Figure 13.6   Backscattering at a soft cylinder. According to Equation (13.70), the PO curve here also displays the backscattering of electromagnetic waves (with E_x-polarization) from a perfectly conducting cylinder.

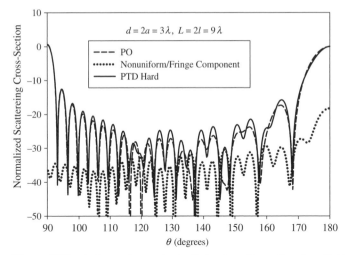

Figure 13.7   Backscattering at a hard cylinder. According to Equation (13.46), the PO curve here also displays the backscattering of electromagnetic waves (with H_x-polarization) from a perfectly conducting cylinder.

13.2 ELECTROMAGNETIC WAVES

The original PTD of electromagnetic waves scattered from a finite perfectly conducting cylinder was published in the work of Ufimtsev (1958a, 1962). Below, we present in brief a revised version based on the concept of EEWs.

13.2.1 E-Polarization

The incident wave is defined as

$$E_x^{\text{inc}} = E_{0x} e^{ik(z\cos\gamma + y\sin\gamma)}, \qquad E_y^{\text{inc}} = E_z^{\text{inc}} = H_x^{\text{inc}} = 0. \qquad (13.42)$$

The uniform component (1.97) of the induced surface current is determined by

$$j_x^{(0)\text{disk}} = 2Y_0 E_{0x} \cos\gamma\, e^{-ikl\cos\gamma} e^{ik\rho\sin\gamma\sin\psi}$$

and

$$j_y^{(0)\text{disk}} = j_z^{(0)\text{disk}} = 0 \qquad (13.43)$$

on the left base of the cylinder (Fig. 13.1), and by

$$j_x^{(0)\text{cyl}} = -2Y_0 E_{0x} \sin\gamma\sin\psi\, e^{ik(z\cos\gamma + a\sin\gamma\sin\psi)},$$

$$j_y^{(0)\text{cyl}} = 2Y_0 E_{0x} \sin\gamma\cos\psi\, e^{ik(z\cos\gamma + a\sin\gamma\sin\psi)},$$

and

$$j_z^{(0)\text{cyl}} = 2Y_0 E_{0x} \cos \gamma \cos \psi e^{ik(z \cos \gamma + a \sin \gamma \sin \psi)} \tag{13.44}$$

on the cylindrical part of the surface ($-l \le z \le l$, $\pi \le \psi \le 2\pi$). Here, $Y_0 = 1/Z_0$ is the admittance of free space (vacuum).

The field $E_x^{(0)}$ generated by the current $\vec{j}^{(0)}$ is found with the help of Equations (1.92) and (1.93), where one should drop off the terms $A_{\varphi,\vartheta}^m$, because $\vec{j}^m = -[\hat{n} \times \vec{E}] = 0$ due to the boundary condition on a perfectly conducting surface. The nonuniform (fringe) currents $\vec{j}^{(1)}$ concentrate near the left ($z = -\ell$) and right ($z = l$) edges. The field $E_x^{(1)}$ radiated by these currents is calculated in accordance with the theory developed in Section 7.8. The total scattered field is the sum

$$E_x = E_x^{(0)\text{disk}} + E_x^{(0)\text{cyl}} + E_x^{(1)\text{left}} + E_x^{(1)\text{right}}. \tag{13.45}$$

One can show that

$$E_x^{(0)\text{disk}} = u_s^{(0)\text{disk}}, \qquad E_x^{(0)\text{cyl}} = u_s^{(0)\text{cyl}}. \tag{13.46}$$

The quantities $u_s^{(0)\text{disk}}$ and $u_s^{(0)\text{cyl}}$ are defined in Section 13.1.1, where one should set $u_0 = E_{0x}$. Therefore, the PO curves in Figures 13.4 and 13.6 for the backscattering of acoustic waves from a soft cylinder also display the backscattering of electromagnetic waves from a perfectly conducting cylinder.

The fields $E_x^{(1)\text{left}}$ and $E_x^{(1)\text{right}}$ are calculated by the integration of EEWs, which are the functions of the local spherical coordinates with the origin at an edge point $x = a \cos \psi$, $y = a \sin \psi$. To avoid the possible confusion with the basic coordinates R, ϑ, φ of the observation point, we re-denote the local coordinates as r, θ, ϕ. The necessary preliminary work is to define the local coordinates in terms of the basic coordinates ϑ, ψ.

First, notice that one can use the following approximations

$$r^{\text{left}} = R + a \sin \vartheta \sin \psi + l \cos \vartheta, \qquad r^{\text{right}} = R + a \sin \vartheta \sin \psi - l \cos \vartheta, \tag{13.47}$$

$$\hat{r}^{\text{left}} \approx \hat{r}^{\text{right}} \approx \hat{R} = -\hat{y} \sin \vartheta + \hat{z} \cos \vartheta \tag{13.48}$$

for the observation point ($x = 0$, $y = -R \sin \vartheta$, $z = R \cos \vartheta$) in the far zone ($R \gg ka^2$, $R \gg kl^2$). Then, we introduce the unit vectors

$$\hat{\theta} = \hat{x}\, \theta_x + \hat{y}\, \theta_y + \hat{z}\, \theta_z, \qquad \hat{\phi} = \hat{x}\, \phi_x + \hat{y}\, \phi_y + \hat{z}\, \phi_z, \tag{13.49}$$

and find their components from the equations

$$\hat{R} \cdot \hat{\theta} = 0, \qquad \hat{\theta} \cdot [\hat{R} \times \hat{i}] = 0, \qquad \hat{i} \cdot \hat{\theta} = -\sqrt{1 - \sin^2 \vartheta \cos^2 \psi}, \qquad \hat{\phi} = \hat{R} \times \hat{\theta}, \tag{13.50}$$

where $\hat{t} = \hat{x}\sin\psi - \hat{y}\cos\psi$ is the tangent to the edge. According to these equations,

$$\theta_x = -\frac{\sin\psi}{\sqrt{1-\sin^2\vartheta\cos^2\psi}}, \qquad \theta_y = \frac{\cos\psi\cos^2\vartheta}{\sqrt{1-\sin^2\vartheta\cos^2\psi}},$$

$$\theta_z = \frac{\cos\psi\sin\vartheta\cos\vartheta}{\sqrt{1-\sin^2\vartheta\cos^2\psi}}, \tag{13.51}$$

$$\phi_x = -\frac{\cos\psi\cos\vartheta}{\sqrt{1-\sin^2\vartheta\cos^2\psi}}, \qquad \phi_y = -\frac{\sin\psi\cos\vartheta}{\sqrt{1-\sin^2\vartheta\cos^2\psi}},$$

$$\phi_z = -\frac{\sin\psi\sin\vartheta}{\sqrt{1-\sin^2\vartheta\cos^2\psi}}. \tag{13.52}$$

The angle θ is defined by the equation

$$\hat{R}\cdot\hat{t} = \cos\theta = \sin\vartheta\cos\psi. \tag{13.53}$$

In order to define the angles ϕ and ϕ_0, one should note that they are measured from the illuminated face of the edge in the plane perpendicular to the tangent $\hat{t}$ to the edge. By projecting the vectors $\hat{R} = -\hat{y}\sin\vartheta + \hat{z}\cos\vartheta$ and $\hat{Q} = -\hat{k}^i = -\hat{y}\sin\gamma - \hat{z}\cos\gamma$ on this plane, one obtains

$$\sin\phi = -\frac{\cos\vartheta}{\sqrt{1-\sin^2\vartheta\cos^2\psi}}, \qquad \cos\phi = \frac{\sin\vartheta\sin\psi}{\sqrt{1-\sin^2\vartheta\cos^2\psi}} \tag{13.54}$$

and

$$\sin\phi_0 = \frac{\cos\gamma}{\sqrt{1-\sin^2\vartheta\cos^2\psi}}, \qquad \cos\phi_0 = \frac{\sin\gamma\sin\psi}{\sqrt{1-\sin^2\vartheta\cos^2\psi}} \tag{13.55}$$

for the left edge ($z = -l$), and

$$\sin\phi = -\frac{\sin\vartheta\sin\psi}{\sqrt{1-\sin^2\vartheta\cos^2\psi}}, \qquad \cos\phi = -\frac{\cos\vartheta}{\sqrt{1-\sin^2\vartheta\cos^2\psi}} \tag{13.56}$$

and

$$\sin\phi_0 = -\frac{\sin\gamma\sin\psi}{\sqrt{1-\sin^2\vartheta\cos^2\psi}}, \qquad \cos\phi_0 = \frac{\cos\gamma}{\sqrt{1-\sin^2\vartheta\cos^2\psi}} \tag{13.57}$$

for the right edge ($z = l$).

Now, according to Section 7.8, one obtains

$$E_x^{(1)\text{left}} = E_{0x}\frac{a}{2\pi}\frac{e^{ikR}}{R}e^{i2kl\cos\vartheta}\int_0^{2\pi}\Big\{\sin\psi\,F_\theta^{(1)}(\psi,\theta,\phi)\cdot\theta_x$$

$$+ \cos\vartheta\cos\psi[G_\theta^{(1)}(\psi,\theta,\phi)\cdot\theta_x + G_\phi^{(1)}(\psi,\theta,\phi)\cdot\phi_x]\Big\}e^{i2ka\sin\vartheta\sin\psi}\,d\psi, \tag{13.58}$$

$$E_{y,z}^{(1)\text{left}} = H_x^{(1)\text{left}} = 0, \tag{13.59}$$

and

$$E_x^{(1)\text{right}} = E_{0x} \frac{a}{2\pi} \frac{e^{ikR}}{R} e^{-i2kl \cos \vartheta} \int_\pi^{2\pi} \left\{ \sin \psi \, F_\theta^{(1)}(\psi, \theta, \phi) \cdot \theta_x \right.$$

$$\left. + \cos \vartheta \cos \psi \, [G_\theta^{(1)}(\psi, \theta, \phi) \cdot \theta_x + G_\phi^{(1)}(\psi, \theta, \phi) \cdot \phi_x] \right\} e^{i2ka \sin \vartheta \sin \psi} \, d\psi,$$

$$(13.60)$$

$$E_{y,z}^{(1)\text{right}} = H_x^{(1)\text{right}} = 0. \tag{13.61}$$

Equations (13.59) and (13.61) show that no cross-polarization takes place in this problem, due to its symmetry with respect to the plane $x = 0$. In the direction $\vartheta = \pi - 0$, which is the focal line for EEWs, these equations predict the field

$$E_x^{(1)\text{left}} = E_{0x} \frac{a}{2} \left[f^{(1)}\left(\frac{\pi}{2}, \frac{\pi}{2}, \frac{3\pi}{2}\right) - g^{(1)}\left(\frac{\pi}{2}, \frac{\pi}{2}, \frac{3\pi}{2}\right) \right] \frac{e^{ikR}}{R} e^{-i2kl}, \tag{13.62}$$

$$E_x^{(1)\text{right}} = E_{0x} \frac{a}{4} \left[f^{(1)}\left(0, 0, \frac{3\pi}{2}\right) - g^{(1)}\left(0, 0, \frac{3\pi}{2}\right) \right] \frac{e^{ikR}}{R} e^{i2kl}. \tag{13.63}$$

The above equations allow the complete calculation of the total scattered field (13.45). They involve elementary functions, Bessel functions, and the one-dimensional integrals (13.58) and (13.60), which can be calculated numerically. However, one can avoid this direct integration by introducing approximations similar to (13.22), (13.23) and (13.26), (13.27).

The idea of these approximations is as follows. Away from the focal line ($ka \sin \vartheta \gg 1$), the asymptotic evaluation of the integrals (13.58) and (13.60) leads to the following ray asymptotics:

$$E_x^{(1)\text{left}} = E_{0x} \frac{a}{2} \frac{1}{\sqrt{\pi ka \sin \vartheta}} [f^{(1)}(1) e^{i2ka \sin \vartheta - i\pi/4}$$

$$+ f^{(1)}(2) e^{-i2ka \sin \theta + i\pi/4}] \frac{e^{ikR}}{R} e^{i2kl \cos \vartheta} \tag{13.64}$$

and

$$E_x^{(1)\text{right}} = E_{0x} \frac{a}{2} f^{(1)}(3) \frac{e^{-i2ka \sin \vartheta + i\pi/4}}{\sqrt{\pi ka \sin \vartheta}} \frac{e^{ikR}}{R} e^{-i2kl \cos \vartheta}. \tag{13.65}$$

These are the electromagnetic versions of the acoustic asymptotics (13.18) and (13.25). They reveal again the equivalence relationships existing between the acoustic and electromagnetic diffracted rays,

$$E_x^{(1)} = u_s^{(1)}, \qquad \text{if } u_0 = E_{0x}. \tag{13.66}$$

As shown above, the focal asymptotics (13.17) and (13.62) for acoustic and electromagnetic waves are different. However, they are small quantities of the order

$(ka)^{-1}$, compared to the basic component $E_{0x}^{(0)\text{disk}}$ scattered from the left base (disk). Therefore, the approximation (13.28) derived for acoustic waves can also be used for calculation of electromagnetic waves, but with the relative error of the order $(ka)^{-1}$. In this case, it is just sufficient to replace in Equation (13.28) the quantity u_0 by E_{0x} and $u_s^{(1)}$ by $E_x^{(1)}$.

13.2.2 H-Polarization

The incident wave

$$H_x^{\text{inc}} = H_{0x}e^{ik(z\cos\gamma + y\sin\gamma)}, \qquad H_{y,z}^{\text{inc}} = E_x^{\text{inc}} = 0 \qquad (13.67)$$

generates the uniform currents

$$j_y^{(0)\text{disk}} = -2H_{0x}e^{-ikl\cos\gamma}e^{ik\rho\sin\gamma\sin\psi}, \qquad j_x^{(0)\text{disk}}j_z^{(0)\text{disk}} = 0 \qquad (13.68)$$

on the left base of the cylinder (Fig. 13.1), and

$$j_z^{(0)\text{cyl}} = -2H_{0x}\sin\psi\,e^{ik(z\cos\gamma + a\sin\gamma\sin\psi)}, \qquad j_x^{(0)\text{cyl}} = j_y^{(0)\text{cyl}} = 0 \qquad (13.69)$$

on its cylindrical part $(-l \le z \le l, \pi \le \psi \le 2\pi)$. One can show that these currents radiate the field

$$H_x^{(0)} = u_h^{(0)} \qquad (13.70)$$

under the condition $u_0 = H_{0x}$, where function $u_h^{(0)}$ is determined in Section 13.1.1 and represents the acoustic field scattered at a rigid cylinder. Therefore, the PO curves in Figures 13.5 and 13.7 for the backscattering of acoustic waves from a rigid cylinder also display the backscattering of electromagnetic waves from a perfectly conducting cylinder.

The nonuniform currents induced in the vicinity of the left $(z = -l)$ and right $(z = l)$ edges radiate the field $H_x^{(1)} = H_x^{(1)\text{left}} + H_x^{(1)\text{right}}$, which is determined according to Section 7.8 as

$$H_x^{(1)\text{left}} = H_{0x}\frac{a}{2}\frac{e^{ikR}}{R}e^{i2kl\cos\vartheta}$$

$$\times \int_0^{2\pi} \{\sin\psi[G_\theta^{(1)}(\psi,\theta,\phi)\cdot\phi_x - G_\phi^{(1)}(\psi,\theta,\phi)\cdot\theta_x]$$

$$- \cos\vartheta\cos\psi\,F_\theta^{(1)}(\psi,\theta,\phi)\cdot\phi_x\}e^{i2ka\sin\vartheta\sin\psi}\,d\psi, \qquad (13.71)$$

and

$$
H_x^{(1)\text{right}} = H_{0x} \frac{a}{2} \frac{e^{ikR}}{R} e^{-i2kl \cos \vartheta}
$$

$$
\times \int_\pi^{2\pi} \{ \sin \psi [G_\theta^{(1)}(\psi,\theta,\phi) \cdot \phi_x - G_\phi^{(1)}(\psi,\theta,\phi) \cdot \theta_x]
$$

$$
- \cos \vartheta \cos \psi F_\theta^{(1)}(\psi,\theta,\phi) \cdot \phi_x \} e^{i2ka \sin \vartheta \sin \psi} \, d\psi. \tag{13.72}
$$

Due to the symmetry of this problem, the cross-polarized component of the backscattered field equals zero, $H_{y,z}^{(1)} = 0$. The local spherical coordinates θ, ϕ of EEWs involved in these integrals are defined in Section 13.2.1.

For the observation point on the focal line ($\vartheta = \pi - 0$), these integrals are calculated in the closed form:

$$
H_x^{(1)\text{left}} = H_{0x} \frac{a}{2} \left[g^{(1)}\left(\frac{\pi}{2},\frac{\pi}{2},\frac{3\pi}{2}\right) - f^{(1)}\left(\frac{\pi}{2},\frac{\pi}{2},\frac{3\pi}{2}\right) \right] \frac{e^{ikR}}{R} e^{-i2kl} \tag{13.73}
$$

and

$$
H_x^{(1)\text{right}} = H_{0x} \frac{a}{4} \left[g^{(1)}\left(0,0,\frac{3\pi}{2}\right) - f^{(1)}\left(0,0,\frac{3\pi}{2}\right) \right] \frac{e^{ikR}}{R} e^{i2kl}. \tag{13.74}
$$

The ray asymptotics of the field $H_x^{(1)}$ are given by

$$
H_x^{(1)\text{left}} = H_{0x} \frac{a}{2} \frac{1}{\sqrt{\pi ka \sin \vartheta}} \left[g^{(1)}(1)e^{i2ka \sin \vartheta - i\pi/4} \right.
$$

$$
\left. + g^{(1)}(2)e^{-i2ka \sin \vartheta + i\pi/4} \right] \frac{e^{ikR}}{R} e^{i2kl \cos \vartheta} \tag{13.75}
$$

and

$$
H_x^{(1)\text{right}} = H_{0x} \frac{a}{2} \frac{1}{\sqrt{\pi ka \sin \vartheta}} g^{(1)}(3)e^{-i2ka \sin \vartheta + i\pi/4} \frac{e^{ikR}}{R} e^{-i2kl \cos \vartheta}. \tag{13.76}
$$

These asymptotics are identical to those found for the acoustic waves $u_h^{(1)}$ in Section 13.1.2.

For electromagnetic waves with this polarization, one can also suggest the approximation similar to Equation (13.29), which follows from Equation (13.29) with the replacements $u_0 \rightarrow H_{0x}$ and $u_h^{(1)} \rightarrow H_x^{(1)}$. The expression obtained in this way exactly transforms into the ray asymptotics (13.75) and (13.76), but it leads to the approximations for the total scattered field, which possess the relative error of the order $(ka)^{-1}$ in the vicinity of the focal line.

PROBLEMS

13.1 Use Equations (13.20) and (13.21) and verify the transition from the focal (13.17) and ray (13.18), (13.19) asymptotics to the uniform approximations (13.22) and (13.23).

13.2 Verify Equation (13.43), use Equations (1.92) and (1.93), and derive the PO approximation $\vec{E}^{(0)\text{disk}}$ for the field backscattered from the left base (disk) of a finite cylinder. Prove that $E_{y,z}^{(0)\text{disk}} = 0$. Compare the results with the acoustic field $u_s^{(0)\text{disk}}$ scattered from a soft cylinder (see Section 13.1.1). Establish the equivalence relationships between acoustic and electromagnetic diffracted waves.

13.3 Verify that the approximations (13.22) and (13.23) exactly transform into the focal asymptotics (13.17) and asymptotically reduce to the ray asymptotics (13.18), (13.19).

13.4 Verify the approximations (13.47) for the distance between the diffraction point on the edge and the observation point in the far zone.

13.5 Use Equation (13.50) and verify the expressions (13.51) and (13.52) for the unit vectors $\hat{\theta}, \hat{\phi}$.

13.6 Derive Equations (13.54) and (13.55) for local coordinates ϕ, ϕ_0 of EEWs diverging from the left edge $(z = -l)$.

13.7 Derive Equations (13.56) and (13.57) for local coordinates ϕ, ϕ_0 of EEWs diverging from the right edge $(z = l)$.

13.8 Use the general Equations (7.130) and (7.131) for EEWs and derive Equation (13.58) for the field scattered by the left edge.

13.9 Use the general Equations (7.130) and (7.131) for EEWs and derive Equation (13.60) for the field scattered by the right edge.

13.10 Use the definitions of the quantities $F_\theta^{(1)}$, $G_{\theta,\phi}^{(1)}$ and prove that

$$F_\theta^{(1)}(\pi - \psi) = F_\theta^{(1)}(\psi), \quad G_\theta^{(1)}(\pi - \psi) = -G_\theta^{(1)}(\psi), \quad G_\phi^{(1)}(\pi - \psi) = G_\phi^{(1)}(\psi).$$

Then, use these relations and prove Equation (13.59), which indicates the absence of cross-polarization in the field backscattered from a finite cylinder.

13.11 Apply the stationary-phase method to the integrals (13.58) and (13.60) and obtain the ray asymptotics (13.64) and (13.65) for the waves diffracted at the left and right edges. Compare them with similar expressions for acoustic waves (see Section 13.1.2). Establish the relationships between the acoustic and electromagnetic edge-diffracted rays.

13.12 Compute and graph the normalized backscattering cross-section for a soft cylinder with parameters $d = 2a = 2\lambda$, $L = 2l = 4\lambda$. Set $\gamma = 45°$. Follow Section 13.1.3.

13.13 Verify Equation (13.68), use Equations (1.92) and (1.93), and derive the field $H_x^{(0)\text{disk}}$ backscattered from the left base of a cylinder. Compare the result with the acoustic field (13.3). Formulate the relationships between acoustic and electromagnetic waves.

13.14 Verify Equation (13.69), use Equations (1.92) and (1.93), and derive the integral expressions for the field $\vec{H}^{(0)\text{cyl}}$ backscattered from a cylinder. Prove that $H_{y,z}^{(0)\text{cyl}} = 0$. Compare $H_x^{(0)\text{cyl}}$ with the acoustic field (13.4). Formulate the relationships between acoustic and electromagnetic waves.

13.15   Use the general Equations (7.130) and (7.131) for EEWs and derive Equation (13.71) for the field scattered by the left edge.

13.16   Use the general Equations (7.130) and (7.131) for EEWs and derive Equation (13.72) for the field scattered by the right edge.

13.17   Use Equations (7.130) and (7.131) and derive the integral expressions for the field components $H_{y,z}^{(1)\text{left}}$ similar to Equation (13.71). Prove that $H_{y,z}^{(1)\text{left}} = 0$ in the backscattering direction.

13.18   Use the general Equations (13.71) and (13.72) and derive the focal asymptotics (13.73) and (13.74).

13.19   Apply the stationary-phase method to the integral (13.71) and derive the ray asymptotic (13.75). Compare it with Equation (13.19) and formulate the equivalence relationships between acoustic and electromagnetic edge-diffracted rays.

Chapter 14

Bistatic Scattering at a Finite-Length Cylinder

14.1 ACOUSTIC WAVES

The geometry of the problem is shown in Figure 14.1. The diameter of the cylinder is $d = 2a$, and its length is $L = 2l$. The incident wave is given by

$$u^{\text{inc}} = u_0 e^{ik(y \sin \gamma + z \cos \gamma)}, \qquad 0 < \gamma < \pi/2. \qquad (14.1)$$

The scattered field is evaluated in the plane $y0z$ ($\varphi = \pi/2$ and $\varphi = 3\pi/2$). It is convenient to indicate the scattering direction by the angle Θ ($0 \leq \Theta \leq 2\pi$),

$$\Theta = \begin{cases} \vartheta, & \text{if } \varphi = \pi/2 \\ 2\pi - \vartheta, & \text{if } \varphi = 3\pi/2, \end{cases} \qquad (14.2)$$

where ϑ ($0 \leq \vartheta \leq \pi$) is the ordinary spherical coordinate of the field point (R, ϑ, φ). One should not confuse this angle Θ with the local angle θ used for description of an EEW diverging from the diffraction point ζ at the edge. The relevant local coordinates r, θ, ϕ were introduced above in Section 13.2.1.

14.1.1 PO Approximation

The incident wave (14.1) generates the uniform component $j_{\text{s,h}}^{(0)}$ of the scattering sources only on the left base (disk) and on the lower lateral part ($\pi \leq \psi \leq 2\pi$) of the cylindrical surface. The scattered field is determined by Equation (1.32). We omit all intermediate calculations and obtain the final expressions for this field

$$u_{\text{s,h}}^{(0)} = u_0 \Phi_{\text{s,h}}^{(0)}(\Theta, \gamma) \frac{e^{ikR}}{R}, \qquad (14.3)$$

Fundamentals of the Physical Theory of Diffraction. By Pyotr Ya. Ufimtsev
Copyright © 2007 John Wiley & Sons, Inc.

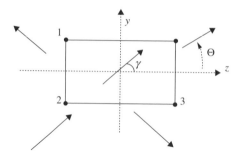

Figure 14.1   Cross-section of the cylinder by the $y0z$-plane. Dots 1, 2, and 3 are the stationary-phase points.

where

$$\Phi_s^{(0)} = ika^2 \cos\gamma \frac{J_1(p)}{p} e^{iq} - \frac{ikal}{\pi} \sin\gamma \frac{\sin q}{q} \int_\pi^{2\pi} e^{ip\sin\psi} \sin\psi \, d\psi, \qquad (14.4)$$

$$\Phi_h^{(0)} = ika^2 \cos\Theta \frac{J_1(p)}{p} e^{iq} - \frac{ikal}{\pi} \sin\Theta \frac{\sin q}{q} \int_\pi^{2\pi} e^{ip\sin\psi} \sin\psi \, d\psi, \qquad (14.5)$$

and

$$p = ka(\sin\gamma - \sin\Theta), \qquad q = kl(\cos\Theta - \cos\gamma). \qquad (14.6)$$

The first terms in functions $\Phi_{s,h}^{(0)}$, which contain the Bessel function $J_1(p)$, relate to the field scattered by the disk, and the second terms describe the scattering at the lateral (cylindrical) surface. These terms were evaluated numerically and their magnitudes are plotted in Figures 14.2 and 14.4.

It follows from the above field expressions that the total scattering cross-section (1.53) is given by

$$\sigma_{s,h}^{tot} = 2A, \qquad (14.7)$$

where

$$A = \pi a^2 \cos\gamma + 4al \sin\gamma \qquad (14.8)$$

is the area of the shadow beam cross-section.

According to Equations (14.4) and (14.5) the normalized scattering cross-section (13.14) is defined as

$$\sigma_{norm}^{(0)}(\Theta, \gamma) = \left(\frac{2}{ka^2} \left| \Phi_{s,h}^{(0)}(\Theta, \gamma) \right| \right)^2. \qquad (14.9)$$

This quantity was calculated numerically and the results are presented in Figures 14.2 to 14.5 for the incident wave direction $\gamma = 45°$. Figures 14.2 and 14.4 show the scattering at the individual parts of the cylinder and demonstrate how the total scattered field is formed.

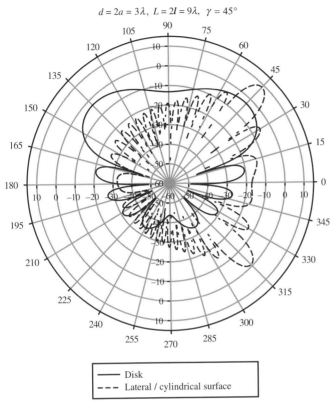

Figure 14.2 Scattering at the individual parts of a soft cylinder. According to Equation (14.62), this figure also demonstrates the PO approximation for electromagnetic waves (with E_x-polarization) scattered from the parts of a perfectly conducting cylinder.

14.1.2 Shadow Radiation as a Part of the Physical Optics Field

In Section 1.3.4 it was shown that the shadow radiation is the constituent part of the PO field. It was noted there that this field concentrates in the vicinity of the shadow region. Now we can verify this property by numerical investigation of the shadow radiation generated by the finite-length cylinder. The most appropriate procedure for doing this work would be the direct application of the shadow contour theorem demonstrated in Section 1.3.5. However, we can facilitate our work by utilizing the relationship given in Equation (1.73),

$$u_{sh} = \tfrac{1}{2}[u_s^{(0)} + u_h^{(0)}], \tag{14.10}$$

and the numerical results obtained in the previous section for the PO fields $u_s^{(0)}$ and $u_h^{(0)}$. Figure 14.6 shows the spatial distribution of the shadow radiation found in this way. The normalized scattering cross-section (14.9) is plotted here in a decibel

$d = 2a = 3\lambda,\ L = 2l = 9\lambda,\ \gamma = 45°$

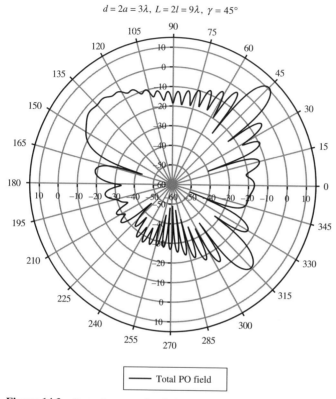

— Total PO field

Figure 14.3   Scattering at a soft cylinder. According to Equation (14.62), this figure also demonstrates the PO approximation for electromagnetic waves (with E_x-polarization) scattered from a perfectly conducting cylinder.

scale. It is clearly seen that the shadow radiation really represents the nature of the scattered field in the forward sector ($0° \leq \Theta \leq 90°$). According to Equations (14.62) and (14.75), this figure also demonstrates the shadow radiation and PO field scattered from a perfectly conducting cylinder.

14.1.3 PTD for Bistatic Scattering at a Hard Cylinder

Here we consider the scattering only at a hard cylinder. This problem is more important from a practical point of view. According to PTD the scattered field is generated by the uniform ($j_h^{(0)}$) and nonuniform ($j_h^{(1)}$) scattering sources induced by the incident wave on the cylinder. The field created by $j_h^{(0)}$ represents the PO field investigated in the previous sections. Now we calculate the field generated by that part of $j_h^{(1)}$ that concentrates near the circular edges of the cylinder and which is also called the fringe source. Then we will combine both components of the scattered field and provide the results of numerical calculation.

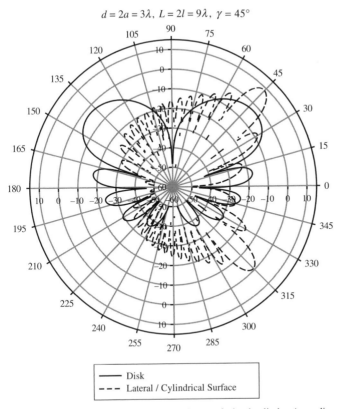

$d = 2a = 3\lambda,\ L = 2l = 9\lambda,\ \gamma = 45°$

Disk
--- Lateral / Cylindrical Surface

Figure 14.4   Scattering at the individual parts of a hard cylinder. According to Equation (14.75), this figure also demonstrates the PO approximation for electromagnetic waves (with H_x-polarization) scattered from the parts of a perfectly conducting cylinder.

The field generated by $j_{\mathrm{h}}^{(1)}$ is found by integrating the elementary edge waves introduced in Chapter 7. This field can be written in the following form:

$$u_{\mathrm{h}}^{(1)} = u_{\mathrm{h}}^{(1)\mathrm{left}} + u_{\mathrm{h}}^{(1)\mathrm{right}} \qquad (14.11)$$

where

$$u_{\mathrm{h}}^{(1)\mathrm{left}} = u_0\frac{a}{2\pi}\frac{e^{ikR}}{R}e^{ikl(\cos\Theta-\cos\gamma)}\int_0^{2\pi} F_{\mathrm{h}}^{(1)\mathrm{left}}(\psi,\theta,\phi)e^{ip\sin\psi}\,\mathrm{d}\psi \qquad (14.12)$$

and

$$u_{\mathrm{h}}^{(1)\mathrm{right}} = u_0\frac{a}{2\pi}\frac{e^{ikR}}{R}e^{-ikl(\cos\Theta-\cos\gamma)}\int_\pi^{2\pi} F_{\mathrm{h}}^{(1)\mathrm{right}}(\psi,\theta,\phi)e^{ip\sin\psi}\,\mathrm{d}\psi. \qquad (14.13)$$

Here, $p = ka(\sin\gamma - \sin\Theta)$ and the angle Θ $(0 \le \Theta \le 2\pi)$ is defined by Equation (14.2). The summands $u_{\mathrm{h}}^{(1)\mathrm{left}}$ and $u_{\mathrm{h}}^{(1)\mathrm{right}}$ represent the components related

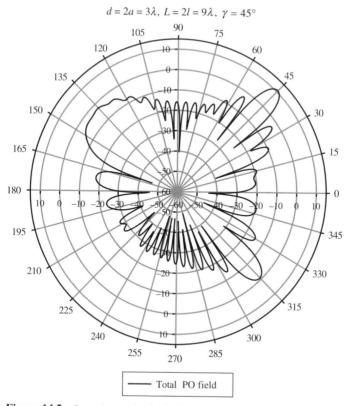

$$d = 2a = 3\lambda, \ L = 2l = 9\lambda, \ \gamma = 45°$$

—— Total PO field

Figure 14.5 Scattering at a hard cylinder. According to Equation (14.75), this figure also demonstrates the PO approximation for electromagnetic waves (with H_x-polarization) scattered from a perfectly conducting cylinder.

to the left and right edges, respectively. Only half of the right edge is illuminated by the incident wave (14.1), which is why the integration limits in Equation (14.13) are π and 2π.

Functions $F_{\mathrm{h}}^{(1)\mathrm{left,right}}(\psi, \theta, \phi)$ are described by Equations (7.91) and (7.92). Section 13.2.1 shows how one defines the local angles γ_0, θ, ϕ, and ϕ_0 for the backscattering direction $\vartheta = \pi - \gamma$. In the same way one can introduce these angles for arbitrary scattering directions in the $y0z$ plane. The relationships

$$\cos \gamma_0 = \sin \gamma \cos \psi, \qquad \cos \theta = -\sin \Theta \cos \psi \qquad (14.14)$$

are valid for both edges. For the left edge one should use the expressions

$$\sin \phi_0 = \frac{\cos \gamma}{\sqrt{1 - \sin^2 \gamma \cos^2 \psi}}, \qquad \cos \phi_0 = \frac{\sin \gamma \sin \psi}{\sqrt{1 - \sin^2 \gamma \cos^2 \psi}}, \qquad (14.15)$$

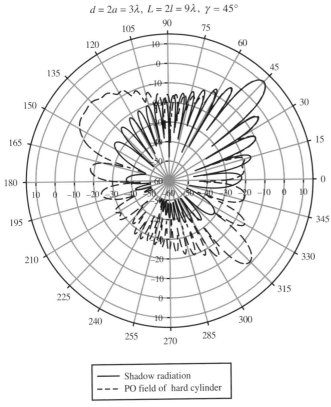

$d = 2a = 3\lambda$, $L = 2l = 9\lambda$, $\gamma = 45°$

—— Shadow radiation
--- PO field of hard cylinder

Figure 14.6   Shadow radiation as a part of the PO field.

$$\sin \phi = -\frac{\cos \Theta}{\sqrt{1 - \sin^2 \Theta \cos^2 \psi}}, \qquad \cos \phi = -\frac{\sin \Theta \sin \psi}{\sqrt{1 - \sin^2 \Theta \cos^2 \psi}}; \qquad (14.16)$$

however, for the right edge one should use the definitions

$$\sin \phi_0 = -\frac{\sin \gamma \sin \psi}{\sqrt{1 - \sin^2 \gamma \cos^2 \psi}}, \qquad \cos \phi_0 = \frac{\cos \gamma}{\sqrt{1 - \sin^2 \gamma \cos^2 \psi}} \qquad (14.17)$$

and

$$\sin \phi = \frac{\sin \Theta \sin \psi}{\sqrt{1 - \sin^2 \Theta \cos^2 \psi}}, \qquad \cos \phi = -\frac{\cos \Theta}{\sqrt{1 - \sin^2 \Theta \cos^2 \psi}}. \qquad (14.18)$$

The numerical results found with these expressions for the normalized scattering cross-section (13.14) are presented below for the two cylinders with parameters $L = 3d = 3\lambda$ and $L = 3d = 9\lambda$. The direction of the incident wave (14.1) is given by the angle $\gamma = 45°$. Figures 14.7 and 14.9 demonstrate the individual contributions

$$d = 2a = \lambda, \quad L = 2l = 3\lambda$$

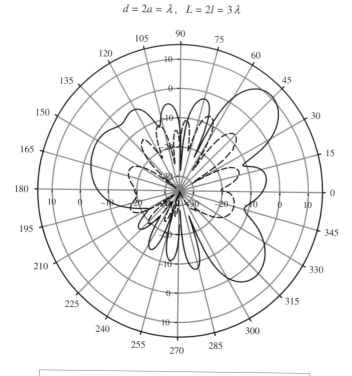

- PO field
- - - - Field generated by nonuniform / fringe sources

Figure 14.7 Bistatic scattering at a hard cylinder. According to Equation (14.75), the PO curve here also demonstrates the scattering of electromagnetic waves (with H_x-polarization) from a perfectly conducting cylinder.

by the PO field and by the field generated by $j_h^{(1)}$. The sum of these fields and its comparison with the PO field are shown in Figures 14.8 and 14.10.

These figures clearly show the influence of the field generated by the nonuniform/fringe scattering sources $j_h^{(1)}$. In particular, this field fills in the deep minima in the PO field. More accurate PTD approximation can be obtained with calculation of the high-order edge waves. However, in contrast to thin dipoles, thick cylinders are not resonant bodies and all high-order edge waves can be neglected when the size of the cylinder exceeds 3–5 wavelengths. The larger the cylinders, the higher the accuracy of the PTD expressions (14.12) and (14.13).

The following comments explain some details of the numeric calculations:

- The functions $F_h^{(1)\text{left,right}}$ are determined using Equations (7.82), (7.87), (7.88), (7.92), and (7.94) through the functions $V_t(\sigma_1, \phi_0)$ and $V_t(\sigma_2, \alpha - \phi_0)$, which

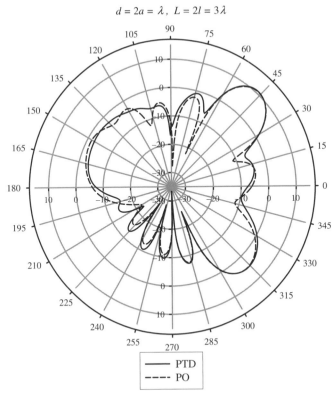

$d = 2a = \lambda, \; L = 2l = 3\lambda$

Figure 14.8 legend:
— PTD
---- PO

Figure 14.8 Scattering at a rigid cylinder. According to Equation (14.75), the PO curve here also demonstrates the scattering of electromagnetic waves (with H_x-polarization) from a perfectly conducting cylinder.

contain the factors $1/\sin \sigma_{1,2}$. These factors become singular when $\sigma_{1,2} \to 0$ or $\sigma_{1,2} \to \pi$. In the case $\sigma_1 \to 0$, the functions V_t remain finite. They can be transformed into the more convenient form

$$V_t(\sigma_1, \phi_0) = \frac{4}{9 \sin^2 \gamma} \frac{\cos \dfrac{\sigma_1}{3}}{1 - \dfrac{4}{3} \sin^2 \dfrac{\sigma_1}{3}} \frac{1}{\cos \dfrac{2\phi_0}{3} - \cos \dfrac{2\sigma_1}{3}}. \tag{14.19}$$

The replacements of σ_1 by σ_2 and ϕ_0 by $\alpha - \phi_0$ in Equation (14.19) lead to the transformed expression for $V_t(\sigma_2, \alpha - \phi_0)$.

- However, these expressions are still singular when $\sigma_{1,2} \to \pi$. In this case one should calculate the products $V_t(\sigma_1, \phi_0) \sin \phi$ and $V_t(\sigma_2, \alpha - \phi_0) \sin(\alpha - \phi)$. They remain finite when $\sigma_{1,2} \to \pi$ because the ratios $\sin \phi / \sin \sigma_1$ and $\sin(\alpha - \phi)/\sin \sigma_2$ are equal to plus or minus unity.

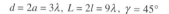

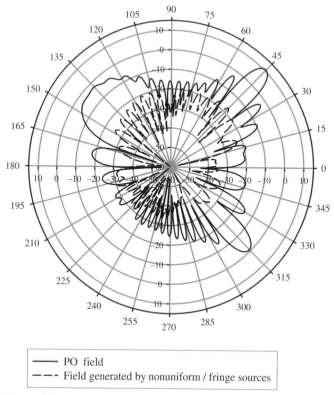

$$d = 2a = 3\lambda, \ L = 2l = 9\lambda, \ \gamma = 45°$$

──── PO field

──── Field generated by nonuniform / fringe sources

Figure 14.9   Scattering at a rigid cylinder. According to Equation (14.75), the PO curve here also demonstrates the scattering of electromagnetic waves (with H_x-polarization) from a perfectly conducting cylinder.

- The function $V(\sigma_2, \alpha - \phi_0)$ related to the left edge is singular at the points $\psi = 0, \pi, 2\pi$ for the observation direction $\Theta = \gamma$. This is the grazing singularity mentioned in Equation (4.21). It is removed by the exclusion of a certain vicinity of the singular points from the integral (14.12). This exclusion is done only for that part of the integral (14.12) that contains the function $V(\sigma_2, \alpha - \phi_0)$. The function $V(\sigma_1, \varphi_0)$ is not singular and is integrated in Equation (14.12) over the entire region $0 \leq \psi \leq 2\pi$. Notice that the theory of EEWs presented in Section 7.9 (which is free from the grazing singularity) cannot treat the above singularity in the direction $\Theta = \gamma$, because this theory is applicable only for objects with planar faces.

- Finally we note that in the case $\sigma_1 \to \phi_0$ or $\sigma_2 \to \alpha - \phi_0$ one should use Equation (7.107) for the functions $V(\phi_0, \phi_0)$ and Equation (7.109) for the function $V(\alpha - \phi_0, \alpha - \phi_0)$.

$d = 2a = 3\lambda$, $L = 2l = 9\lambda$, $\gamma = 45°$

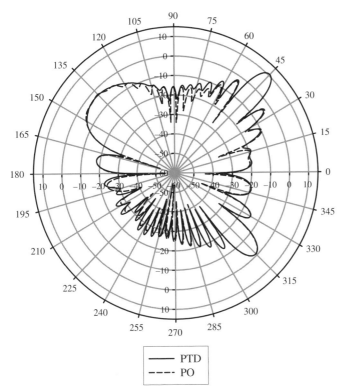

Figure 14.10 Scattering at a rigid cylinder. According to Equation (14.75), the PO curve here also demonstrates the scattering of electromagnetic waves (with H_x-polarization) from a perfectly conducting cylinder.

14.1.4 Beams and Rays of the Scattered Field

The previous section provides a numerical investigation of the scattered field. Here we consider its physical structure and present simple high-frequency asymptotics for the directivity pattern $\Phi(\Theta, \gamma)$ defined by the equation

$$u^{sc} = u_0 \Phi(\Theta, \gamma) \frac{e^{ikR}}{R}. \tag{14.20}$$

From the physical point of view, the scattered field consists of the following basic components:

- The reflected beam in the vicinity of the direction $\Theta = \pi - \gamma$. This beam appears due to the transverse diffusion of the field in ordinary rays reflected from the left base of the cylinder (Fig. 14.1). It is described by the first terms in Equations (14.4) and (14.5), which contain the Bessel function $J_1(p)$. Exactly in this direction its value equals

$$\Phi_s^{\text{beam 1}} = -\Phi_h^{\text{beam 1}} = \frac{ika^2}{2}\cos\gamma e^{-i2kl\cos\gamma}. \qquad (14.21)$$

- The reflected beam in the vicinity of the direction $\Theta = 2\pi - \gamma$. It appears due to the transverse diffusion of the field in ordinary rays reflected from the lower lateral part of the cylinder. This beam is described by the high-frequency asymptotics for the second terms in Equations (14.4) and (14.5):

$$\Phi_s^{\text{beam 2}} = ikal\sin\gamma\frac{\sin q}{q}\sqrt{\frac{2}{\pi p}}e^{-ip}e^{i\pi/4}, \qquad (14.22)$$

$$\Phi_h^{\text{beam 2}} = ikal\sin\Theta\frac{\sin q}{q}\sqrt{\frac{2}{\pi p}}e^{-ip}e^{i\pi/4}. \qquad (14.23)$$

Exactly in the direction $\Theta = 2\pi - \gamma$ its value equals

$$\Phi_s^{\text{beam 2}} = -\Phi_h^{\text{beam 2}} = l\sqrt{\frac{ka\sin\gamma}{\pi}}e^{-i2ka\sin\gamma}e^{i3\pi/4}. \qquad (14.24)$$

We note that $p = ka(\sin\gamma - \sin\Theta)$ and $q = kl(\cos\Theta - \cos\gamma)$.

- The beam of the shadow radiation in the vicinity of the direction $\Theta = \gamma$. It is described by both terms in Equations (14.4) and (14.5). Exactly in this direction its value equals

$$\Phi_s^{\text{shad.beam}} = \Phi_h^{\text{shad.beam}} = \frac{ika^2}{2}\cos\gamma + \frac{i2kal}{\pi}\sin\gamma. \qquad (14.25)$$

- The beam of edge-diffracted rays generated by the fringe scattering sources, which are located near the left edge. It propagates in the direction $\Theta = \pi - \gamma$ and supplements the reflected beam (14.21). This beam is described by Equation (14.12). In the case of the soft cylinder, the corresponding expression follows from Equation (14.12) with the obvious replacement of $F_h^{(1)}$ by $F_s^{(1)}$. Exactly in the direction $\Theta = \pi - \gamma$, it is determined by

$$\Phi_s^{\text{fr.beam 1}} = \frac{a}{2\pi}e^{-i2kl\cos\gamma}\int_0^{2\pi} f^{(1)\text{left}}(\psi, \pi - \gamma)d\psi, \qquad (14.26)$$

$$\Phi_h^{\text{fr.beam 1}} = \frac{a}{2\pi}e^{-i2kl\cos\gamma}\int_0^{2\pi} g^{(1)\text{left}}(\psi, \pi - \gamma)d\psi. \qquad (14.27)$$

- The beam of edge-diffracted rays generated by the fringe scattering sources, which are located near both edges. It propagates in the direction $\Theta = \gamma$ and supplements the shadow beam. This beam is described by Equation (14.11). Exactly in this direction it is described by

$$\Phi_s^{\text{fr.shad.beam}} = \frac{a}{2\pi} \left[\int_0^{2\pi} f^{(1)\text{left}}(\psi, \gamma) d\psi + \int_\pi^{2\pi} f^{(1)\text{right}}(\psi, \gamma) d\psi \right],$$
(14.28)

$$\Phi_h^{\text{fr.shad.beam}} = \frac{a}{2\pi} \left[\int_0^{2\pi} g^{(1)\text{left}}(\psi, \gamma) d\psi + \int_\pi^{2\pi} g^{(1)\text{right}}(\psi, \gamma) d\psi \right].$$
(14.29)

To avoid the grazing singularity, one should exclude a certain vicinity of the points $\psi = 0, \pi, 2\pi$ in the integral over the left edge.

Away from these beams, the scattered field contains the three edge-diffracted rays generated by the total surface current $j_{s,h}^{(0)} + j_{s,h}^{(1)}$. They can be determined by the asymptotic estimation of the field $u_{s,h}^{(0)} + u_{s,h}^{(1)}$ described by Equations (14.4), (14.5), and (14.11). However, a simpler way is to apply the modified asymptotics (8.12) and (8.13), where one should replace the functions $f^{(1)}$, $g^{(1)}$ with f, g. As a result, one obtains the following expressions for these rays.

- *Ray 1:*

$$\Phi_s^{\text{ray}}(1) = \frac{a}{\sqrt{2\pi |p|}} f(1) e^{i(p+q)} e^{\mp i\pi/4}$$
(14.30)

and

$$\Phi_h^{\text{ray}}(1) = \frac{a}{\sqrt{2\pi |p|}} g(1) e^{i(p+q)} e^{\mp i\pi/4}$$
(14.31)

propagates from the stationary point 1 (Fig. 14.1) and exists in the regions $0 \le \Theta < \gamma, \gamma < \Theta < \pi - \gamma$, and $\pi - \gamma < \Theta \le 3\pi/2$. Factor $\exp(-i\pi/4)$ is taken for positive values of p and factor $\exp(+i\pi/4)$ is valid for the negative values of p.

- *Ray 2:*

$$\Phi_s^{\text{ray}}(2) = \frac{a}{\sqrt{2\pi |p|}} f(2) e^{i(q-p)} e^{\pm i\pi/4}$$
(14.32)

and

$$\Phi_h^{\text{ray}}(2) = \frac{a}{\sqrt{2\pi |p|}} g(2) e^{i(q-p)} e^{\pm i\pi/4}$$
(14.33)

propagates from the stationary point 2 and exists in the region $\pi/2 \le \Theta < \pi - \gamma, \pi - \gamma < \Theta < 2\pi - \gamma$, and $2\pi - \gamma < \theta \le 2\pi$. Factor $\exp(+i\pi/4)$ is taken for positive values of p and factor $\exp(-i\pi/4)$ is valid for negative values of p.

• *Ray* 3:

$$\Phi_s^{\mathrm{ray}}(3) = \frac{a}{\sqrt{2\pi|p|}} f(3)e^{-i(q+p)}e^{\pm i\pi/4} \qquad (14.34)$$

and

$$\Phi_h^{\mathrm{ray}}(3) = \frac{a}{\sqrt{2\pi|p|}} g(3)e^{-i(q+p)}e^{\pm i\pi/4} \qquad (14.35)$$

propagates from the stationary point 3 (Fig. 14.1) and exists in the regions $0 \le \Theta < \gamma, \gamma < \Theta \le \pi/2, \pi \le \Theta \le 2\pi - \gamma$, and $2\pi - \gamma < \Theta \le 2\pi$. Factor $\exp(+i\pi/4)$ is taken for positive values of p and factor $\exp(-i\pi/4)$ is valid for negative values of p.

In the above expressions, functions f, g and $f^{(1)}$, $g^{(1)}$ are defined according to Chapters 2, 3, and 4:

$$f(m) = f(\phi_m, \phi_{0m}, \alpha), \qquad g(m) = g(\phi_m, \phi_{0m}, \alpha), \qquad (14.36)$$

$$f^{(1)}(m) = f^{(1)}(\phi_m, \phi_{0m}, \alpha), \qquad g^{(1)}(m) = g^{(1)}(\phi_m, \phi_{0m}, \alpha), \qquad (14.37)$$

with $m = 1, 2, 3$. Here, $\alpha = 3\pi/2$ and

$$\phi_1 = 3\pi/2 - \Theta, \qquad \phi_{01} = \pi/2 - \gamma, \qquad (14.38)$$

$$\phi_2 = \Theta - \pi/2, \qquad \phi_{02} = \pi/2 + \gamma, \qquad (14.39)$$

$$\phi_3 = \Theta - \pi, \qquad \text{if } \pi \le \Theta \le 2\pi, \qquad (14.40)$$

$$\phi_3 = \pi + \Theta, \qquad \text{if } 0 \le \Theta \le \pi/2, \qquad (14.41)$$

$$\phi_{03} = \gamma. \qquad (14.42)$$

The angles ϕ, ϕ_0 for the functions $f^{(1)\mathrm{left}}[\phi(\psi), \phi_0(\psi), \alpha]$, $g^{(1)\mathrm{left}}[\phi(\psi), \phi_0(\psi), \alpha]$ are determined by Equations (14.15) and (14.16), and for the functions $f^{(1)\mathrm{right}}[\phi(\psi), \phi_0(\psi), \alpha]$, $g^{(1)\mathrm{right}}[\phi(\psi), \phi_0(\psi), \alpha]$ they are found with Equations (14.17), (14.18).

All diffracted rays undergo a phase shift equal to $\pm\pi/2$ when they cross the focal lines $\Theta = \gamma$ and $\Theta = \pi - \gamma$, which are the axes of the field beams.

14.1.5 Refined Asymptotics for the Specular Beam

We refer again to Figure 14.1 and focus on the beam reflected from the lower lateral surface of the cylinder and propagating in the specular direction $\Theta = 2\pi - \gamma$. In the first-order approximation, this beam was evaluated in the previous section. Here we consider some fine features of the theory, which are beyond the first approximation. The final results of this section were published in the paper by Ufimtsev (1989).

According to PTD the scattered field is generated by the uniform $j^{(0)}$ and non-uniform $j^{(1)}$ components of the scattering sources induced by the incident wave on the object. Up to now we have calculated the field radiated only by the basic part of the component $j^{(1)}$ that is caused by sharp bending (edges). This is the so-called fringe component $j^{(1)\mathrm{fr}}$. The other component $j^{(1)\mathrm{sm}}$ is caused by the smooth bending of the scattering surface and it is asymptotically small compared to $j^{(0)}$ and $j^{(1)\mathrm{fr}}$. In particular on a circular cylinder, the ratio $j^{(1)\mathrm{sm}}/j^{(0)}$ is of the order $1/ka$. That is why the component $j^{(1)\mathrm{sm}}$ is usually neglected for thick cylinders. However, in our papers (Ufimtsev, 1979, 1981, 1989), it was shown that this small component distributed over the entire generatrix ($-l \leq z \leq l$, $\psi = 3\pi/2$, $kl \gg 1$) creates in the specular direction $\Theta = 2\pi - \gamma$ the co-phased radiation of the same order $[(ka)^{-1/2}]$ as the field generated by $j^{(1)\mathrm{fr}}$. For this reason one should include this additional radiation in the beam field. This is the first fine feature of the theory.

It was also shown (Ufimtsev, 1979, 1981, 1989) that the second term of the asymptotic expansion for the PO field (in the specular direction) is also a quantity of the order $(ka)^{-1/2}$ and it also should be included in the beam field. Usually, the high-order terms in the PO field are considered incorrect. However, in the framework of PTD, the PO field is the constituent part of the scattered field. Therefore one should incorporate into the field expression the high-order asymptotic terms of the PO field, which are of the same order of magnitude as those taken from the asymptotic expansion of the field generated by the nonuniform sources $j_{\mathrm{s,h}}^{(1)}$. This is the second fine and important feature of PTD.

Here, these observations are demonstrated in the analytic form for the directivity pattern $\Phi(\Theta, \gamma)$ introduced by Equation (14.20). The scattered field is evaluated in the vicinity of the specular direction $\Theta = 2\pi - \gamma$. The PO field generated by $j^{(0)}$ is described by Equations (14.4) and (14.5). The first term there relates to the field scattered by the left base (disk) of the cylinder. Its contribution to the field in the region $3\pi/2 < \Theta < 2\pi$ is created by the vicinity of point 2 (Fig. 14.1). By asymptotic evaluation of this contribution, the expressions (14.4) and (14.5) can be written as

$$\Phi_s^{(0)} = -\frac{ae^{i\pi/4}}{\sqrt{2\pi p}}\frac{\cos\gamma}{\sin\gamma - \sin\Theta}e^{i(q-p)} - \frac{ikal}{\pi}\sin\gamma\frac{\sin q}{q}\int_\pi^{2\pi}e^{ip\sin\psi}\sin\psi\,d\psi,$$

$$(14.43)$$

and

$$\Phi_h^{(0)} = -\frac{ae^{i\pi/4}}{\sqrt{2\pi p}}\frac{\cos\Theta}{\sin\gamma - \sin\Theta}e^{i(q-p)} - \frac{ikal}{\pi}\sin\Theta\frac{\sin q}{q}\int_\pi^{2\pi}e^{ip\sin\psi}\sin\psi\,d\psi,$$

$$(14.44)$$

where $p = ka(\sin\gamma - \sin\Theta)$ and $q = kl(\cos\Theta - \cos\gamma)$. Here, in accordance with the above discussion, we retain the two first terms in the asymptotic expansion for the integrals and obtain

$$\Phi_s^{(0)} = -\frac{a}{\sqrt{2\pi p}}\frac{\cos\gamma}{\sin\gamma - \sin\Theta}e^{i(q-p)+i\pi/4}$$

$$- \frac{ikal}{\pi}\sin\gamma\frac{\sin q}{q}\sqrt{\frac{2\pi}{p}}\left(-1 + i\frac{3}{8p}\right)e^{-ip+i\pi/4} \qquad (14.45)$$

and

$$\Phi_h^{(0)} = -\frac{a}{\sqrt{2\pi p}}\frac{\cos\Theta}{\sin\gamma - \sin\Theta}e^{i(q-p)+i\pi/4}$$

$$-\frac{ikal}{\pi}\sin\Theta\frac{\sin q}{q}\sqrt{\frac{2\pi}{p}}\left(-1+i\frac{3}{8p}\right)e^{-ip+i\pi/4}. \tag{14.46}$$

The field radiated by $j^{(1)\text{fr}}$ is determined in accordance with Section 14.1.4 as

$$\Phi_s^{(1)\text{fr}} = \frac{a}{\sqrt{2\pi p}}[f^{(1)}(2)e^{iq}+f^{(1)}(3)e^{-iq}]e^{-ip+i\pi/4} \tag{14.47}$$

and

$$\Phi_h^{(1)\text{fr}} = \frac{a}{\sqrt{2\pi p}}[g^{(1)}(2)e^{iq}+g^{(1)}(3)e^{-iq}]e^{-ip+i\pi/4}. \tag{14.48}$$

Functions $f^{(1)} = f - f^{(0)}$, $g^{(1)} = g - g^{(0)}$ are introduced in Chapters 2, 3, and 4. Their arguments are defined in Equations (14.38) to (14.42).

The nonuniform component $j^{(1)\text{sm}}$ caused by the smooth bending of the cylindrical surface is found by the extension of the results of the paper by Franz and Galle (1955) to the case of the oblique direction of the incident wave. The asymptotic approximations found in this way are

$$j_s^{(1)\text{sm}} = u_0\frac{1}{a\sin^2\psi}e^{ik(z\cos\gamma+a\sin\gamma\sin\psi)} \tag{14.49}$$

and

$$j_h^{(1)\text{sm}} = u_0\frac{i}{ka\sin\gamma\sin^3\psi}e^{ik(z\cos\gamma+a\sin\gamma\sin\psi)}. \tag{14.50}$$

According to Equations (1.16), (1.17), and (14.20), the field radiated by these sources is described by

$$\Phi_s^{(1)\text{sm}} = -\frac{l}{2\pi}\frac{\sin q}{q}\int_\pi^{2\pi}\frac{e^{ip\sin\psi}}{\sin^2\psi}d\psi \sim -\frac{l}{2\pi}\frac{\sin q}{q}\sqrt{\frac{2\pi}{p}}e^{-ip+i\pi/4} \tag{14.51}$$

and

$$\Phi_h^{(1)\text{sm}} = \frac{l}{2\pi}\frac{\sin\Theta}{\sin\gamma}\frac{\sin q}{q}\int_\pi^{2\pi}\frac{e^{ip\sin\psi}}{\sin^2\psi}d\psi \sim \frac{l}{2\pi}\frac{\sin\Theta}{\sin\gamma}\frac{\sin q}{q}\sqrt{\frac{2\pi}{p}}e^{-ip+i\pi/4}, \tag{14.52}$$

under the condition $p \gg 1$.

The total field equals

$$\Phi_{s,h}^{\text{tot}} = \Phi_{s,h}^{(0)} + \Phi_{s,h}^{(1)\text{fr}} + \Phi_{s,h}^{(1)\text{sm}}. \tag{14.53}$$

Away from the specular direction this field contains the terms of the order $(ka)^{-1/2}$ and $(kl\sqrt{ka})^{-1}$. Exactly in the specular direction $\Theta = 2\pi - \gamma$, its components are equal to

$$\Phi_s^{(0)} = \frac{a}{\sqrt{\pi ka \sin \gamma}} \left(ikl \sin \gamma + \frac{3}{16} \frac{l}{a} - \frac{1}{4} \cot \gamma \right) e^{-i2ka \sin \gamma + i\pi/4}, \tag{14.54}$$

$$\Phi_h^{(0)} = -\frac{a}{\sqrt{\pi ka \sin \gamma}} \left(ikl \sin \gamma + \frac{3}{16} \frac{l}{a} + \frac{1}{4} \cot \gamma \right) e^{-i2ka \sin \gamma + i\pi/4}, \tag{14.55}$$

$$\Phi_s^{(1)\text{fr}} = -\frac{a}{\sqrt{\pi ka \sin \gamma}} \left\{ \frac{1}{\sqrt{3}} \left[\frac{1}{2} + \cos \left(\frac{2(\pi - 2\gamma)}{3} \right) \right]^{-1} \right.$$
$$\left. + \frac{1}{3\sqrt{3}} - \frac{1}{4} \cot \gamma \right\} e^{-i2ka \sin \gamma + i\pi/4}, \tag{14.56}$$

$$\Phi_h^{(1)\text{fr}} = -\frac{a}{\sqrt{\pi ka \sin \gamma}} \left\{ \frac{1}{\sqrt{3}} \left[\frac{1}{2} + \cos \left(\frac{2(\pi - 2\gamma)}{3} \right) \right]^{-1} \right.$$
$$\left. - \frac{1}{3\sqrt{3}} - \frac{1}{4} \cot \gamma \right\} e^{-i2ka \sin \gamma + i\pi/4}, \tag{14.57}$$

$$\Phi_s^{(1)\text{sm}} = \Phi_h^{(1)\text{sm}} = -\frac{a}{\sqrt{\pi ka \sin \gamma}} \frac{l}{2a} e^{-i2ka \sin \gamma + i\pi/4}. \tag{14.58}$$

By the summation of these components one obtains the total field in the specular direction

$$\Phi_s = \frac{a}{\sqrt{\pi ka \sin \gamma}} \left\{ ikl \sin \gamma + \frac{3}{16} \frac{l}{a} - \frac{1}{\sqrt{3}} \left[\frac{1}{2} + \cos \left(\frac{2(\pi - 2\gamma)}{3} \right) \right]^{-1} \right.$$
$$\left. - \frac{1}{3\sqrt{3}} - \frac{l}{2a} \right\} e^{-i2ka \sin \gamma + i\pi/4} \tag{14.59}$$

and

$$\Phi_h = \frac{a}{\sqrt{\pi ka \sin \gamma}} \left\{ -ikl \sin \gamma - \frac{3}{16} \frac{l}{a} - \frac{1}{\sqrt{3}} \left[\frac{1}{2} + \cos \left(\frac{2(\pi - 2\gamma)}{3} \right) \right]^{-1} \right.$$
$$\left. + \frac{1}{3\sqrt{3}} - \frac{l}{2a} \right\} e^{-i2ka \sin \gamma + i\pi/4}. \tag{14.60}$$

The origination and meaning of each term here is clear from the preceding expressions for the field components. The first and second terms in the braces relate respectively to the first and second terms in the asymptotic expansion for the PO field. The third and fourth terms represent the contribution by the fringe sources $j_{s,h}^{(1)\text{fr}}$, and the last term shows the contribution by the sources $j_{s,h}^{(1)sm}$ caused by the smooth bending of the cylindrical surface.

Notice that the primary edge waves propagating along the cylinder base from point 1 (Fig. 14.1) to point 2 undergo edge diffraction and generate the secondary diffracted rays in the region $3\pi/2 < \Theta \leq 2\pi$. These secondary rays can be calculated by application of the theory developed in Chapter 10. Their contributions to the field in the specular direction are of the order $(ka)^{-3/2}$ for the soft cylinder and of the order $(ka)^{-1}$ for the hard cylinder. They are small compared to the field (14.59), (14.60) and can be neglected.

The counterpart of the present theory developed for electromagnetic waves scattered at a perfectly conducting cylinder of finite size has been published in the paper by Ufimtsev (1981) and is considered below in Section 14.2.3.

14.2 ELECTROMAGNETIC WAVES

14.2.1 *E-Polarization*

On the surface of a perfectly conducting cylinder (Fig. 14.1), the incident wave

$$E_x^{inc} = E_{0x}e^{ik(z\cos\gamma + y\sin\gamma)}, \qquad E_{y,z}^{inc} = H_x^{inc} = 0 \qquad (14.61)$$

generates the uniform currents given by Equations (13.43) and (13.44). The PO field radiated by these currents is calculated according to Equations (1.92) and (1.93). In the plane $y0z$ ($\varphi = \pi/2$ and $\varphi = 3\pi/2$), this field is described by the expressions totally identical to Equations (14.3) and (14.4) derived above for acoustic waves:

$$E_x^{(0)} = [E_{0x}/u_0] \cdot u_s^{(0)}. \qquad (14.62)$$

In particular, this relationship means that the PO curves in Figures 14.2 and 14.3 for the acoustic waves *scattered from a soft cylinder* also demonstrate the electromagnetic waves with *E*-polarization *scattered from a perfectly conducting cylinder* of the same size.

The field radiated by the nonuniform current $\vec{j}^{(1)}$ is calculated according to Section 7.8 as

$$E_x^{(1)} = E_x^{(1)left} + E_x^{(1)right}, \qquad (14.63)$$

where the function

$$E_x^{(1)left} = E_{0x}\frac{a}{2\pi}\frac{e^{ikR}}{R}e^{iq}\int_0^{2\pi}\{\sin\psi F_\theta^{(1)left}(\psi,\theta,\phi)\cdot\theta_x - \cos\gamma\cos\psi$$
$$\times [G_\theta^{(1)left}(\psi,\theta,\phi)\cdot\theta_x + G_\phi^{(1)left}(\psi,\theta,\phi)\cdot\phi_x]\}e^{ip\sin\psi}\,d\psi \qquad (14.64)$$

describes the field scattered from the left edge, and the function

$$E_x^{(1)\text{right}} = E_{0x} \frac{a}{2\pi} \frac{e^{ikR}}{R} e^{-iq} \int_\pi^{2\pi} \{\sin\psi \, F_\theta^{(1)\text{right}}(\psi,\theta,\phi) \cdot \theta_x - \cos\gamma\cos\psi$$

$$\times [G_\theta^{(1)\text{right}}(\psi,\theta,\phi) \cdot \theta_x + G_\phi^{(1)\text{right}}(\psi,\theta,\phi) \cdot \phi_x]\} e^{ip\sin\psi} \, d\psi \quad (14.65)$$

represents the field scattered from the right edge. Here, $p = ka(\sin\gamma - \sin\Theta)$, $q = kl(\cos\Theta - \cos\gamma)$. In addition, Equations (14.15) to (14.18) define the local angles θ, ϕ, ϕ_0 (different for the left and right edges), but Equations (13.51) and (13.52) determine the unit vectors $\hat{\theta}, \hat{\phi}$ (the same for the left and right edges).

For the forward scattering direction $\Theta = \gamma$ (belonging to the diffraction cone), these expressions reduce to

$$E_x^{(1)\text{left}} = E_{0x} \frac{a}{2\pi} \frac{e^{ikR}}{R} \int_0^{2\pi} \left\{ \frac{\sin^2\psi}{\sin^2\gamma_0} f^{(1)}(\pi+\phi_0,\phi_0,3\pi/2) + \frac{\cos^2\gamma\cos^2\psi}{\sin^2\gamma_0} \right.$$

$$\left. \times g^{(1)}(\pi+\phi_0,\phi_0,3\pi/2) + \frac{\sin\gamma\cos\gamma\cos^2\psi\sin\psi}{\sin^2\gamma_0} \right\} d\psi \quad (14.66)$$

and

$$E_x^{(1)\text{right}} = E_{0x} \frac{a}{2\pi} \frac{e^{ikR}}{R} \int_\pi^{2\pi} \left\{ \frac{\sin^2\psi}{\sin^2\gamma_0} f^{(1)}(\pi+\phi_0,\phi_0,3\pi/2) + \frac{\cos^2\gamma\cos^2\psi}{\sin^2\gamma_0} \right.$$

$$\left. \times g^{(1)}(\pi+\phi_0,\phi_0,3\pi/2) + \frac{\sin\gamma\cos\gamma\sin\psi\cos^2\psi}{\sin^2\gamma_0} \right\} d\psi. \quad (14.67)$$

We emphasize again that the local angles θ, ϕ, ϕ_0 in functions $E_x^{(1)\text{left}}$ and $E_x^{(1)\text{right}}$ are different.

Notice also that in the case of scattering at the disk, the expression similar to Equation (14.66) (but with the wedge angle equal to $\alpha = 2\pi$) exactly transforms into the function

$$E_x^{(1)} = E_{0x} \frac{a}{\pi\sin^2\gamma} \left[2\cos\gamma K\left(\frac{\pi}{2},\sin\gamma\right) - \frac{1+\cos^2\gamma}{\cos\gamma} E\left(\frac{\pi}{2},\sin\gamma\right) \right] \frac{e^{ikR}}{R}$$

$$\quad (14.68)$$

derived earlier by Ufimtsev (1962) and Butorin et al. (1988). Here,

$$K(\pi/2,x) = \int_0^{\pi/2} \frac{d\psi}{\sqrt{1-x^2\cos^2\psi}}, \qquad E(\pi/2,x) = \int_0^{\pi/2} \sqrt{1-x^2\cos^2\psi} \, d\psi$$

$$\quad (14.69)$$

are complete elliptic integrals (Gradshteyn and Ryzhik, 1994).

In order to simplify the comparative analysis of the electromagnetic and acoustic waves scattered at a finite cylinder, let us represent the total scattered field in the form

$$E_x = E_x^{(0)} + E_x^{(1)} = E_{0x}[\Phi_{ex}^{(0)} + \Phi_{ex}^{(1)}] \frac{e^{ikR}}{R} = E_{0x}\Phi_{ex}(\Theta,\gamma) \frac{e^{ikR}}{R}, \quad (14.70)$$

similar to Equation (14.20) for acoustic waves. According to Equation (14.62), all of the beam and ray asymptotics of the electromagnetic field $E_x^{(0)}$ are identical to those presented in Section 14.1.4 for the acoustic field $u_s^{(0)}$:

$$\Phi_{ex}^{(0)}(\Theta, \gamma) = \Phi_s^{(0)}(\Theta, \gamma). \tag{14.71}$$

The ray asymptotics of the field $E_x^{(1)}$ are also identical to those of the acoustic field $u_s^{(1)}$ shown in Section 14.1.4:

$$\Phi_{ex}^{(1)}(\Theta, \gamma) = \Phi_s^{(1)}(\Theta, \gamma). \tag{14.72}$$

However, the beam asymptotics associated with the field generated by the nonuniform (fringe) components are different for electromagnetic and acoustic waves:

$$\Phi_{ex}^{(1)beam}(\Theta, \gamma) \neq \Phi_s^{fr.beam}(\Theta, \gamma). \tag{14.73}$$

14.2.2 H-Polarization

On the surface of a perfectly conducting cylinder (Fig. 14.1), the incident wave

$$H_x^{inc} = H_{0x}e^{ik(z\cos\gamma + y\sin\gamma)}, \qquad H_{y,z}^{inc} = E_x^{inc} = 0 \tag{14.74}$$

generates currents with the uniform components (13.68) and (13.69). The field radiated by these currents is calculated according to Equations (1.92) and (1.93). In the plane $y0z$ ($\varphi = \pi/2$ and $\varphi = 3\pi/2$), it is described by the expressions identical to Equations (14.3) and (14.5) for acoustic waves:

$$H_x^{(0)} = [H_{0x}/u_0] \cdot u_h^{(0)}. \tag{14.75}$$

In particular, this relationship means that the PO curves in Figures 14.4, 14.5, 14.7, 14.8, and 14.9 plotted for the acoustic waves *scattered from a rigid cylinder* also demonstrate the electromagnetic waves with H_x-polarization *scattered from a perfectly conducting cylinder* of the same size.

The field radiated by the nonuniform component of the current $(\vec{j}^{(1)})$ is calculated according to Section 7.8. The expression

$$H_x^{(1)left} = H_{0x}\frac{a}{2\pi}\frac{e^{ikR}}{R}e^{iq}\int_0^{2\pi}\left\{\cos\gamma\cos\psi\, F_\theta^{(1)left}(\psi,\theta,\phi)\cdot\phi_x + \sin\psi\right.$$
$$\left. \times [G_\theta^{(1)left}(\psi,\theta,\phi)\cdot\phi_x - G_\phi^{(1)left}(\psi,\theta,\phi)\cdot\theta_x]\right\}e^{ip\sin\psi}d\psi \tag{14.76}$$

describes the field scattered from the left edge, and the expression

$$H_x^{(1)right} = H_{0x}\frac{a}{2\pi}\frac{e^{ikR}}{R}e^{-iq}\int_\pi^{2\pi}\left\{\cos\gamma\cos\psi\, F_\theta^{(1)left}(\psi,\theta,\phi)\cdot\phi_x + \sin\psi\right.$$
$$\left. \times [G_\theta^{(1)left}(\psi,\theta,\phi)\cdot\phi_x - G_\phi^{(1)left}(\psi,\theta,\phi)\cdot\theta_x]\right\}e^{ip\sin\psi}d\psi \tag{14.77}$$

represents the field scattered from the right edge. Here, $p = ka(\sin \gamma - \sin \Theta)$ and $q = kl(\cos \Theta - \cos \gamma)$. In addition, Equations (14.15) to (14.18) define the local angles ϕ, ϕ_0 (different for the left and right edges), and Equations (13.51) and (13.52) determine the unit vectors $\hat{\theta}$, $\hat{\phi}$ (the same for the left and right edges).

For the forward scattering direction $\Theta = \gamma$ (belonging to the diffraction cone), these expressions reduce to

$$H_x^{(1)\text{left}} = H_{0x}\frac{a}{2\pi}\frac{e^{ikR}}{R}\int_0^{2\pi}\left\{\frac{\cos^2\gamma\cos^2\psi}{\sin^2\gamma_0}f^{(1)}(\pi+\phi_0,\phi_0,3\pi/2)\right.$$
$$\left.+\frac{\sin^2\psi}{\sin^2\gamma_0}g^{(1)}(\pi+\phi_0,\phi_0,3\pi/2)-\frac{\sin\gamma\cos\gamma\sin\psi\cos^2\psi}{\sin^2\gamma_0}\right\}d\psi$$

(14.78)

and

$$H_x^{(1)\text{right}} = H_{0x}\frac{a}{2\pi}\frac{e^{ikR}}{R}\int_\pi^{2\pi}\left\{\frac{\cos^2\gamma\cos^2\psi}{\sin^2\gamma_0}f^{(1)}(\pi+\phi_0,\phi_0,3\pi/2)\right.$$
$$\left.+\frac{\sin^2\psi}{\sin^2\gamma_0}g^{(1)}(\pi+\phi_0,\phi_0,3\pi/2)-\frac{\sin\gamma\cos\gamma\sin\psi\cos^2\psi}{\sin^2\gamma_0}\right\}d\psi.$$

(14.79)

We emphasize again that the local angles ϕ, ϕ_0 in (14.78) and (14.79) are different.

Notice also that in the case of scattering at the disk, the expression similar to Equation (14.78) (but with the wedge angle equal to $\alpha = 2\pi$) exactly transforms into the known function

$$H_x^{(1)} = H_{0x}\frac{a}{\pi\sin^2\gamma}\left[\frac{1+\cos^2\gamma}{\cos\gamma}E\left(\frac{\pi}{2},\sin\gamma\right)-2\cos\gamma K\left(\frac{\pi}{2},\sin\gamma\right)\right]\frac{e^{ikR}}{R}$$

(14.80)

derived earlier by Ufimtsev (1962) and Butorin et al. (1988). Here, functions $E(\pi/2, x)$ and $K(\pi/2, x)$ are the complete elliptic integrals (14.69).

To continue the comparative analysis of electromagnetic and acoustic waves scattered at a finite cylinder, it is convenient to represent the total scattered field in the form

$$H_x = H_x^{(0)} + H_x^{(1)} = H_{0x}[\Phi_{hx}^{(0)} + \Phi_{hx}^{(1)}]\frac{e^{ikR}}{R} = H_{0x}\Phi_{hx}(\Theta,\gamma)\frac{e^{ikR}}{R},$$

(14.81)

similar to Equation (14.20) for acoustic waves. Due to Equation (14.75), all of the beam and ray asymptotics for the electromagnetic field $H_x^{(0)}$ are the same as those for the acoustic field $u_h^{(0)}$ (presented in Section 14.1.4):

$$\Phi_{hx}^{(0)}(\Theta,\gamma) = \Phi_h^{(0)}(\Theta,\gamma).$$

(14.82)

It turns out that the ray asymptotics for the electromagnetic field $H_x^{(1)}$ and those (found in Section 14.1.4) for the acoustic field $u_h^{(1)}$ are also identical:

$$\Phi_{hx}^{(1)}(\Theta, \gamma) = \Phi_h^{(1)}(\Theta, \gamma). \tag{14.83}$$

However, the beam asymptotics for the fields $H_x^{(1)}$ and $u_h^{(1)}$ generated by the nonuniform (fringe) components $j^{(1)}$ (in the scattering directions $\Theta = \gamma$, $\Theta = \pi - \gamma$, and $\Theta = 2\pi - \gamma$) are different,

$$\Phi_{hx}^{(1)\text{beam}}(\Theta, \gamma) \neq \Phi_h^{(1)\text{beam}}(\Theta, \gamma), \tag{14.84}$$

although they are of the same order of magnitude. Besides, the examination of Figures 14.7 and 14.9 reveals the following situations:

- Already in the case of the cylinder with diameter $d = \lambda$ and length $L = 3\lambda$, the quantity $\Phi_h^{(1)\text{beam}}$ is about 18 dB less compared to the PO beams $\Phi_h^{(0)\text{beam}}$ in the directions $\Theta = \gamma$, $\Theta = \pi - \gamma$, and about 25 dB less in the direction $\Theta = 2\pi - \gamma$.

- In the case of the cylinder with diameter $d = 3\lambda$ and length $L = 9\lambda$, the quantity $\Phi_h^{(1)\text{beam}}$ is about 28 dB less compared to the PO beams $\Phi_h^{(0)\text{beam}}$ in the directions $\Theta = \gamma$, $\Theta = \pi - \gamma$, and about 35 dB less in the direction $\Theta = 2\pi - \gamma$.

These observations clearly show that, for cylinders of such size and larger, the difference between acoustic and electromagnetic scattering (in the plane $y0z$ (Fig. 14.1, and for electromagnetic waves with E_x- or H_x-polarization) is practically negligible.

14.2.3 Refined Asymptotics for the Specular Beam Reflected from the Lateral Surface

This section is the electromagnetic version of Section 14.1.5. It studies the scattered field in the vicinity of the specular direction $\Theta = 2\pi - \gamma$ in the plane $\varphi = 3\pi/2$. This study is based on the paper by Ufimtsev (1981).

According to Sections 14.2.1 and 14.2.2, the following asymptotic relationships exist between the electromagnetic and acoustic scattered waves:

$$E_x = E_x^{(0)} + E_x^{(1)\text{fr}} = [E_{0x}/u_0] \cdot u_s = [E_{0x}/u_0] \cdot [u_s^{(0)} + u_s^{(1)\text{fr}}] \tag{14.85}$$

and

$$H_x = H_x^{(0)} + H_x^{(1)\text{fr}} = [H_{0x}/u_0] \cdot u_h = [H_{0x}/u_0] \cdot [u_h^{(0)} + u_h^{(1)\text{fr}}]. \tag{14.86}$$

They are valid under the conditions $ka \sin \gamma \gg 1$ and $kl \gg 1$. Here the fields with the superscript "0" are generated by the uniform components of the surface scattering sources ($j^{(0)}$) and represent the PO fields. The fields with the superscript "(1)fr" are generated by the nonuniform/fringe components of the surface scattering sources

$(j^{(1)\mathrm{fr}})$, which concentrate in the vicinity of the edges. Besides, one should also include into the scattered field the contributions $E_x^{(1)\mathrm{sm}}$ and $H_x^{(1)\mathrm{sm}}$ generated by that part of the nonuniform component $j^{(1)\mathrm{sm}}$ that is caused by the smooth bending of the cylindrical surface. As shown in Section 14.1.5, these contributions (in the vicinity of the specular direction) are the quantities of the same order of magnitude as those radiated by the fringe currents. To calculate them is a main objective of the present section.

First it necessary to determine the nonuniform currents $\vec{j}^{(1)\mathrm{sm}}$. We use the results of the paper by Franz and Galle (1955), which are also reproduced in the work of Bowman et al. (1987). This paper contains the high-frequency asymptotics for the surface field induced on an infinite circular cylinder by the incident plane wave. It is assumed there that the incident wave propagates in the direction perpendicular to the cylinder axis. A quite subtle procedure is to extend the results of this paper to the oblique incidence and to obtain correct formulas for the currents $\vec{j}^{(1)\mathrm{sm}}$. Eventually one obtains

$$j_x^{(1)\mathrm{sm}} = -Y_0 E_{0x} \frac{i}{ka \sin^2 \psi} e^{ik(z \cos \gamma + a \sin \gamma \sin \psi)}, \tag{14.87}$$

$$j_y^{(1)\mathrm{sm}} = Y_0 E_{0x} \frac{i \cos \psi}{ka \sin^3 \psi} e^{ik(z \cos \gamma + a \sin \gamma \sin \psi)}, \tag{14.88}$$

$$j_z^{(1)\mathrm{sm}} = Y_0 E_{0x} \frac{i \cos \gamma \cos \psi}{ka \sin \gamma \sin^3 \psi} e^{ik(z \cos \gamma + a \sin \gamma \sin \psi)} \tag{14.89}$$

in the case of E-polarization, and

$$j_z^{(1)\mathrm{sm}} = H_{0x} \frac{i}{ka \sin \gamma \sin^2 \psi} e^{ik(z \cos \gamma + a \sin \gamma \sin \psi)}, \qquad j_{x,y}^{(1)\mathrm{sm}} = 0 \tag{14.90}$$

in the case of H-polarization. Then one applies Equations (1.92) and (1.93) and calculates the retarded vector-potential. The integrals over the variable ζ are calculated in closed form. Integrals over the variable ψ are evaluated asymptotically by the stationary-phase technique. As a result one finds

$$E_x^{(1)\mathrm{sm}} = E_{0x} \Phi_{\mathrm{ex}}^{(1)\mathrm{sm}}(\Theta, \gamma) \frac{e^{ikR}}{R} \tag{14.91}$$

and

$$H_x^{(1)\mathrm{sm}} = H_{0x} \Phi_{\mathrm{hx}}^{(1)\mathrm{sm}}(\Theta, \gamma) \frac{e^{ikR}}{R}, \tag{14.92}$$

where

$$\Phi_{\mathrm{ex}}^{(1)\mathrm{sm}}(\Theta, \gamma) = l \frac{\sin q}{q} \frac{e^{-ip+i\pi/4}}{\sqrt{2\pi p}} \tag{14.93}$$

and

$$\Phi_{\mathrm{hx}}^{(1)\mathrm{sm}}(\Theta, \gamma) = -l \frac{\sin \Theta}{\sin \gamma} \frac{\sin q}{q} \frac{e^{-ip+i\pi/4}}{\sqrt{2\pi p}} \tag{14.94}$$

with $p = ka(\sin \gamma - \sin \Theta)$ and $q = kl(\cos \Theta - \cos \gamma)$. The comparison of these quantities with their acoustic counterparts (14.51), (14.52) shows that they differ only in sign.

Now one can calculate the total scattered field,

$$E_x^{\text{tot}} = E_{0x} \Phi_{\text{ex}}^{\text{tot}}(\Theta, \gamma) \frac{e^{ikR}}{R}, \tag{14.95}$$

$$H_x^{\text{tot}} = H_{0x} \Phi_{\text{hx}}^{\text{tot}}(\Theta, \gamma) \frac{e^{ikR}}{R}, \tag{14.96}$$

where

$$\Phi_{\text{ex}}^{\text{tot}} = \Phi_{\text{ex}}^{(0)} + \Phi_{\text{ex}}^{(1)\text{fr}} + \Phi_{\text{ex}}^{(1)\text{sm}} \tag{14.97}$$

and

$$\Phi_{\text{hx}}^{\text{tot}} = \Phi_{\text{hx}}^{(0)} + \Phi_{\text{hx}}^{(1)\text{fr}} + \Phi_{\text{hx}}^{(1)\text{sm}}. \tag{14.98}$$

According to Equations (14.85) and (14.86),

$$\Phi_{\text{ex}}^{(0)} = \Phi_{\text{s}}^{(0)}, \qquad \Phi_{\text{ex}}^{(1)\text{fr}} = \Phi_{\text{s}}^{(1)\text{fr}}, \tag{14.99}$$

$$\Phi_{\text{hx}}^{(0)} = \Phi_{\text{h}}^{(0)}, \qquad \Phi_{\text{hx}}^{(1)\text{fr}} = \Phi_{\text{h}}^{(1)\text{fr}}, \tag{14.100}$$

where $\Phi_{\text{s,h}}^{(0)}$, $\Phi_{\text{s,h}}^{(1)\text{fr}}$ are the acoustic functions defined in Section 14.1.5. Exactly in the specular direction $\Theta = 2\pi - \gamma$, the scattered field is determined as

$$\Phi_{\text{ex}}^{\text{tot}} = \frac{a}{\sqrt{\pi ka \sin \gamma}} \left\{ ikl \sin \gamma + \frac{3}{16} \frac{l}{a} - \frac{1}{\sqrt{3}} \left[\frac{1}{2} + \cos\left(\frac{2(\pi - 2\gamma)}{3}\right) \right]^{-1} \right.$$

$$\left. - \frac{1}{3\sqrt{3}} + \frac{l}{2a} \right\} e^{-i2ka \sin \gamma + i\pi/4}, \tag{14.101}$$

$$\Phi_{\text{hx}}^{\text{tot}} = \frac{a}{\sqrt{\pi ka \sin \gamma}} \left\{ -ikl \sin \gamma - \frac{3}{16} \frac{l}{a} - \frac{1}{\sqrt{3}} \left[\frac{1}{2} + \cos\left(\frac{2(\pi - 2\gamma)}{3}\right) \right]^{-1} \right.$$

$$\left. + \frac{1}{3\sqrt{3}} + \frac{l}{2a} \right\} e^{-i2ka \sin \gamma + i\pi/4}. \tag{14.102}$$

The origination and meaning of each term here is clear. The first and second terms in the braces relate respectively to the first and second terms in the asymptotic expansion for the PO field. The third and fourth terms represent the contribution by the fringe sources $j_{\text{s,h}}^{(1)\text{fr}}$, and the last term shows the contribution by the sources $j_{\text{s,h}}^{(1)\text{sm}}$ caused by the smooth bending of the cylindrical surface.

Thus, the calculation of the specular reflected beam of electromagnetic waves has been completed.

PROBLEMS

14.1   Start with expressions (1.33) and (1.34) and derive the directivity pattern (14.4) of acoustic waves scattered at a soft cylinder. See details in Sections 6.2 and 13.1.1.

14.2   Start with expressions (1.33) and (1.34) and derive the directivity pattern (14.5) of acoustic waves scattered at a hard cylinder. See details in Sections 6.2 and 13.1.1.

14.3   Use Equations (1.53) and (14.4) and derive the total cross-section (14.7) of a soft cylinder.

14.4   Use Equations (1.53) and (14.5) and derive the total cross-section (14.7) of a hard cylinder.

14.5   Use expression (14.4) and compute the normalized scattering cross-sections (14.9) for the individual parts of a soft cylinder with diameter $d = 2a = \lambda$ and length $L = 2l = 3\lambda$. Set $\gamma = 45°$. Represent the results in the form of a diagram as in Figure 14.2. Write the expression for the electromagnetic wave incident on a perfectly conducting cylinder, whose PO scattering cross-section is identical to that plotted by you.

14.6   Use expression (14.5) and compute the normalized scattering cross-sections (14.9) for the individual parts of a hard cylinder with diameter $d = 2a = \lambda$ and length $L = 2l = 3\lambda$. Set $\gamma = 45°$. Represent the results in the form of a diagram as in Figure 14.4. Write the expression for the electromagnetic wave incident on a perfectly conducting cylinder, whose PO scattering cross-section is identical to that plotted by you.

14.7   Use Equations (14.4) and (14.5) and compute the directivity pattern (14.9) of the shadow radiation (14.10) for the cylinder with diameter $d = 2a = \lambda$ and length $L = 2l = 3\lambda$. Set $\gamma = 45°$. Represent the results in the form of a diagram as in Figure 14.6. Write the expression for the electromagnetic wave incident on a perfectly conducting cylinder and generating the same shadow radiation as that plotted by you.

14.8   The incident wave (14.1) hits a hard cylinder (Fig. 14.1). Use the asymptotic expression (7.90) and derive the function (14.12), which describes the field radiated by the scattering sources $j_h^{(1)}$ induced near the left edge. Write the explicit expressions for the function $F_h^{(1)}$ in terms of functions $V_t(\sigma_1, \varphi_0)$, $V_0(\beta_1, \varphi_0)$, $V_t(\sigma_2, \alpha - \varphi_0)$, $V_0(\beta_2, \alpha - \varphi_0)$ and define the parameters $\sigma_{1,2}$, $\beta_{1,2}$.

14.9   Elementary edge waves are the functions of the local angles γ_0, θ. Derive Equations (14.14), which define these angles. Use the dot products between the unit vectors $\hat{R}, \hat{t}$ and $\hat{t}, \hat{k}^i$ where $\hat{R}$ shows the scattering direction in the plane $y0z$, $\hat{t}$ is the tangent to the edge, and $\hat{k}^i$ shows the direction of the incident wave.

14.10   Elementary edge waves are the functions of the local angles ϕ, ϕ_0. Derive Equations (14.15) and (14.16), which define these angles. Hint: project the unit vectors $\hat{R} = \hat{y}\sin\Theta + \hat{z}\cos\Theta$ and $\hat{Q} = -\hat{k}^i = -\hat{y}\sin\gamma - \hat{z}\cos\gamma$ on the plane normal to the tangent $\hat{t}$. Recall the note box following Equation (7.131).

14.11   Apply the stationary-phase technique to the second terms in Equations (14.4) and (14.5), and derive asymptotics (14.22), (14.23) for the beam scattered from the lateral/cylindrical surface.

14.12   Use Equation (14.12), apply the stationary-phase technique, and derive the asymptotic expression (similar to Equation (14.31)) for the edge-diffracted ray diverging from the stationary point 1 (Fig. 14.1).

14.13   Verify that the incident wave (13.42) generates the surface current (13.43) on the left base (disk) of a perfectly conducting cylinder (Fig. 13.1). Use Equations (1.91), (1.92), and (1.93) and derive the explicit expression for the electromagnetic field in the plane $y0z$. Compare the E_x-component with the acoustic field (14.3), (14.4) scattered from a soft disk. Confirm the statement (14.62).

14.14 Verify that the incident wave (13.42) generates the surface current (13.44) on the lateral/cylindrical surface of a perfectly conducting cylinder (Fig. 13.1). Use Equations (1.91), (1.92), and (1.93) and derive the explicit expression for the electromagnetic field in the plane $y0z$. Compare the E_x-component with the acoustic field (14.3), (14.4) scattered from a soft cylindrical surface. Confirm the statement (14.62).

14.15 Use Equations (13.50) and verify the expressions (13.51) and (13.52) for the unit vectors $\hat{\theta}, \hat{\phi}$.

14.16 Use Equations (7.130), (7.131), recall the note box following Equation (7.131), apply the local coordinates ϕ, ϕ_0, θ (introduced in Section 13.2.1), and derive the function (14.64) for the field scattered by the left edge of a perfectly conducting cylinder.

14.17 Use Equation (14.64) and verify expression (14.66) for the diffracted beam in the forward direction.

14.18 Verify that the incident wave (13.67) generates the surface current (13.68) on the left base (disk) of a perfectly conducting cylinder (Fig. 13.1). Use Equations (1.91), (1.92), and (1.93) and derive the explicit expression for the electromagnetic field in the plane $y0z$. Compare the H_x-component with the acoustic field (14.3), (14.5) scattered from a hard disk. Confirm the statement (14.75).

14.19 Verify that the incident wave (13.67) generates the surface current (13.69) on the lateral/cylindrical surface of a perfectly conducting cylinder (Fig. 13.1). Use Equations (1.91) to (1.93) and derive the explicit expression for the electromagnetic field in the plane $y0z$. Compare the H_x-component with the acoustic field (14.3), (14.5) scattered from a hard cylindrical surface. Confirm the statement (14.75).

14.20 Use Equations (7.130) and (7.131), recall the note box following Equation (7.131), apply the local coordinates ϕ, ϕ_0, θ (introduced in Section 13.2.1), and derive the function (14.76) for the field scattered by the left edge of a perfectly conducting cylinder.

14.21 Use Equation (14.76) and verify the expression (14.78) for the diffracted beam in the forward direction.

Conclusions

The PTD developed in this book clarifies the scattering physics. Via the shadow radiation, it elucidates the nature of the Fresnel diffraction and forward scattering, as well as the optical theorem. It also establishes the diffraction limit for the reduction of the total power scattered by large (compared to the wavelength) objects covered by absorbing materials. This theory shows that, even with the application of perfectly absorbing coatings on perfectly reflecting objects, their total scattered power can be reduced solely by a factor of two. This means that, against bistatic sonar and radar, it is impossible to completely mask the scattering object with any absorbing materials (Ufimtsev, 1996).

As a source-based theory, PTD allows the calculation of contributions to the scattered field, which are generated by individual elements of the scattering surface. Such data are valuable in the design of antennas and objects with given characteristics of radiation and scattering.

PTD is a flexible theory amenable to further development and generalization. In combination with other analytic and numeric approaches, it can be used to create efficient hybrid techniques for the solution of complex diffraction problems. Some examples are presented in the papers listed in the section "Additional References Related to the PTD Concept: Applications, Modifications and Developments".

Fundamentals of the Physical Theory of Diffraction. By Pyotr Ya. Ufimtsev
Copyright © 2007 John Wiley & Sons, Inc.

References

M. ABRAMOWITZ AND I.A. STEGUN (1972): *Handbook of Mathematical Functions*, Dover Publication, Inc., New York.

J.S. ASVESTAS (1985): Line integrals and Physical Optics. Part 1: The transformation of the solid-angle surface integral to a line integral. *J. Opt. Soc. Am.*, 2(6), 891–895.

J.S. ASVESTAS (1985): Line integrals and Physical Optics. Part II: The conversion of the Kinchhoff surface integral to a line integral. *J. Opt. Soc. Am.*, 2(6), 896–902.

J.S. ASVESTAS (1986): The Physical Optics fields of an aperture on a perfectly conducting screen in terms of line integrals. *IEEE Trans Antennas Propagat.*, AP-34(9), 1155–1158.

J.S. ASVESTAS (1995): The Physical Optics integral and computer graphics. *IEEE Trans Antennas Propagat.*, 43(12), 1459–1460.

B.B. BAKKER AND E.T. COPSON (1939): *The Mathematical Theory of Huygen's Principle*, Oxford, University Press.

C.A. BALANIS (1989): *Advanced Engineering Electromagnetics*, John Wiley & Sons, New York.

H. BATEMAN (1955): *The Mathematical Analysis of Electrical and Optical Wave-Motion on the Basis of Maxwell's Equations*, Dover Publications Inc., New York, pp. 90–94.

J. BOERSMA AND Y. RAHMAT-SAMII (1980): Comparison of two leading uniform series of edge diffraction with the exact uniform asymptotic expansion. *Radio Sci.*, 15, 1179–1194.

V.A. BOROVIKOV (1966): *Diffraction at Polygons and Polyhedrons*, Nauka, Moscow.

V.A. BOROVIKOV AND B.E. KINBER (1994): *Geometrical Theory of Diffraction*, Institution of Electrical Engineering, London, UK.

M. BORN AND E. WOLF (1980): *Principles of Optics*, Pergamon Press, London, New York.

J.J. BOWMAN, T.B.A. SENIOR, AND P.L.E. USLENGHI, Eds (1987): *Electromagnetic and Acoustic Scattering by Simple Shapes*, Hemisphere Publishing Corp., New York.

O. BREINBJERG (1992): Higher order equivalent edge currents for fringe wave radar scattering by perfectly conducting polygonal plates. *IEEE Trans Antennas Propagat.*, 40(12), 1543–1554.

D. BRILL AND G.C. GAUNAURD (1993): Approximate description of the sound fields scattered by insonified, submerged, ribbed, flat-ended cylindrical structures. *J. Acoust. Soc. Am.*, 93(1), 71–79.

I.N. BRONSHTEIN AND K.A. SEMENDYAEV (1985): *Handbook of Mathematics*, Van Norstrand Reinhold Company, New York.

M.W. BROWNE (1991a): Two rival designers led the way to stealthy warplanes. *New York Times*, Sci. Times Sec., May 14, 1991.

M.W. BROWNE (1991b): Lockheed credits Soviet theory in design of F-117. *Aviation Week Space Technol.*, p. 27, December 1991.

D.I. BUTORIN AND P. YA. UFIMTSEV (1986): Explicit expressions for an acoustic edge wave scattered by an infinitesimal edge element. *Soviet Physics–Acoustics*, 32(4), 283–287.

D.I. BUTORIN, N.A. MARTYNOV, AND P. YA. UFIMTSEV (1987): Asimptoticheskie vyrazheniya dlya elementarnoi kraevoi volny. *Radiotekhnika i elektronika*, 32, 1818–1828 [English translation, Asymptotic expressions for the elementary edge wave. *Soviet Journal of Communications Technology and Electronic*, 1988, 33(1), 17–26].

Fundamentals of the Physical Theory of Diffraction. By Pyotr Ya. Ufimtsev
Copyright © 2007 John Wiley & Sons, Inc.

P.C. CLEMMOW (1950): Some extensions to the method of integration by steepest descents. *Quart. J. Mech. Appl. Math.*, 3(2), 241–256.

C. CHESTER, B. FRIEDMAN, AND F. URSELL (1957): An extension of the method of steepest descents, *Proc. Camb. Phil. Soc.*, 54, 599–611.

E.T. COPSON (1965): *Asymptotic Expansions*, University Press, Cambridge.

L.B. FELSEN (1955): Backscattering from wide-angle and narrow-angle cones. *J. Appl. Phys.*, 26(3), 138–151.

V.A. FOCK (1965): *Electromagnetic Diffraction and Propagation Problems*, Pergamon Press, London.

W. FRANZ AND R. GALLE (1955): Semiasymptotische Reihen fur die Beugung einer ebenen Welle am Zylinder. *Z. Naturforschung*, 10a(5), 374–378.

J.I. GLASER (1985): Bistatic RCS of complex objects near forward scatter. *IEEE Trans Aerosp. Electron. Syst.*, AES-21(1), 70–78.

W.B. GORDON (1994): High frequency approximations to the Physical Optics scattering integral. *IEEE Trans. Antennas Propagat.*, 42(3), 427–432.

W.B. GORDON (2003): Calculating scatter from surface with zero curvature. *IEEE Trans Antennas Propagat.*, 51(9), 2506–2508.

W.B. GORDON AND H.J.B. BILOW (2002): Reduction of surface integrals to contour integrals. *IEEE Trans Antennas Propagat.*, 50(3), 308–311.

I.S. GRADSHTEYN AND I.M. RYZHIK (1994): *Tables of Integrals, Series, and Products*, Academic Press, Inc., New York.

G.L. JAMES (1980): *Geometrical Theory of Diffraction for Electromagnetic Waves*, Institution of Electrical Engineers, Peter Peregrinus Ltd., Stevenage, UK, and New York.

P.M. JOHANSEN (1996): Uniform physical theory of diffraction equivalent edge currents for truncated wedge strips. *IEEE Trans Antennas Propagat.*, AP-44(7), 989–995.

A. KALASHNIKOV (1912): The Gouy–Sommerfeld diffraction. *Zhurnal Russkogo Fiziko-Khemicheskogo Obshchestva, Fizicheskyi Otdel* [*Journal of the Russian Physical-Chemical Society, Physical Division*], 44(3), 137–144.

S.N. KARP AND J.B. KELLER (1961): Multiple diffraction in a hard screen. *Optica Acta*, 8(1), 61–72.

J.B. KELLER (1962): Geometrical theory of diffraction. *J. Opt. Soc. Am.*, 52(2), 116–130.

L.E. KINSLER, A.R. FREY, A.B. COPPENS, AND J.V. SANDERS (1982): *Fundamentals of Acoustics*, John Wiley & Sons, New York.

F. KOTTLER (1923): Elektromagnetische Theorie der Beugung an schwarzen Schirmen. *Annalen der Physik*, 71(15), 457–508.

R.G. KOUYOUMJIAN AND P.H. PATHAK (1974): A uniform theory of diffraction for an edge in a perfectly conducting surface. *Proc. IEEE*, 62(11), 1448–1461.

E.F. KNOTT AND T.B.A. SENIOR (1973): Equivalent currents for a ring discontinuity. *IEEE Trans Antennas Propagat.*, AP-21(9), 698–696.

E.F. KNOTT (1985): A progression of high-frequency RCS prediction techniques. *Proc. IEEE*, 73(2), 252–264.

YU.A. KRAVTSOV AND YU.I. ORLOV (1983): Caustics, catastrophes and wave fields. *Uspekhi Fizicheskikh Nauk*, 141(4), 591–627 (in Russian) [English translation: *Sov. Phys. Usp.*].

H.M. MACDONALD (1902): *Electric Waves*, The University Press, Cambridge, England, pp. 186–198.

H.M. MACDONALD (1912): The effect produced by an obstacle on a train of electric waves. *Phil. Trans. Royal Soc. London, Series A, Math. Phys. Sci.*, 212, 299–337.

G.A. MAGGI (1888): Sulla propagazione libra e perturbata delle onde luminose in un mezzo isotropo. *Annali di Matematica*, 16(2), 21–48.

D.A. MCNAMARA, C.W.I. PISTORIUS, AND J.A.G. MALHERBE (1990): *Introduction to the Uniform Geometrical Theory of Diffraction*, Artech House, Boston–London.

P. MEINCKE, O. BREINBERG, AND E. JORGENSON (2003): An exact line integral representation of the magnetic field Physical Optics scattered field. *IEEE Trans. Antennas Propagat.*, 51(6), pp. 1395–1398.

P. MENOUNOU, M.R. BAILEY, AND D.T. BLACKSTOCK (2000): Edge wave on axis behind an aperture or disk having a ragged edge. *J. Acoust. Soc. Am.*, 107(1), pp. 103–111.

C.A. MENTZER, L. PETERS JR., AND R.C. RUDDUCK (1975): Slope diffraction and its application to horns. *IEEE Trans. Antennas Propagat.*, 23(2), 153–159.

A. MICHAELI (1986): Elimination of infinities in equivalent edge currents. *IEEE Trans. Antennas Propagat.*, AP-34(7), 912–918.

A. MICHAELI (1987): Equivalent currents for second order diffraction by the edges of perfectly conducting polygonal surfaces. *IEEE Trans. Antennas Propagat.*, AP-35(2), 183–190.

K.M. MITZNER (1974): Incremental length diffraction coefficients. Technical Report AFAL-TR-73-26, Northrop Corporation, Aircraft Division.

F.A. MOLINET (2005): Edge-excited rays on convex and concave structures: A review, *IEEE Antennas Propagat. Mag.*, 47(5), 34–46.

B.T. MORSE (1964): Diffraction by polygonal cylinders. *J. Math. Phy.*, 5(2), 199–214.

P.J. MOSER, H. UBERALL, AND J.R. YUAN (1993): Sound scattering from a finite cylinder with ribs. *J. Acoust. Soc. Am.*, 94(6), 3342–3351.

J.D. MURRAY (1984): *Asymptotic Analysis*, Springer-Verlag, New York.

P.H. PATHAK (1988): "Techniques for high-frequency problems," Ch. 4 in, Eds., Y.T. LO AND S.W. LEE, *Antenna Handbook*, Van Nostrand Reinhold Company, New York.

W. PAULI (1938): On asymptotic series for functions in the theory of diffraction of light. *Phys. Rev.*, 54(2), 924–931.

G. PELOSI, S. SELLERI, AND P. YA. UFIMTSEV (1998): Newton's observations of diffracted rays. *IEEE Antennas Propagat. Mag.*, 40(2), 7–14.

A.D. PIERCE (1994): *Acoustics, Introduction to Its Physical Concepts and Applications*, Acoustical Society of America, New York.

Y. RAHMAT-SAMII AND R. MITTRA (1978): Analysis of high-frequency diffraction of an arbitrary incident field by a half-plane – Comparison with four asymptotic techniques. *Radio Sci.*, 13(1), 31–48.

B. RICH (1994): Inside the top secret skunk works. *Popular Sci.*, October, 61–81.

B. RICH AND L. JANOS (1994): *Skunk Works*, Little Brown, Boston.

A. RUBINOWICZ (1917): Die Beugungswelle in der Kichhoffschen Theorie der Beugungserscheinungen. *Annalen der Physik*, IV Folge, Band 53, Heft 12, 257–278.

A. RUBINOWICZ (1924): Zur Kirchhoffschen Beugungstheorie. *Annalen derPhysik*, Folge 4, Band 73, 339–364.

A. RUBINOWICZ (1965): Darstellung der Sommerfeldschen Beugungswelle in einer Gestalt, die Beitrage der einzelnen Elemente der beugende Kante zur gesamten Beugungswelle erkennen last. *Acta Physica Polonica*, 28, 6(12) 811–860.

G.T. RUCK, D.E. BARRICK, W.D. STUART, AND C.K. KIRCHBAUM (1970): *Radar Cross Section Handbook*, Vols. 1 and 2, Plenum Press, New York.

C.E. SCHENSTED (1955): Electromagnetic and acoustic scattering by a semi-infinite body of revolutions. *J. Appl. Phys.*, 26(3), 306–308.

K. SCHWARZSCHILD (1902): Die Beugung und Polarisation des Lichts durch einen Spalt. *Mathematische Annalen*, 55, 177–247.

T.B.A. SENIOR AND P.L.E. USLENGHI (1972): Experimental detection of the edge-diffraction cone. *Proc. IEEE*, PROC-60, 1448.

A. SOMMERFELD (1896): Mathematische Theorie der Diffraction. *Mathematische Annalen*, 47, 317–374.

A. SOMMERFELD (1935): "Theorie der Beugung," Ch. 20 in, F. FRANK AND R.V. MIZES. Eds., *Die Differential- und Integralgleichungen der Mechanik und Physik*, Vol. 2, *Physical Part*. Friedr. Vieweg & Sohn, Braunschweig, Germany. (American Publications, New York, 1943, 1961).

R. TIBERIO AND S. MACI (1994): An incremental theory of diffraction: Scalar Formulation. *IEEE Trans. Antennas Propagat.*, 42(5), 600–612.

R. TIBERIO, S. MACI, AND A. TOCCAFONDI (1995): An incremental theory of diffraction: Electromagnetic formulation. *IEEE Trans. Antennas Propagat.*, 43(1), 87–96.

R. TIBERIO, A. TOCCAFONDI, A. POLEMI, AND S. MACI (2004): Incremental theory of diffraction: A new improved formulation. *IEEE Trans. Antennas Propagat.*, 52(9), 2234–2243.

M. TRAN VAN NHIEU (1995): Diffraction by plane screens. *J. Acoust. Soc. Am.*, 97(2), 796–806.

M. TRAN VAN NHIEU (1996): Diffraction by the edge of a three-dimensional object. *J. Acoust. Soc. Am.*, 99(1), 79–87.

P. YA. UFIMTSEV (1957): "Diffraktsiya na kline i lente", part I of "Priblizhennyi raschet diffraktsii ploskikh electromagnitnykh voln na nekotorykh metallicheskikh telakh" (Diffraction at a wedge and a strip, part I of "Approximate computation of the diffraction of plane electromagnetic waves at certain metallic objects"). *Zhurnal Tekhnicheskoi Fiziki*, 27(8), 1840–1849. (English translation published by *Soviet Physics–Technical Physics.*)

P. YA. UFIMTSEV (1958a): "Diffraktsiya na diske i konechnom tsilindre," part II of "Priblizhennyi raschet diffraktsii ploskikh elektromagnitnykh voln na nekotorykh metallicheskikh telakh." (Diffraction at a disk and a finite cylinder, part II of "Approximate computation of the diffraction of plane electromagnetic waves at certain metallic objects.") *Zhurnal Tekhnicheskoi Fiziki*, 28(11), 2604–2616. (English translation published by Soviet *Physics–Technical Physics.*)

P. YA. UFIMTSEV (1958b): Secondary diffraction of electromagnetic waves by a strip. *Soviet Physics–Technical Physics*, 3(3), 535–548.

P. YA. UFIMTSEV (1958c): Secondary diffraction of electromagnetic waves by a disk. *Soviet Physics–Technical Physics*, 3(3), 549–556.

P. YA. UFIMTSEV (1961): Symmetrical illumination of finite bodies of revolution. *Radio Eng. Electron. Phys.*, 6(4), 492–500.

P. YA. UFIMTSEV (1962): *Metod Kraevykh Voln v Fizicheskoi Teorii Diffraktsii (Method of Edge Waves in the Physical Theory of Diffraction)*. Moscow, Sovetskoe Radio, 243 pp. (Machine translated into English by the U.S. Air Force, Foreign Technology Division (National Air Intelligence Center), Wright-Patterson AFB, OH, 1971. Technical Report AD 733203, Defense Technical Information Center of USA, Cameron Station, Alexandria, VA, 22304-6145, USA.)

P. YA. UFIMTSEV (1968): Diffraction of electromagnetic waves at black bodies and semitransparent plates. *Radiophys. Quantum Electron.*, 35(6), 527–538 (translated by Consult. Bureau, New York).

P. YA. UFIMTSEV (1969): Asymptotic investigation of the problem of diffraction on a strip. *Radio Eng. Electron. Phys.*, 14(7), 1014–1025.

P. YA. UFIMTSEV (1970): Asymptotic solution to the problem of diffraction from a strip using Dirichlet boundary conditions. *Radio Eng. Electron. Phys.*, 15(5), 782–757.

P. YA. UFIMTSEV (1979): Uniform asymptotic theory of diffraction by a finite cylinder. *SIAM*, 37(3), 459–466.

P. YA. UFIMTSEV (1981): Reflection of electromagnetic waves from a finite cylinder. *Radio Eng. Electron. Phys.*, 26(2), 59–65.

P. YA. UFIMTSEV (1989): Theory of acoustical edge waves. *J. Acoust. Soc. Amer.*, 86(2), 463–474.

P. YA. UFIMTSEV (1990): Black bodies and shadow radiation, *Soviet J. Commun. Technol. Electron.*, 35(5), 108–116 (translated by Scripta Technica).

P. YA. UFIMTSEV (1991): Elementary edge waves and the physical theory of diffraction. *Electromagnetics*, 11(2), 125–160.

P. YA. UFIMTSEV (1995): Rubinowicz and the modern theory of diffracted rays. *Electromagnetics*, 15(5), 547–565.

P. YA. UFIMTSEV AND Y. RAHMAT-SAMII (1995): Physical theory of slope diffraction. Special issue on Radar Cross Section of Complex Objects. *Annales des Telecommunications*, *(Annals of Telecommunications)* 50(5–6), 487–498.

P. YA. UFIMTSEV (1996): Comments on diffraction principles and limitations for RCS reduction techniques, *Proc. IEEE*, 84(12), 1828–1851.

P. YA. UFIMTSEV (1998): Fast convergent integrals for nonuniform currents on wedge faces. *Electromagnetics*, 18(3), 289–313. Corrections in *Electromagnetics*, 19(5), 473 (1999).

P. YA. UFIMTSEV (1999): Backscatter, in *Wiley Encyclopedia of Electrical and Electronics Engineering*, John Wiley & Sons, Inc., New York.

P. YA. UFIMTSEV (2003): *Theory of Edge Diffraction in Electromagnetics*, Tech Science Press, Encino, California.

P. YA. UFIMTSEV (2006a): Improved theory of acoustic elementary edge waves. *J. Acoust. Soc. Amer.*, 120(2), 631–635.

P. YA. UFIMTSEV (2006b): Improved physical theory of diffraction: Removal of the grazing singularity, *IEEE Trans. Antennas Propagat.*, 54(10), 2698–2702.

N.J. WILLIS (1991): *Bistatic Radars*, Artech House, Boston–London.

H.H. WITTE AND K. WESTPFAHL (1970): Hochfrequente Schallbeugung an der Kreisblende: numerische Ergebnisse. *Annalen der Physik*, Folge 7, Band 25, Heft 4, 375–382.

P. WOLF (1967): A new approach to edge diffraction, *SIAM J. Appl. Math.*, 15(6), 1434–1469.

ADDITIONAL REFERENCES RELATED TO THE PTD CONCEPT: APPLICATIONS, MODIFICATIONS, AND DEVELOPMENTS

A. ALTINTAS, O.M. BUYUKDURA, AND P.H. PATHAK (1994): An extension of the PTD concept for aperture radiation problems. *Radio Sci.*, 29(6), 1403–1407.

D.J. ANDERSH, M. HAZLETT, S.W. LEE, D.D. REEVES, D.P. SULLIVAN, AND Y. CHU (1984): *X-PATCH*: A high-frequency electromagnetic-scattering code and environment for complex three-dimensional objects. *IEEE Antennas Propagat. Mag.*, 36(1), 65–69.

T. AKASHI, M. ANDO, AND T. KINOSHITA (1989): Effects of multiple diffraction in PTD analysis of scattered field from a conducting disk. *Trans. IEICE*, E72(4), 259–261.

M. ANDO (1990): Modified physical theory of diffraction, in E. YAMASHITA, Ed., *Analysis Methods for EM Wave Problems*, Artech House, Boston–London.

M. ANDO AND T. KINOSHITA (1989): PO and PTD analysis in polarization prediction for plane wave diffraction from large circular disk. *Digests of 1989 IEEE AP/S International Symposium*, June 26–30, 1989, San Jose, California.

M. ANDO AND T. KINOSHITA (1989): Accuracy comparison of PTD and PO for plane wave diffraction from large circular disk. *Trans. IEICE*, E72(11), 1212–1218.

J.S. ASVESTAS (1995): A class of functions with removable singularities and their application in the physical theory of diffraction. *Electromagnetics*, 15(2), 143–155.

P. BALLING (1995): Fringe-currents effects on reflector antenna crospolarization. *Electromagnetics*, 15(1), 55–69.

S.S. BOR, S.Y. YANG, S.M. YETH, S.R. HWANG, AND C.C. HWANG (1996): Electromagnetic backscattering of helicopter rotor. *Electromagnetics*, 16(1), 63–74.

D.P. BOUCHE, J.J. BOUQUET, H. MANENE, AND R. MITTRA (1992): Asymptotic computation of the RCS of low observable axisymmetric objects at high frequency. *IEEE Trans. Antennas Propagat.*, 40(10), 1165–1174.

D.P. BOUCHE, F.A. MOLINET, AND R. MITTRA (1995): Asymptotic and hybrid techniques for electromagnetic scattering. *Proc. IEEE*, 81(12), 1658–1684.

M. BOUTILLIER AND M.A. BLONDELL-FOURNIER (1995): CAD based high-frequency RCS computing code for complex objects: Sermat. Special issue on Radar Cross Section of Complex Objects. *Annales des Telecommunications*, 50(5–6), 536–539.

O. BREINBJERG AND E. JORGANSEN (1999): Slope diffraction in the geometrical and physical theories of diffraction. *USNC/URSI Meeting*, July 11–16, Orlando, Florida, Digests, p. 90.

R.T. BROWN (1984): Treatment of singularities in the physical theory of diffraction. *IEEE Trans. Antennas Propagat.*, AP-32(6), 640–641.

C.C. CHA, J. MICHELS AND E. STARCZEWSKI (1988): An analysis of airborne vehicles dependence on frequency and bistatic angle. *Proc. 1988 IEEE National Radar Conference*, pp. 214–219. April 20–21, 1988. University of Michigan, Ann Arbor.

P. CORONA, A. DE BONITATIBUS, G. FERRARA, AND C. GENNARELLI (1993): Accurate evaluation of backscattering by 90° dihedral corners. *Electromagnetics*, 13(1), 23–36.

M.G. COTE, M.B. WOODWORTH, AND A.D. YAGHJIAN (1988): Scattering from perfectly conducting cube. *IEEE Trans. Antennas Propagat.*, 36(9), 1321–1329.

M. DOMINGO, R.P. TORRES, AND M.F. CATEDRA (1994): Calculation of the RCS from the interaction of edges and faces. *IEEE Trans. Antennas Propagat.*, 42(6), 885–898.

M. DOMINGO, F. RIVES, J. PEREZ, R.P. TORRES, AND M.F. CATEDRA (1995): Computation of the RCS of complex bodies modeled using NURBS surfaces. *IEEE Trans. Antennas Propagat.*, 37(6), 36–47.

D.W. DUAN, Y. RAHMAT-SAMII, AND J.P. MAHON (1991): Scattering from a circular disk: Comparative study of PTD and GTD techniques. *Proc. IEEE*, 79(10), 1472–1480.

D.D. GABRIELYAN, O.M. TARASENKO, AND V.V. SHATSKYI (1991): Ispol'zovanie predstavleniya kraevykh voln v sochetanii s metodom integral'nykh uravnenyi pri reshenii zadach difraktsii na ideal'no provodyaschikh telakh slozhnoi formy [Usage of the edge-wave representation combined with the method of integral equations to solve problems of diffraction by ideally conducting bodies with a complicated shape]. *Radiotekhnika I Elektronika*, 36(6), 1159–1163 [English translation *J. Commun. Technol. Electron.*].

J.L. GUIRAUD (1983): Une approche spectrale de la theorie physique de la diffraction. *Annales des Telecommunications*, 38(3–4), 145–157.

T.B. HANSEN AND R.A. SHORE (1998): Incremental length diffraction coefficients for the shadow boundary of a convex cylinder. *IEEE Trans. Antennas Propagat.*, 46(10), 1458–1466.

K. HONGO AND H. KOBAYASHI (2001): Evaluation of the surface field scattered by an impedance polygonal cylinder. *Electromagnetics*, 21, 319–339.

M. IDEMEN AND A. BUYUKAKSOY (1984): High-frequency surface currents induced on a perfectly conducting cylindrical reflector. *IEEE Trans. Antennas Propagat.*, AP-32(5), 501–507.

S.K. JENG (1998): Near-field scattering by PTD and shooting and bouncing rays. *IEEE Trans. Antennas Propagat.*, AP-46(4), 551–558.

P.M. JOHANSEN (1996): Uniform physical theory of diffraction equivalent edge currents for truncated wedge strips. *IEEE Trans. Antennas Propagat.*, 44(7), 989–995.

P.M. JOHANSEN (1999): Time-domain version of the PTD. *IEEE Trans. Antennas Propagat.*, 47(2), 261–270.

E. JORGANSEN, A. TOCCAFONDI, AND S. MACI (1999): Integral equation for truncated slab structures by using a fringe current formulation. *IEEE APS Intern. Meeting*, July 11–16, Orlando, Florida. Digests, 4, pp. 2546–2549.

E. JORGANSEN, S. MACI, AND A. TOCCAFONDI (2001): Fringe integral equation method for a truncated grounded dielectric slab. *IEEE Trans. Antennas Propagat.*, 49(8), 1210–1217.

J.J. KIM AND O.B. KESLER (1996): Hybrid scattering analysis (PTD + UFIM) for large airframe with small details. *USNC/URSI Radio Science Meeting*, Digests, July 21–26 1996, Baltimore, MD, p. 263.

S.Y. KIM, J.W. RA, AND S.Y. SHIN (1991): Diffraction by an arbitrary-angled dielectric wedge: Part II—Corrections to physical optics solution. *IEEE Trans. Antennas Propagat.*, 39(9), 1282–1292.

H. KOBAYASHI AND K. HONGO (1997): Scattering of electromagnetic plane waves by conducting plates. *Electromagnetics*, 17(6), 573–587.

I.L. LANDSBERG (1974): O polarizatsionnoi structure izlucheniya osesymmetrichnogo zerkala vblizi osi [Polarization structure of radiation by an axisymmetric reflector close to the symmetry axis]. *Radiotekhnika i Elektronika,* 19(9), 1817–1623 [English translation: *Radio Engineering and Electronic Physics*].

I.L. LANDSBERG (1979): Scattering of a plane wave at a metallic cone close to its symmetry axis (in Russian). *Radiotekhnika i Elektronika*, 24(5), 886 [English translation: *Radio Engineering and Electronic Physics*].

S.W. LEE (1977): Comparison of uniform asymptotic theory and Ufimtsev's theory of electromagnetic edghe diffraction. *IEEE Trans. Antennas Propagat.*, AP-25(2), 162.

M. MARTINEZ-BURDALO, A. MARTIN, AND R. VILLAR (1993): Uniform PO and PTD solution for calculating plane wave backscattering from a finite cylindrical shell of arbitrary cross section. *IEEE Trans. Antennas Propagat.*, 41(9), 1336–1339.

F.A. MOLINET (1991): Modern high frequency techniques for RCS computation: A comparative analysis. Special issue on RCS. *ACS J.*, September 1991.

A. MICHAELI (1985): A new asymptotic high-frequency analysis of electromagnetic scattering by a pair parallel wedges: closed form results. *Radio Sci.*, 20, 1537–1548.

A. MICHAELI (1995): Incremental diffraction coefficients for the extended physical theory of diffraction. *IEEE Trans. Antennas Propagat.*, 43(7), 732–734.

N. MORITA (1971): Diffraction by arbitrary cross-sectional semi-infinite conductor. *IEEE Trans. Antennas Propagat.*, AP-19(5), 358–364.

J.D. MURRAY (1984): Asymptotic Analysis, Springer-Verlag, New York Inc.

P.K. MURTHY AND G.A. THIELE (1986): Non-uniform currents on a wedge illuminated by a TE-plane wave. *IEEE Trans. Antennas Propagat.*, AP-34(8), 1038–1045.

G. PELOSI, S. MACI, R. TIBERIO, AND A. MICHAELI (1992): Incremental length diffraction coefficients for an impedance wedge. *IEEE Trans. Antennas Propagat.*, 40(10), 1201–1210.

A.C. POLYCARPOU, C.A. BALANIS, AND C.R. BITCHER (1995): Radar cross section of trihedral corner reflectors using PO and MEC. Special issue on Radar Cross Section of Complex Objects. *Annales des Telecommunications*, 50(5–6), 510–516.

J.M. Ruis, M. Ferrando, and L. Jofre (1993): *GRECO:* High-frequency RCS of complex radar targets in real-time. *IEEE Trans. Antennas Propagat.*, 41(9), 1308–1319.

J.M. Ruis, M. Ferrando, AND L. Jofre (1993): *GRECO:* Graphical electromagnetic computing for RCS prediction in real-time. *IEEE Trans. Antennas Propagat.*, 35(2), 7–17.

J.M. Rius, M. Vall-Lossera, AND A. Cardama (1995): *GRECO:* Graphical processing methods for high-frequency RCS prediction. Special issue on Radar Cross Section of Complex Objects. *Annales des Telecommunications*, 50(5–6), 551–556.

S.S. Skyttemyr (1986): Cross polarization in dual reflector antennas–A PO and PTD analysis. *IEEE Trans. Antennas Propagat.*, AP-34(6), 849–853.

R.A. Shore and A.D. Yaghjian (1993): Application of incremental length diffraction coefficients to calculate the pattern effects of the rim and surface cracks of the reflector antenna. *IEEE Trans. Antennas Propagat.*, 41(1), 1–11.

R.A. Shore and A.D. Yaghjian (2004): A comparison of high-frequency scattering determined from PO, enhanced with alternative ILDC's. *IEEE Trans. Antennas Propagat.*, 52(1), 336–341.

V.A. Somov and Vyaz'mitinova (1990): Primenenie metoda kraevykh voln pri chislennom analize zerkal'nykh antenn. [Application of the edge wave method for the numeric analysis of reflector antennas]. *Padiotekhnika*, no. 1, pp. 69–71 [English translation in *Telecommunication and Radio Engineering*, 45(2), 99–102, published by Scripta Technica, Inc.].

S.J. Schretter and D.M. Bolle (1969): Surface currents on a wedge under plane wave illumination: an approximation. *IEEE Trans. Antennas Propagat.*, AP-17, 246–248.

H.H. Syed and J.L. Volakis (1996): PTD analysis of impedance structures. *IEEE Trans. Antennas Propagat.*, 44(7), 983–988.

H.B. Tran and T.J. Kim (1989): The interior wedge scattering. Ch. 4 in *Monostatic and Bistatic RCS Analysis*, Vol. 1: *The High-Frequency Electromagnetic Scattering Theory*, Northrop Corporation, Aircraft Division, Report NOR-82-215, Dec. 1989.

E.N. Vasil'ev, V.V. Solodukhov, AND A.I. Fedorenko (1991): The integral equation method in the problem of electromagnetic waves diffraction by complex bodies. *Electromagnetics*, 11(2), 161–182.

S. Vermersch, M. Sesques, AND D. Bouche (1995): Computation of the RCS of coated objects by a generalized PTD approach. Special issue on Radar Cross Section of Complex Objects. *Annales des Telecommunications*, 50(5–6), 563–572.

D.S. Wang, AND L.N. Medgyesi-Mitschang (1985): Electromagnetic scattering from finite circular and elliptic cone. *IEEE Trans. Antennas Propagat.*, AP-33(5), 488–497.

S.Y. Wang AND S.K. Jeng (1998): A compact RCS formula for a dihedral corner reflector at arbitrary aspect angles. *IEEE Trans. Antennas Propagat.*, AP-46(7), 1112–1113.

A.D. Yaghjian AND R.V. McGahan (1985): Broadside RCS of the perfectly conducting cube. *IEEE Trans. Antennas Propagat.*, AP-33(3), 321.

A.P. Yarygin (1972): Primenenie metoda kraevykh voln v zadachakh difraktsii na telakh nakhodyaschikhsya v plavno neodnorodnoi srede. [Application of the edge waves method to problems of diffraction from bodies placed in smoothly inhomogeneous medium.] *Radiotekhnika i Elektronika*, 17(10), 1601–1609 [English translation: *Radio Engineering and Electronic Physics*].

N.N. Youssef (1989): Radar cross section of complex targets. *Proc. IEEE*, 77(5), 722–734.

Index

Acoustic
  diffraction problems, 5–31
  EEW, grazing singularity removal and,
      199–203
  waves, 169–210
    axially symmetric scattering and,
        115–166
    backscattering
      nonuniform component $j^{(1)}$, 273–277
      PO approximation, 269–272
      total field of, 277–279
        hard fringe, 278
        soft fringe, 278
    Bessel functions, 128
    bodies of revolution, 115–166
    caustic asymptotics and, 221–225
    finite length cylinder
      backscattering, 269–279
      bistatic scattering and, 287–304
    grazing diffraction and, 230–234
    ray asymptotics and, 213–217
    slope diffraction and, 236–238,
        240–243
Acoustically
  hard surface, diffraction interaction and,
      248–250
  soft surface, diffraction interaction and,
      250–252
Asymptotic approximations, cones focal
    field and, 134–138
Asymptotic expression, steepest descent
    method, 44
Asymptotics, 33–56
  EEW and, 183–187
  electromagnetic waves and, 308–310
  first order, 145
  nonuniform component radiation and,
      71–76
  paraboloids backscattering and,
      145–147

Pauli, 47–51
ray, 67–68, 156–159
Sommerfeld ray, 44–47
specular beam and, 300–304
uniform, 51–55
Axially symmetric
  bistatic scattering, 155–156
    bodies and revolution and, 155–156
    PO field, 156–159
    PTD field, 160–161
    ray asymptotics, 156–159
  scattering
    acoustic waves, 115–166
    backscattered focal fields, 141–155
    bodies of revolution, 155–166
    cones, focal field, 134–141
    diffraction, canonical conic surface,
        115–126

Backscattered focal fields, bodies of
    revolution, 141–155
Backscattering
  acoustic waves
    nonuniform component $j^{(1)}$, 273–277
    PO approximation, 269–272
    total field of, 277–279
  cones focal fields, numerical analysis,
      134–141
  cross section, 9
  finite length cylinder, 269–284
    acoustic waves, 269–279
    electromagnetic waves, 279–284
      E-polarization, 279–283
      H-polarization, 283–284
  first order PTD approximation, 107–109
  PO approximation and, 102–104
Beams
  bistatic scattering and, 297–300
  specular, 300–304

Fundamentals of the Physical Theory of Diffraction. By Pyotr Ya. Ufimtsev
Copyright © 2007 John Wiley & Sons, Inc.

Bessel
  function, 35, 40, 128
  interpolations, PO field and, 159–160
  interpolations, PTD field and, 160–161
Bistatic
  cross section, 9
    backscattering, 9
    geometrical acoustics, 9–10
    monostatic, 9
    smooth convex, 9
  scattering
    axially symmetric, 155–156
    beams, 297–300
    finite length cylinder and, 287–310
      acoustic waves, 287–304
      acoustic waves
        PO approximation, 287–288
    finite length cylinder and
      electromagnetic waves, 304–310
    physical optics field, shadow radiation,
        289–290
    PTD, 290–296
    rays, 297–300
    specular beam, 300–304
Bodies of revolution
  axially symmetric bistatic scattering,
      155–156
  backscattered focal fields and, 141–155
  first-order PTD asymptotics, 145
  nonzero Gaussian curvature, 141–156
  paraboloids backscattering, 145–151
  PO approximation, 143–144
  spherical segment backscattering,
      151–155
Branched wave functions, 33

Canonical
  conic surface
    diffraction and, 115–126
      disk scattering, 126–127
      field $u_{s,h}^{(1)}$ Bessel interpolations,
          125–126
      focal fields, 124–125
      ray asymptotics, 118–124
      scattered field integrals and,
          117–118
  form conversion
    Cauchy residue theorem, 63

physical optics integrals to, 61–67
  wedge, elementary strips and, 170–171
Cauchy
  residue theorem, 63
  theorem, 41
Caustic asymptotics, 220–226
  acoustic waves, 221–225
  electromagnetic waves, 225–226
  edge diffracted waves and, 213–226
Cones
  axially symmetric scattering and, focal
      field, 134–141
  focal field
    asymptotic approximations, 134–138
    backscattering, numerical analysis,
        138–141
Convex body of revolution
  diffraction, 255–260
  multiple acoustic edge waves diffraction,
      255–260
Cylinders, polygonal, 83–112

Diffracted
  field, physical optics, 67–68
  ray, origin of, 46
Diffraction
  axially symmetric scattering and,
      115–126
  canonical conic surface, 115–126
    scattered field integrals, 117–118
  cone, EEW and, 190–191
  convex body of revolution, 255–260
  first order, 83–112
  formulation of, 5–7
  interaction
    acoustically
      hard surface, 248–250
      soft surface, 250–252
    electromagnetic waves, 252–254
    neighboring edges, 247–254
  multiple
    hard, 256–258
    soft, 258–260
  part, 26
  problems, 5–31
    electromagnetic waves, 27–31
    induced surface field, 25–27
    physical optics, 11–25
  slope, 229–245

strip, 83–99
total scattered field and, 87–92
Dirichlet condition, 6
Disk
  diffraction, multiple edge waves and,
      261–268
  scattering, 126–127
    nonuniform, field generated, 130–132
    physical optics approximation, 127–130
    total scattered field, 132–134

Edge
  diffracted waves
    caustic asymptotics, 213–226
    ray asymptotics, 213–226
  waves
    electromagnetic, 169–210
    multiple diffraction, 229–245
      grazing incidence, 229–245
      grazing, 230–236
      slope, 229–245
EEW (elementary edge waves)
  analytic properties of, 187–191
    diffraction cone, 190–191
  electromagnetic, 194–198
  general asymptotics, 183–187
  grazing singularity
    nonuniform component $\vec{j}^{(1)}$, 204–205
    removal of, 198–208
  high frequency asymptotics, 169
  history of, 209–210
  important points, 175
  other interpretations of, 209–210
  nonuniform, 192–193
  numerical calculations of, 191–194
  one dimensional integrals, 178–183
  triple integrals, 175–177
  transformation of, 178–183
Electromagnetic
  diffraction problems, 5–31
  EEW, 194–198
    grazing singularity removal and,
        203–208
  waves, 27–31, 129–130
    backscattering finite length cylinder,
        279–284
    caustic asymptotics and, 225–226
    diffraction interaction and, 252–254

edge, 169–210
  E-polarization, 279–283, 304–306
  finite length cylinder bistatic scattering
      and, 304–310
  grazing diffraction and, 234–236
  H-polarization, 283–284, 306–308
  multiple diffraction, 267–268
  ray asymptotics and, 218–219
  Shadow Contour Theorem, 30
  shadow radiation, 16–21
  slope diffraction and, 238–240,
      243–245
  specular beam, refined asymptotics,
      308–310
  uniform component, 28
Elementary edge waves (EEWs), 169
Elementary strips
  canonical wedge, 170–171
  $j_{s,h}^{(1)}$ integrals, 171–175
E-polarization, 279–283, 304–306
Exact solution, 33–56
Extended Kirchoff Approximation (KA). *See*
    scalar physical optics.

Far zone, 7–10
  bistatic cross-section, 9
Field generated nonuniform scattering
    sources, 130–132
Field $u_{s,h}^{(1)}$ Bessel interpolations, 125–126
Finite length cylinder
  backscattering, 269–284
    acoustic waves, 269–279
    electromagnetic waves, 279–284
  bistatic scattering, 287–310
    acoustic waves, 287–304
    electromagnetic waves, 304–310
First order
  diffraction
    polygonal cylinders, 83–112
    strips, 83–112
    triangular cylinder, 99–112
  PTD
    approximation
      backscattering and, 107–109
      symmetric scattering and,
          104–107
    asymptotics, bodies of revolution and,
        145
    Hankel function, 96

First order (*Continued*)
  PTD (*Continued*)
    strip diffraction and, 95–99
    TED approximation, 99
    truncated scattering sources, 95–99
Focal field
  cone scattering and, 134–141
  canonical conic surface diffraction and, 124–125
Fock, theory of, 10
Fresnel integral, 48, 50, 53
Fringe waves, 26
Functions
  $f^{(1)}$ integrals, 76–78
  $g^{(1)}$ integrals, 76–78

Geometrical acoustics, 9–10
Geometrical optics. *See* uniform component.
GO boundary, PTD field and, 161
Gradshteyn and Ryzhik table formula, 60
Grazing
  diffraction, 230–236
    acoustic waves, 230–234
    electromagnetic waves, 234–236
    edge waves multiple diffraction and, 229–245
    EEW, nonuniform component $\vec{j}^{(1)}$, 204–205
  singularity, removal of,
    acoustic EEW, 199–203
    EEW and, 198–208
    electromagnetic EEW, 203–208
    uniform component $\vec{j}^{(0)} \equiv \vec{j}^{(hp)}$, 205–208
Green theorem, 36, 39

Hankel function, 35, 80, 96
  integral forms, 60
Hard
  diffraction, multiple, 261–264
  fringe field, acoustic wave backscattering and, 278
Helmholtz
  equivalency theorem, 17
  integral expressions, 7
High frequency asymptotics, elementary edge waves and, 169
H-polarization, 283–284, 306–308

Incident plane wave, 40
Incomplete asymptotic expansion, multiple soft diffraction and, 258–260
Induced surface field
  nonuniform component, 25–27
  uniform component, 25–27
Integrals
  functions $f^{(1)}$, 76–78
  functions $g^{(1)}$, 76–78
  nonuniform components and, 71

$j_{s,h}^{(1)}$ integrals, elementary strips and, 171–175

Laplacian operator, 34

Monstatic cross section, 9
Multiple
  acoustic edge waves, diffraction convex body of revolution, 255–260
  diffraction
    edge waves and, 229–245
    electromagnetic waves, 267–268
  edge waves
    disk diffraction, 261–268
    hard diffraction, 261–264
    hard diffraction, 256–264
    soft diffraction, incomplete asymptotic expansion, 257–260

Neighboring edges diffraction interaction, ruled surface, 247–254
Neuman condition, 6
Nonuniform
  component, 25–27
    $\vec{j}^{(1)}$, grazing singularity, EEW and, 204–205
    $j^{(1)}$, backscattering acoustic waves and, 273–277
  radiation
    asymptotics, 71
    integrals, 71
      functions $f^{(1)}$, 76–78
      functions $g^{(1)}$, 76–78
  surface sources, 71–82
  diffraction part, 26
  fringe waves, 26
  physical optics diffracted field and, 68

disk scattering, field generated, 130–132
EEW, 192–193
Nonzero Gaussian curvature
    axially symmetric bistatic scattering,
        155–156
    bodies of revolution and, 141–156
Numerical analysis
    paraboloids backscattering and, 147–151
    scattered field and, 92–95, 110–111
    spherical segment backscattering and,
        153–155

Oblique incidence
    Hankel function, 80
    plane wave and, 78–82
One dimensional integrals, EEW and,
    178–183
Optical theorem, 15

Paraboloids backscattering, 145–151
    asymptotics, 145–147
    numerical analysis, 147–151
Pascul, pressure unit, 7
Pauli
    asymptotics, 47–51
        Fresnel integral, 48
        Sommerfeld ray, 51
        Taylor series, 47
    technique, uniform asymptotoic, 51–55
        Fresnel integrals, 53
        Taylor series, 52
Physical optics (PO)
    approximation, 101–102, 127–130
        acoustic waves, 127–129
            backscattering and, 269–272
        backscattering, 102–104
        bistatic scattering and, 287–288
        bodies of revolution and, 143–144
        electromagnetic waves, 129–130
    canonical form conversion, 61–67
        Cauchy residue theorem, 63
    definition of, 11–13
    diffracted field
        nonnuniform components, 68
        ray asymptotics, 67–68
    diffraction problems and, 11–25
    equations, 13

field
    axially symmetric bistatic scattering
        and, 156–159
    Bessel interpolations, 159–160
    bistatic scattering and, 289–290
    ray
        asymptotics, 156–159
        region, 161–165
    reflected rays, shadow region, 165–166
    shadow radiation, 289–290
    transition region, 161–165
    wedge diffraction and, 59–68
geometrical optics, 11
optical theorem, 15
original integrals, 59–61
    Gradshteyn and Ryzhik table formula,
        60
    Hankel function, 60
    Sommerfeld formula, 61
part, strip diffraction and, 85–87
properties of, 24–25
scalar, 13
shadow
    contour theorem, 21–24
    radiation, 16–21
total scattering cross section, 13–15,
    21–24
uniform components, 12, 59
uniform components, 59
Physical theory of diffraction (PTD)
    approximation, first order, 104–109
    asymptotics, first order, 145
    bistatic scattering, 290–296
    field
        axially symmetric bistatic scattering
            and, 160–161
        Bessel interpolations, 160–161
        GO Boundary, 161
    first order, 95–99
    physical theory of diffraction, 5–7
    scattered field, 7–10
Plane wave
    excitation transition, 38–40
        Green theorem, 39
        incident plane wave, 40
        Sommerfeld radiation condition, 39
    oblique incidence, 78–82
PO. *See* physical optics.
Polygonal cylinders, 83–112

Poynting vector for electromagnetic waves, 5
Pressure unit, Pascul, 7
PTD. *See* physical theory of diffraction.

Radiation
  nonuniform component, surface sources,
    71–82
  wedge diffraction and, 71–82
Ray
  asymptotics
    acoustic waves, 213–217
    diffraction, canonical sonic surface and,
      118–124
    edge diffracted waves and, 213–226
    electromagnetic waves, 218–219
    physical optics diffracted field and,
      67–68
    PO field and, 156–159
  bistatic scattering and, 297–300
  component. *See* uniform component.
  region, PO field and, 161–165
Reflected rays, shadow region, 165–166
Ruled surface, neighboring edges diffraction
    interaction, 247–254
Ryzhik and Gradshteyn table formula, 60

Scalar physical optics, 13
Scattered
  field
    far zone, 7–10
    Helmholtz integral expressions, 7
    integrals, canonical conic surface and,
      117–118
    numerical analysis, 92–95
    PTD and, 7–10
    strip diffraction and, physical optics
      part, 85–87
    theory of Fock, 10
    total, 132–134
  sources, first order PTD and, 95–99
Series solution, 40–44
  conversion to Sommerfeld integrals, 40
Shadow
  contour theorem, 21–24
  radiation, 12–21
    Helmholtz equivalency theorem, 17
    physical optics field and, 289–290
  region, reflected rays and, 165–166

Slope diffraction, 236–240
  acoustic waves, 236–238, 240–243
  edge waves and, 229–245
  electromagnetic waves, 238–240, 243–245
Smooth convex, 9
Soft
  diffraction, multiple, 261–264
  fringe field, acoustic wave backscattering
    and, 278
Sommerfeld
  formula, 61
  integrals
    Bessel function, 40
    Cauchy theorem, 41
    series solution conversion to, 40
  radiation condition, 39
    classical form, 40
  ray asymptotics
    diffracted ray, origin of, 46
    Pauli, 47–51
    wedge diffraction and, 44–47
Specular beam
  bistatic scattering and, 300–304
  electromagnetic waves and, refined
    asymptotics, 308–310
  refined asymptotics, 300–304
Spherical segment backscattering
  bodies of revolution and, 151–155
  numerical analysis, 153–155
Steepest descent method, 44
Strip diffraction, 83–99
  first order PTD, 95–99
  scattered field, physical optics part, 85–87
  total scattered field, 87–92
Strips, first order diffraction, 83–112
Surface sources, nonuniform component
    radiation and, 71–82
Symmetric scattering
  first-order PTD approximation, 104–107
  PO approximation, 101–102

Taylor series, 47, 52
TED approximation 92, 99
Theory of Fock, 10
Total
  field, backscattering acoustic waves and,
    277–279
    hard fringe, 278
    soft fringe, 278

scattered field, 132–134
  strip diffraction and, 87–92
scattering cross section, 13–15, 21–24
Transition region, PO field and, 161–165
Triangular cylinder
  diffraction, scattered field, numerical
    analysis, 110–111
  first order diffraction and, 99–112
  symmetric scattering, 101–102
Triple integrals, EEW and, 175–177
  transformation of, 178–183
Truncated scattering sources, 95–99

Uniform
  asymptotics, 51–55
    extension of Pauli technique, 51–55
      Fresnel integrals, 53
      Taylor series, 52
  components, 11, 12, 25, 28, 59
  $\vec{j}^{(0)} \equiv \vec{j}^{hp}$, grazing singularity and,
    205–208

Wedge diffraction, 33–56
  asymptotics, 33–56
  Bessel function, 35
  branched wave functions, 33
  classical studies, 33–38
  exact solution, 33–56
  Green theorem, 36
  Hankel function, 35
  Laplacian operator, 34
  perfectly conducting wedge,
    34
  physical optics field, 59–68
  plane wave
    excitation transition, 38–40
    oblique incidence, 78–82
  radiation, 71–82
  series solution, 40–44
  Sommerfeld ray asymptotics,
    44–47
  uniform asymptotics, 51–55
  Wronskian equation, 37